Prof Ste
CREATOR OF THE **CHIMP** I

A **Path** through the **Jungle**

A Psychological Health
and Wellbeing Programme to:
**Develop Robustness
and Resilience**

Graphic design:
Jeff Battista

mind*field*
M E D I A

Published in 2021 by Mindfield Media Limited.
Company number: 07228776

For contact information, please direct all enquiries through
www.chimpmanagement.com

Paperback ISBN 9781998991105
eBook ISBN 9781998991112

Copyright © Prof Steve Peters 2021

Prof Steve Peters is identified as the author of this Work in accordance with the Copyright, Designs and Patents Act 1988

All rights reserved. No part of this publication may be reproduced, stored in a retrieval system, or transmitted in any form or by any means, electronic, mechanical, photocopying, recording or otherwise, without the prior permission of the copyright owner.

A CIP Catalogue record for this book is available from the British Library.

Designed by Jeff Battista
Graphic material copyright © Jeff Battista 2021

Printed and bound by:
Page Bros Group Ltd, Mile Cross Lane, Norwich, NR6 6SA, UK

Copies are available at special rates for bulk orders through
www.chimpmanagement.com

Every reasonable effort has been made to trace copyright holders of material reproduced in this book, but if any have been inadvertently overlooked the publishers would be glad to hear from them.

The author and publisher shall have no liability or responsibility to any person or entity regarding any loss, damage or injury incurred, or alleged to have incurred, directly or indirectly, by the information contained in this book.

Prof Steve Peters

Consultant Psychiatrist and CEO of Chimp Management Ltd.

MBBS; MRCPsych; D.Sc.; PhD; BA; MEd.; PGCE; Dip. Sports Med.

Professor Steve Peters is a medical doctor and consultant psychiatrist, specialising in mental health and the functioning of the human mind. He has dedicated his working life to help people get the best out of themselves and to be in a good place. His current and past experiences include: clinical director of Mental Health Services within the NHS at a district hospital, Consultant Forensic Psychiatrist and Undergraduate Dean at Sheffield Medical School. He has spent 20 years as an examination panel member at the Royal College of Psychiatry and has been an expert advisor to World-Anti Doping Agency.

As an author he has written 4 books including the best selling self-help book of all time in the UK, The Chimp Paradox, which has sold over a million copies. - as per Nielsen TCM Chart 2020.

Other achievements include attending Downing Street for winning the senate award twice for his excellence in teaching. He has been a consultant to over 20 Olympic and national sporting teams and organisations over the course of his career.

Chimp Management Ltd.

Our aim is to help you get the best out of yourself and others. We offer a variety of programmes, courses, keynotes, workshops, individual mentoring, conferences and retreats. Whether you are an individual, a team or a whole organisation we will find a way to help.

I'm proud of the team we have. They are all likeable, compassionate people who are here to help. It can be daunting to get in touch but please do, without any obligation, to see if we can help.

Please visit or contact us:
www.chimpmanagement.com
enquiries@chimpmanagement.com

Join 'The Troop' for access to free videos and discussion forums at: *www.thetroop.chimpmanagement.com*

Psychological skills mentoring

Contents

Introduction – *Welcome everybody* VI

Stage 1 – **Understanding your mind**

Explanations will be given of the neuroscience of brain structure and functioning, the rules of the mind and how information enters and is processed by the mind. An explanation and application of the Chimp model will be introduced with an emphasis on the development of self.

1. The mind and Chimp Model introduced 2
2. The 'rules' of the mind 24
3. Developing yourself 45

Stage 2 – **Emotional management**

This stage will help you to understand and manage the emotionally based system within your mind. It will explain how all emotions can be helpful and protective.

4. Managing emotional reactions 64
5. Your mind in harmony 80
6. How to nurture your Chimp 94
7. Managing your drives 106

Stage 3 – **Working with emotions**

When emotions are understood, we can then appreciate how to use and work with all emotions to effect positive changes. Panic attacks, emotional scars and ghost emotions will be looked at in detail and how to manage them.

8. Emotion – learning an internal language 126
9. Disguised emotional messages explained 137
10. Expressing emotion with insight and change 153

Stage 4 – **Changing habits and managing life events**

The science of habit formation is explained and how we can change habits, beliefs and behaviours, including resistant habits. The grief reaction and loss are used as examples to demonstrate how we come to terms with life events.

11. Changing habits, beliefs and behaviours 167
12. Processing and managing life events 184
13. Managing significant life events 197

Stage 5 – **The two main stabilisers of the mind**

By using the two main stabilisers of the mind we can establish a basis for happiness and peace of mind. How to find and implement your values along with perspective is reviewed and how to create a rigorous support system.

14. Working with reality and truths _____ 216
15. Establishing peace of mind and happiness _____ 235
16. Keeping events in perspective _____ 249
17. External support _____ 266

Stage 6 – **Creating a stress-free lifestyle**

This stage explains the neuroscience of stress and how to work with the stress reaction. We look at how to prevent stress from occurring and how managing our environment can help with this. Lifestyle choices and recuperation, including the neuroscience of sleep and its applications, are addressed.

18. Managing stress _____ 273
19. Preventing stress from occurring _____ 289
20. Managing your environment and lifestyle _____ 300
21. Recuperation _____ 308

Stage 7 – **Optimising interactions with others**

We will consider how to get the best out of someone and review some common problem scenarios involving people interactions. We will then look at how to optimise relationships and how to manage past experiences and relationships.

22. Getting the best out of others _____ 325
23. A basis for relationships _____ 345
24. Optimising relationships _____ 354
25. Communication _____ 367

Stage 8 – **Pulling it all together**

The final stage will pull the previous stages together to explain how robustness and resilience can be attained. Some common blockers to attaining this will be addressed in the troubleshooting section.

26. Robustness and resilience _____ 376
27. Troubleshooting _____ 392

Going forward – *a few lines to encourage you and wish you well* _____ 408

References – *followed by the index* _____ 409

Introduction
– developing robustness and resilience

Welcome everybody

Welcome to this programme, that will take you on a path through the jungle of life to robustness and resilience. It will give an understanding of how the mind works and how to get the best out of yourself and others. Robustness and resilience are based on such things as good self-esteem, great relationships, confidence, a successful outlook, happiness and peace of mind. All of these things and more will be covered as the programme progresses.

I wrote the Chimp Paradox book to introduce the concept of how the mind is structured and works. The first stage in this book gives a summary of this. The remaining seven stages will form the path to robustness and resilience. The course is written following requests to have a structured practical programme that will help people to apply the Chimp Model and to develop emotional skills for life. By undertaking the programme, it will enable the reader to acquire and maintain psychological health and wellbeing.

I've had the privilege of spending a lifetime supporting people. Many have had struggles, or are going through a rough patch, which just about includes us all. The one thing that stands out to me, is that no matter what someone faces, if **they are in a good place** they will cope and thrive. A major pitfall when dealing with problems and troubles is to engage with them, **before first getting yourself into a good place**, and then finding solutions and ways forward.

With these thoughts in mind, I have constructed the programme to help you to focus on yourself, empower you, develop robustness and resilience and support you getting into a good place.

I sincerely hope that you will find the book helpful, and that the jungle of life will become much more manageable, as you find your own inner strength. This book is for you and about you.

How to use this book

It is best to work through the book, stage by stage, and do the exercises at the end of each unit. Working through the programme with others can be very helpful. As you progress through the book, you will be building the foundations for robustness and resilience. These foundations will be pulled together in the final two units. Stage one gives a summary of the Chimp Model. It is the basis for going forward and might be challenging, so please persevere!

The eight stages are divided into units. Each unit has a section of information with examples and a reminder of important points at the end of it. The science boxes are there only for those who are interested, but they are not essential to the unit. Following each unit, there are practical exercises to try. I strongly recommend that you do these, as they will help to bring to life the key points.

For each topic covered, it is important to reflect and think about how this applies to you. When we reflect or revisit ideas, we often find more ways to integrate and apply them into our lives. Therefore, it is helpful to go steadily and at your own pace. You are unique, so some examples and themes will resonate much more with you than others; work with those that do. Some themes might spark your own ideas, which would be great to work with.

As you are changing your way of thinking and behaving, it is very important to practice these changes. Please be encouraged, because you can learn the skill of managing your thinking, behaviour and emotions. Over time, these changes will become noticeable to you and others.

The Mind Management Skills for Life course

A Path through the Jungle is the handbook for our online course 'Mind Management Skills for Life'. The course is facilitated by one of our Chimp Management Mentors and is available for individuals, teams and organisations.

An independent evaluation of the programme as an intervention was conducted for workplace wellbeing. It concluded that the intervention is effective in improving occupational burnout and wellbeing.

More information about the research can be found in The Journal of Mental Health:
https://doi.org/10.1080/09638237.2023.2182423

If you think the programme may be beneficial for yourself or your organisation please get in touch via a phone call, email or website. We would love to speak with you:

www.chimpmanagement.com
enquiries@chimpmanagement.com

Unit 1
The mind and Chimp Model introduced

STAGE 1: will explain the structure and rules of the mind and how to find yourself.

Unit 1: covers the structure and functioning of the mind and introduces the Chimp Model.

The mind and the rules by which it operates

The mind is usually considered to be the part of the brain dealing with behaviours, thinking and emotions.

We all understand that our physical body is like a machine and has rules by which it operates. Breaking those rules has consequences. For example, if we eat too much, then we put on weight. If we are not used to physical exercise and we over exercise, we are likely to suffer, usually with a lot of muscle aches.

The mind also works like a machine and it also has rules by which it operates. If we break those rules then there are consequences. Therefore, if we understand the mind and the rules by which it operates, then we will be able to run this machine correctly.

Most of us don't understand how the mind works or the rules by which it operates, so we inadvertently break those rules. When we break the rules, the mind usually reacts by giving us negative emotions, such as anger, frustration or anxiety. These negative emotions can be seen as messages from the mind telling us to stop and work within the rules. For example, one rule of the mind is that we must accept the reality of what is in front of us and then work with it. If we refuse to accept the facts of a situation then our mind will react and give us unwelcome emotions.

The Mind produces:
- Thoughts
- Behaviours
- Emotions

The Mind operating

This is breaking the rules!

FACTS → Refusing to accept the facts →
- Frustration
- Anger
- Upset

Breaking the rules of the mind is similar to trying to operate a machine using the wrong fuel; it might operate, but it will complain, and eventually it is likely to become damaged.

How the mind functions

Two functions of the mind

The mind functions in two different ways:

1. It uses **inbuilt drives** to keep the body alive and produce the next generation
2. It perceives, interprets and **interacts** with the world around it

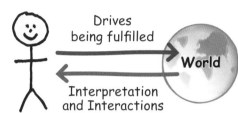

1. Inbuilt drives

The drives we have, which trigger actions such as eating, finding a mate, establishing a territory or finding a secure place to live, are all fixed within us from birth. If we don't fulfil drives, we can begin to feel uncomfortable.

Drives
- Scientific points

Drives are natural forces that help us to survive and perpetuate the species. [1] *Examples include: the drive to eat, have sex, quench our thirst and protect our young.* **Inbuilt drives are formed and managed by multiple areas of the brain.** The hypothalamus (H) contains many centres, called nuclei, which promote fundamental survival drives, such as energy regulation and eating.[2] We have to learn how to manage our drives, and this isn't easy, as nature intends to make them compulsive!

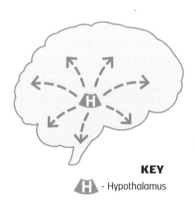

KEY
H - Hypothalamus

2. Interacting with the world

We can interact with the world around us by using instincts or by learning.

Interacting with the world using instincts

Instincts are inbuilt automatic reactions to given stimuli. For example: fight, flight or freeze, is an instinctive reaction to a threat and this immediate reaction can help us to stay safe. We use this instinct in every day life all of the time.

If someone confronts us and we feel threatened, our instinct will be to either, fight (confront them), flight (get away from them as quickly as possible), or freeze (don't react and hope they go away).

Instincts
- Scientific points

Instincts are inbuilt reactions to specific stimuli.
For example, a typical inbuilt reaction to seeing a snake is to set off an alarm system in our mind, which results in the flight, fight or freeze reaction (FFF). The alarm is started in the orbitofrontal cortex (OFC), which sends a message to the amygdala (A), along with other structures. [3][4]

The amygdala works with the FFF reaction. When the amygdala has received the message, it sends back advice to the OFC, forming a two-way conversation. The OFC then makes a decision about what to do.

KEY: OFC - Orbitofrontal cortex
A - Amygdala

Interaction with the world by learning

We can use instincts to react to situations but we have an alternative way of reacting and this is to learn a response. When it comes to learning how to interpret and interact with our experiences, we now find a problem!

Two very different systems in our mind are trying to interpret our experiences, interact with them, and form a plan of how to respond.

Two systems interacting with our world
- Scientific points

When the brain receives information, it is first sent to the thalami (T). The thalami are two walnut sized structures at the centre of the brain and act as relay stations. [5][6] Each thalamus relays the information to two different systems:

1. An emotionally based limbic system, containing the orbitofrontal cortex (OFC): This first system is reactive, impulsive and outside of our direct control and acts immediately. [3][7]

2. A rationally based system, termed the dorsolateral prefrontal cortex (DLPFC): This system works with executive functioning, for example, using logic, organisational skills and perspective. [8][9][10]
   It acts only after the first system has reacted but it can influence or manage the first system.

The conjoined twins who share the same thalami:
Two girls were born joined at the head and shared the same thalami. Whatever one girl saw, the other would be able to describe. It showed how the thalami are the relay stations that first receive messages before sending them out to the other structures within the brain. [11][12][13]

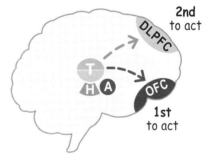

KEY:
DLPFC - Dorsolateral prefrontal cortex
OFC - Orbitofrontal cortex
T - Thalamus
H - Hypothalamus
A - Amygdala

We now have a picture of how the mind functions. It is programmed by nature to use drives and instincts but it also has the ability to think and interpret. Therefore, the way to understand and manage the mind is to accept and work with the drives and instincts we have been given, and then learn how to manage the two systems that interpret our experiences.

Making things easier

This picture is still quite complex; so to make working with the mind easier, we can simplify things by recognising that there are three different 'teams' that operate within the mind.

The three teams operating within the mind

Team one:
- Genetically determined
- Operates with drives
- Reacts to experiences with instincts
- Thinks from an emotional basis
- Interprets with feelings
- Its agenda is to perpetuate the species and help us to survive

This team is ***not within our control***, although we can influence it. This team can run our lives and it acts without our permission! The team leader is the orbitofrontal cortex (OFC). [3] [7]

Team two:
- Thinks from a factual basis
- Responds to experiences with rationality
- Interprets with logic

We have complete control over this system and can choose how to work with it.[8] The team leader is the dorsolateral prefrontal cortex (DLPFC). [9] [10]

Team three:
This team is very similar to a computer.

It has two functions:
- Storing memory, facts and experiences
- Performing automatic behaviours

Teams 1 and 2 programme this computer. [14] [15] [16]

Memory storage

Team 1 stores emotional memory (mainly in the amygdala) and team 2 stores factual memory (mainly in the hippocampal formation). Emotional memory gives feelings to events or situations. [17] [18] Factual memory recalls events in detail, such as times, places and people. The two usually go together but can work independently. Many other types of memory also exist; examples include memory for actions, recognition of faces and musical memory. There is no obvious team leader when it comes to memory. The parietal lobe (P) and the ventromedial prefrontal cortex (VMPFC) are also key memory storage areas.

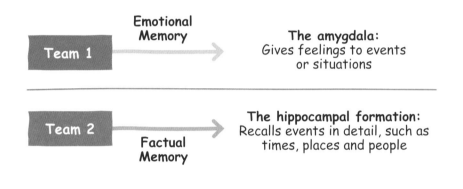

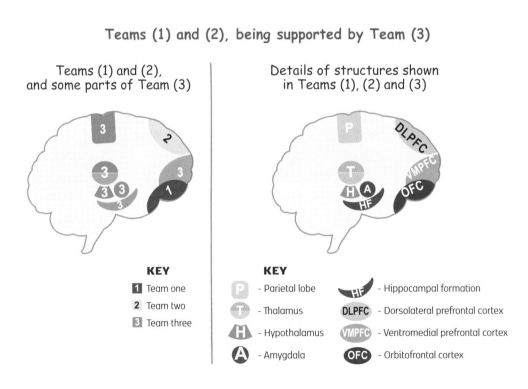

The Blue square experiment
- Scientific points

Memory storage in the brain is very complex. This is because many different types of memory are stored in different places. **Emotional memory** means that we record feelings but not facts. **Factual memory** records facts but not feelings.

The blue square experiment [19] [20] demonstrates the difference between emotional memory and factual memory. People are shown images of different coloured squares on a screen. When a blue square appears they receive a mild but unpleasant electric shock. Once the subjects have had chance to register that a blue square means an electric shock, they then watch the Computer screen. When the blue square appears, people who have damage to their emotional memory banks (amygdala) will recognise the blue square with their factual memory (hippocampal formation) but will not show any fear before they receive the shock. [21] [22] [14] People who have damage to their factual memory banks will experience fear but not be able to explain why they are afraid. Occasionally, a healthy brain can work with only one of the memory banks; either the emotional or the factual. This can explain why we sometimes don't experience emotions when we expect to, and at other times, we experience emotions but can't understand why.

We now have a simplified picture of how the mind works when it is learning, which looks like this:

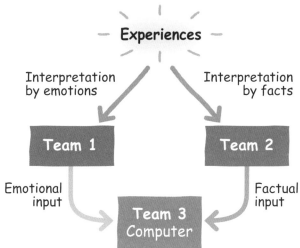

These teams represent different structures and regions that work together to perform different functions. Some structures are in more than one team. The teams are trying to work simultaneously. The problem is that they frequently have different agendas, different ways of thinking and interpreting, and different values to operate from.

It is very important to recognise just how differently these three teams work.
As we progress in our understanding of the mind, we will see how crucial it is *to choose* the appropriate team to work with, if you want to achieve your aims. You can change teams instantly, once you learn the skill of how to do this.

The Chimp Model

The complexities of the neuroscience of the mind can be difficult to follow. For this reason, I gave names to the three operating teams. In order to bring them to life, I gave them the names: Human, Chimp and Computer.[23] [24] Here is why:

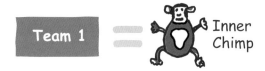

Human beings are biologically classified in a group called the hominids. These are the great apes and include the human, chimpanzee, gorilla, bonobo and orang-utan. The first team of structures in the mind is very similar to those of the other great apes (and many other animals). In particular, this first team or system in our mind functions very similarly to that in the mind of the chimpanzee. [25] [26] Therefore, it is no surprise that when it takes over our mind, or the chimpanzee's mind, we often display similar types of behaviours, emotions and thinking. This first system can therefore be named the 'inner Chimp'. Clearly, there are differences between our mind and that of the other apes, but this set of structures is very similar.

The inner Chimp represents this inbuilt genetically determined team of structures. It has survival and perpetuation of the species as its agenda, working with drives and instincts, and uses emotion as its basis for learning.

The Chimp system developing
- Scientific points

The first team begins to develop at just eight weeks into foetal life. It takes on character traits, such as how quickly it will react to stress. For example, some foetuses react if the mother's heart rate goes up but other foetuses remain calm. Research shows that six months into life as an infant, the same traits remain constant for each baby; some are still reacting to the world, while others remain calm. [27] These characteristics remain in this system throughout life. Effectively, we have been given personality traits of **our unique Chimp system**, but **not our** personality traits. These personality traits are genetically determined, but we can modify their expression by managing them. Therefore, we are **predisposed** to present a certain disposition to the world, but we are not **predetermined.** We can choose.

The second team represents the part of the mind that **you are in control of**. It is effectively you. Therefore, it is easiest to depict this as a Human. The Human is you, as an individual. You operate with areas of the brain that you can control. You learn by using logic and rationality.

The Human system
- Scientific points

The Human circuits represent various areas in the brain: the most important being the dorsolateral prefrontal cortex. These areas develop very slowly during foetal life but accelerate into operation in the first few years of childhood. They continue to mature and can take as long as thirty years to become fully matured and fully functional. The Human brings rational judgement and perspective to our thinking, which can be disturbingly frail during our teenage years, while our brain is still developing. [28] [29]

As stated earlier, the Computer not only stores information and memories, but it can also be programmed to carry out automatic behaviours and automatic thinking. [30] We can programme our inner Computer by putting in beliefs or behaviours that we find constructive. We can practise behaviours that we want to perform and turn them into habits, and we can think through and establish beliefs, in order to reinforce these habits. This Computer system is extremely fast when it operates. The Computer always advises the Human and Chimp when they are trying to make a decision. Therefore, it is very important to check what beliefs the Human or the Chimp has stored in it.

The Computer system
- Scientific points

It could be argued that the entire brain, including the mind, is one big computer system. Please remember that the Chimp model is exactly that, a model to describe the mind in order to access and work with it. In this model, I have separated the brain into areas of active originators of thinking and decision-making from those areas of the mind that assist with decision-making and approaches to situations. These support areas react to any stimuli they receive and then relay their 'advice' back to the two originator areas to influence them before they act.

Many of these advising areas of the brain also act automatically. They do this by mentally and physically repeating patterns of thinking or behaviour, and thereby eliminate the need for any originality. Hence, these supporting areas and automatic functioning become the 'Computer' system within the mind.

You, sharing your mind with a machine

The teams in action

Here is a table to demonstrate the three teams in action in two different scenarios.

Two examples, where the Human and Chimp might think and act differently:

Choosing a meal and discussing a heated topic.

Choose a Meal	Healthy option	Pleasurable option	Advice
Discuss a heated topic	Find a solution	Win an argument	Advice

In both of these examples, your unique Chimp or Human may have different responses to those shown. I have chosen common responses. The Computer's response will depend on what the Human or the Chimp has put into it. For example, your Chimp could have put in a belief that says, "We can always begin the diet tomorrow", and your Human might have put in "I will only regret a poor choice and I really want to be healthy". The Computer can only feed back to the Human and Chimp what beliefs are stored in it. If both Human and Chimp have put beliefs into the Computer, then the Computer will feed back both suggestions because it hasn't been tidied up. The Chimp will select the suggestion it likes best, and this is likely to be to wait until tomorrow to start the diet. Therefore, it is important to make sure the advice and beliefs fed back to the Human and Chimp are helpful.

In other words, *the Computer must have helpful beliefs in place ready to respond to a situation before it meets that situation*. The Computer always needs tidying up, by addressing and removing unhelpful Chimp suggestions. We will learn how to do this in due course, when we cover the rules of the mind.

Every day there is the potential for a clash between the teams within your mind, much like a heated debate going on. The Human will want to find a solution, but the Chimp will want to win, and it will therefore only look at how it can achieve a win. It is common to observe these two very different approaches during group discussions, when people alternate between working from their Human system and working from their Chimp system.

Key Point

The Human looks for solutions:

The Chimp looks to win.

Recognising the Chimp at work

Try to recognise when your Chimp is operating. Remember that the Chimp system is impulsive, has an agenda, which you might not agree with, and will overpower you if necessary.

> **Key Point**
>
> The way to distinguish between the Human and Chimp working is to ask the question:
>
> *"Do I want this thought, behaviour or feeling?"*

If the answer to the Key Point opposite is "no", then the Chimp is hijacking you. If the answer is "yes" then it is you, the Human, who is in charge (or it could be the Chimp is in charge, and you are just agreeing with it).

Sometimes, we agree with what the Chimp is doing and then there is no problem!

If you are still unsure about which system is operating, then another way to find out is to ask yourself: *"How will I feel in an hour's time?"*. When we act, while we are in Chimp mode, we often later regret it.

Example: *Stella and the argument*

Stella works in an office and one of her colleagues, Ben, has just criticised some of her work unjustifiably. Stella's Chimp has now become angry and has taken over. Stella's Chimp is about to raise her voice at Ben and tell him his behaviour is unacceptable.

The two questions for Stella are:
- "Do I want to raise my voice?"
- "Do I want to let him know his behaviour is unacceptable?"

The answer from her Chimp to both questions is "YES"!

The answer from her Human is, "No", to raising her voice, but "Yes" to letting him know his behaviour is unacceptable.

In this example, Stella can see that she disagrees with her Chimp about raising her voice but agrees with her Chimp about letting Ben know.

If Stella doesn't learn how to distinguish between the two teams and manage them, then it is likely that her Chimp will win. This will probably result in Stella feeling bad and apologising to Ben for raising her voice.

Example: *Archie's choice*

Let's say that Archie has had a fall-out with a close friend. If he allows the Chimp system to operate his mind, it will experience emotions that are probably unhelpful. The Chimp will be agitated and focus on the problem. In an attempt to remove the problem, it might simply decide not to speak to his friend again. This isn't a solution; it is just removing the problem. If Archie were to go into Human mode then the Human in Archie will focus on finding a solution, which is likely to be a discussion with apologies or some reconciliation.

Example: *Lisa and the exam*

Lisa has failed an exam at night school and now needs to take it again. Her Chimp might well decide to remove the problem by dropping out, even though the Chimp might want to pass the exam. Her Chimp will focus on things such as how badly she has done and what everyone will think. However, if Lisa changes to the Human system then she is going to focus on the solution. She will work out how to improve for the re-sit and prevent herself from failing again.

Logic and emotion as the basis for action

The Human **DOES NOT** represent logic and the Chimp **DOES NOT** represent emotion. Both teams possess logic and emotion. The difference is how they use these and where their starting point is. The Chimp bases its thinking on emotion and the Human bases its thinking on logic. **They both operate with logic and emotion.**

Our basis for working is different

Ways of working and shift between the two Modes

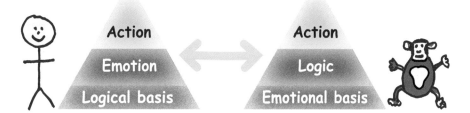

Key Point

The Human bases its emotion on logic and the Chimp bases its logic on emotion (making it potentially less reliable).

The Human team begins with establishing the facts and the logic of a situation and puts the information together rationally.[31] When the Human has done this, these facts evoke emotion. The final action taken is still based on facts but receives some energy from the emotion.

The Chimp begins with a basis of emotion.[32] This means it works with *feelings* and goes on its *intuition* about what is right and what is happening. It doesn't use facts but could be influenced by them. Starting with this emotional basis, it tries to work out what is happening by using an emotionally based logic. The final action taken is based on emotion with some emotionally based logic. This can give rise to unhelpful actions.

Example: *Claudette and the accusation*

Here is an example of how the two teams operate differently in a given situation. I will make the distinction clear by deliberately exaggerating the responses each team gives.

Claudette has strong ethical values and sees herself as a moral person who upholds her values. She has been accused of falsifying some documents for financial gain. Here are some typical responses from her Human and Chimp that have both recognised that an injustice has taken place.

The Human begins by establishing the facts. These are:

- That she is innocent
- She has the evidence to support her case
- The truth is likely to come out
- She is likely to receive an apology
- All she can do is explain the truth of the situation
- She cannot control what someone else believes

The emotions evoked by these facts are:
- Sadness
- Disappointment
- Calmness

The action taken, based on these facts and driven by the emotions evoked, is to explain and hopefully resolve the situation.

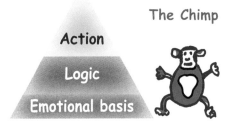

The Chimp begins with an emotional reaction. Such as:
- Anger
- Frustration
- Outrage
- Indignation

These emotions make Claudette's Chimp feel insulted and ready to fight her corner. The logic based on these emotions is:

- I need to win in this situation
- I need to express my feelings emotionally and make sure I am listened to
- I need to change the minds of other people

The resulting action is typically one of hostility and confrontation with a view to winning.

Therefore, the pathway that each team takes to try to resolve the situation is somewhat different.

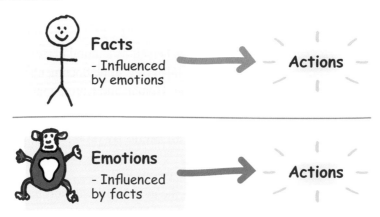

Even though there might be a measure of agreement between the Human and Chimp, the difference in approach makes the Chimp's pathway much more likely to result in a negative effect on others and a less favourable outcome. The Chimp might not accept the facts, distort them or not even understand them; therefore they are only an influence on it. The Human can reject emotion or be persuaded by it.

The contents of the Computer

The contents of the Computer are so important that a brief explanation is needed at this point, but they will be looked at in detail as we work through this course.

We have established that the Computer essentially has two functions:

1. To store factual and emotional beliefs or memory from the Human and Chimp, and then to remind or advise them of these before they act. [33]
2. To act automatically by being programmed by either Human or Chimp. [34] [35]

Stored behaviours or beliefs: *Autopilots, Gremlins and Goblins*

Helpful or constructive automatic behaviours or beliefs we will call **Autopilots**.

Unhelpful or destructive automatic behaviours or beliefs that can be removed we will call **Gremlins**.

Unhelpful or destructive automatic behaviours or beliefs *that are extremely resistant to being removed* we will call **Goblins**.

Examples include, a belief that you:
- Can only do your best is a constructive Autopilot
- Can always learn from any outcome is a helpful Autopilot
- Are not as good as everybody else can be a destructive Gremlin
- Are better than others can also be a destructive Gremlin!
- Will fail at what you do is an unhelpful Gremlin
- Are not a leader could be a Goblin

You might ask why the last example of not being a leader is a Goblin. Leadership skills can be learnt. However, research shows that being a leader is partly genetic. Therefore, a belief that you cannot become a leader, might well have its roots in your genes. [36] Whether someone can become a leader if they lack the leadership gene, clearly depends on the individual. This would determine if the belief is a Goblin or not.

We have thousands of learned programmes (Autopilots or Gremlins) in our Computers, some are behavioural, some are belief-based and some are emotionally based. There are lots of ways in which we can programme and also re-programme the Computer to learn how to react to certain stimuli.

Gremlins and Autopilots in action

Example: *Robin and procrastination*

Robin has decided to sort out his garage because it has become very messy and overloaded with unwanted things. His Human has decided to begin today. Robin's Chimp is concerned about how he is going to manage the task because his Chimp sees the task as being too formidable. As Robin opens the garage door and looks at the mess, his Chimp panics and turns to the Computer for advice. The outcome now depends on what is in the Computer. If the Computer has a Gremlin in it that says to the Chimp, "You will never get through this today and there's just no point in starting" then Robin is likely to feel overwhelmed and fail to begin. Another Gremlin might also be saying, "I don't know where to start, so why don't I sit down and make a plan". This can be a clever way to avoid starting!

However, if the Computer has an Autopilot in it that is programmed to deal with overwhelming situations, then there might be a very different outcome. For example, an Autopilot could say to the Chimp, "Don't look at the whole task, but instead just take a small first step, that is easy to manage quite quickly, and see that as a successful start". Then it is likely that the Chimp will take the small step and begin the process. Alternatively, there could be an Autopilot in the Computer that says, "I am a person who loves a challenge, and getting started is the first part of the challenge". This is again a helpful belief and is much more likely to result in some positive action rather than further avoidance and procrastination.

Autopilots and Gremlins are unique to you. Your beliefs might differ from those of other people and what might be a Gremlin to them could be an Autopilot to you. Unhelpful beliefs, behaviours or habits that are stored in the Computer but have not been considered, tend to sabotage our functioning. They advise both the Chimp and the Human. This advice can be changed and improved by working on what you are storing in your Computer.

Goblins

The term Goblin represents a behaviour or belief that has been put into the Computer and is resistant to being removed. It could be thought of as a virus in the system or as some irreparable damage to the system.

For example, if we have a turbulent time and have some damaging experiences while the brain is developing during childhood, these experiences can make changes to the way the brain functions later in life. These changes in the brain can be accepted and managed. Sometimes traumatic events in our adult life can also effect changes to our mind that we need to accept and manage. These events often result in certain beliefs or behaviours that become ingrained and difficult to change. For example, a child who is rejected by a parent might find they experience feelings or beliefs of a fear of rejection in future relationships and these can be very difficult, if not impossible, to change. However, they can be managed. Another example is someone who has gone through a very traumatic divorce and this could alter the way they see themselves or future partners.

The reason I make a distinction between a Gremlin and a Goblin is to help people not to put unreasonable pressure on themselves in trying to change the impossible. Of course, we will always try to remove unhelpful beliefs or behaviours but we need to recognise when to stop and instead of removing them, learn to manage them.

The ventromedial prefrontal cortex (VMPFC)
- Scientific points

The VMPFC, as part of the Computer, develops significantly during early childhood. However, if we have a childhood where stability and security is compromised, the VMPFC fails to reach its developmental potential. It literally doesn't grow to full size. The window for development is only during early childhood. Therefore, when we reach our adult life, this area cannot function as well as it might have done.

One of its functions is to settle down the amygdala (A) from overreacting to situation.
[37] [38] The lack of development of the VMPFC results in some adults, who have had poor childhood experiences, being emotionally more reactive to stress. [39] Hence, a potential Goblin has been formed. Thankfully, there are ways of managing and compensating for this to bring resilience, despite the Goblin, and we shall address this as we work through the course.

The ventromedial prefrontal cortex (VMPFC)
- **Scientific points** - continued

One function is to settle the amygdala from overreacting

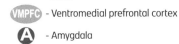

KEY
VMPFC - Ventromedial prefrontal cortex
A - Amygdala

The Stone of Life: *A deeper level of operating*

The 'Stone of Life' is an extremely important **reference point** in our Computers. It can be the ultimate reference point that stabilises the whole of the mind. It contains three aspects: reality (Truths), values and perspective. We will be covering each of these aspects in great detail later in the course.

The Computer can work on two different levels: specific situations and all situations.

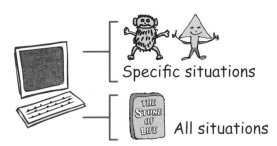

The first level is where specific programmes are developed to deal with **specific situations**, such as, how to manage fear of flying or giving a talk to an audience. This level involves Gremlins and Autopilots. Although specific to a situation, removing Gremlins and replacing them with Autopilots can result in excellent outcomes. This level forms the basis for all the therapies that have a 'cognitive' aspect to them.

The second level is much deeper and is arguably far more influential. It is what we have termed the 'Stone of Life'. **It is more influential because it can be applied to all situations.**

Please think carefully about who you are
Remember that you and your machine are different.

Unfortunately, you do not have complete control of your machine. What you present to the world is you, but this presentation is modified by your Chimp's hijacks and your Computer's influence and take-overs. **The world might never see the real you. This doesn't mean that YOU can't see the real you.**

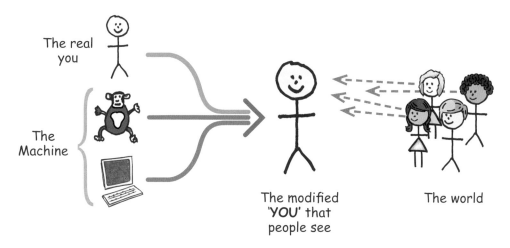

It is so important to know who you are because this will help you to recognise yourself as a genuine person. This will lift your self-esteem and your self-respect.

Unit 1

We can think of the mind as being composed of three teams or systems:

The Human system, which represents you:
- *Works with facts*
- *Interprets logically*

The Chimp system:
- *Works with feelings*
- *Interprets emotionally*

The Computer System:
- *Stores information and beliefs*
- *Is programmed to act automatically*
- *Is managed by both Human and Chimp*
- *Autopilots are helpful, constructive beliefs and behaviours*
- *Gremlins are unhelpful, removable beliefs and behaviours*
- *Goblins are fixed, or extremely difficult to remove, beliefs and behaviours*
- *The Stone of Life is the main stabiliser of the mind*

You, sharing your mind with a machine

You | The Machine

Unit 1

Focus points and reflective exercises

1. *Knowing who you are, and being able to separate yourself from your machine*
2. *Recognising your Chimp in action*

Focus 1: *Knowing who you are, and being able to separate yourself from your machine*

"**Who are you?**" This is the most important question to answer in the whole of this course. The answer is the starting point to finding yourself and not muddling yourself up with a machine.

You | The Machine

The concept of 'how to find the real you' is very important. I cannot emphasise this enough because it is always the fall back position in helping you to both respect and love yourself as a person.

It is crucial that you do not muddle yourself up with the machine. Once you have separated yourself from your machine, it then becomes apparent that *the task ahead is to learn how to manage the machine.* When the machine is managed it will not interfere, but instead enhance you, and you will present the real you to the world.

Once someone has grasped this concept then you will no longer hear them saying such things as, "I am a worrier" or "I get angry so easily". These are not scientifically valid statements because the truth is that their machine hijacks them. It is the machine that is becoming worried or angry. The machine then presents to the world a person who worries a lot or a person who seems to have a short fuse and easily becomes angry.

Your friends or a partner might know who the real you is, because when the Chimp goes quiet and the Computer remains silent, your Human appears to them.

If, at times, you do not present yourself to the world as a pleasant individual, then the machine has hijacked you.

Exercise 1: *Finding the real you*

There is a very simple way to find out who you are. The easiest way is to take a blank piece of paper and write down the characteristics of the ideal or perfect person that you would like to be.

So, for example, you might write:
- Honest
- Calm
- Sense of humour
- Thoughtful
- Compassionate
- Industrious
- Friendly

Hint: think about how you would like to be in your workplace, your home life, in relationships with people and in your approach to life itself. What behaviours do you want to display?

When your list is complete you can write your name above it, because you have now described the real you. What you have written is the way that you would present to the world *if your Chimp and Computer did not interfere or influence you*. If we could surgically remove both of these systems from your mind, then we would be left with just you. **The list that you have composed is therefore you.**

You might argue, "Surely everyone writes the same list?". In fact they don't. When I first asked people to do this, so that I could find out who they were, the lists varied quite a lot. I asked various types of people, including those people who had been diagnosed with psychopathic personalities. I can assure you that they did not include characteristics such as compassion or honesty!

Key Point

Please remember the Chimp model is NOT an excuse model. We are all fully responsible and accountable for managing our Chimps.

Focus 2: *Recognising your Chimp in action*

Once you have found out who you are, it isn't a big step to recognise when your Chimp is hijacking you.

The golden question to ask is: "**Do I want...?**" and complete the sentence.

For example, "Do I want to feel like this?", "Do I want to behave like this?" or "Do I want to think like this?". If the answer is "no", then a hijack is occurring. However, there is more to the Chimp than just a hijack. When we are operating with the Chimp system it has a completely different approach to the Human system. For example, when our mind is operating with the Chimp system we cannot search for solutions to problems. What the Chimp system does is to react to a problem and then try to remove the problem, rather than solve the problem. The Human system will look for solutions and will not focus on the problem.

The mind and Chimp Model introduced

The Chimp:
- Is likely to continually focus on the problem
- Will be unable to look for solutions
- Might try to just ignore or remove the problem

The Human:
- Is likely to focus on finding a solution
- Will want to resolve the problem
- Is likely to keep searching until a solution is found

Exercise 2: *Recognising whether you are in Chimp mode or Human mode*
This exercise is based on the following key point.

> **Key Point**
> *It is important to be able to recognise exactly which mode you are operating in, if you want to increase your chances of successful outcomes.*

To help you to distinguish between Chimp and Human functioning, try to imagine situations that you know are likely to provoke an emotional response in you. First, predict how your Chimp would react, and how it might think and behave. Then think how you *would like to respond* to the situation, as this will show you *what* would happen if you were in Human mode.

By visualising your Chimp's reaction and then your Human's response to various scenarios, you will begin to develop the skill of recognising the difference between the two modes of operating. You will also come to appreciate how they operate from a different basis: the Human from a logical basis and the Chimp from an emotional basis. When you next meet a setback or problem, try to recognise the two systems in action.

Unit 2
The 'rules' of the mind

Unit 2: *will focus on how the Human, Chimp and Computer interact, and how we can programme and tidy up the Computer. We will see the rules of the mind in operation and how to begin working within these rules.*

How the Computer, Human and Chimp interact

Before we look at how the Computer can come to our rescue, we will first consider how it functions. Our Computers tend to run our lives by being programmed for most situations. The Computer merely recognises a situation and then follows a pattern for dealing with it. For example, when we meet people we know, we usually follow a pre-programmed welcome by asking how they are and offering encouraging comments. Often these welcomes are not thought through, they are pleasant automatic social functioning. However, if we meet an unfamiliar situation or a situation where there is a need for interpretation, then the Human and Chimp systems come into play. Consider meeting your new partner's family or new colleagues for the first time. The Computer will stop if it does not know how to advise the Human or Chimp and hasn't been programmed to act in any particular way.

Whenever we meet unusual, different or threatening situations, the Computer will stop, while the Human and Chimp each make an instantaneous interpretation of what is happening. The Human uses rational processes and the Chimp uses emotional processes. When the Chimp and Human have finished interpreting any new situation or experience, they act on their interpretation and then store their beliefs in the Computer for future reference.

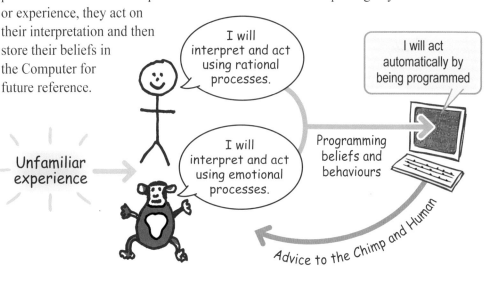

How the Computer is programmed

Example: *The dog bite*

A dog has just bitten Jenny. It approached her, and without warning or provocation, bit her leg. The Human and Chimp within Jenny both have to interpret what happened and what it means in the future when seeing another dog. There are many ways that the Human or Chimp could interpret this, so I will choose two common interpretations. The Chimp might generalise and say that all dogs will bite me and therefore, whenever I see a dog, I will react by trying to get away as soon as possible. Here, two Gremlins have been formed by the Chimp and placed in the Computer; the first Gremlin is a belief, and the second Gremlin is a behaviour. The belief is that all dogs will bite me, and the behaviour is to run away when you see a dog.

However, Jenny's Human might say that this particular dog was disturbed when it attacked me, which had little to do with me. Therefore, whenever I meet a dog in the future, I need to be able to assess what its intentions are, because most dogs are friendly. The Human has placed two Autopilots into the Computer. The first is a belief that most dogs are friendly. The second is a behaviour, which is to assess the dog and then act appropriately.

The Computer now has conflicting beliefs and advice in it and needs tidying up. Otherwise, when Jenny next sees a dog, she is more likely to take the Chimp's advice, in order to make sure she is protected and takes little risk.

The good news is that we can change programmes in our Computer by *changing the way we think about something and establishing what is true and what is helpful.* We can replace Gremlins with Autopilots.

The Computer running a programme

When the Chimp and Human interpret situations, they always check with the Computer for advice. Paradoxically, the Computer is only going to give them the advice that they put into it in the first place! It is more of a reminder. This 'reminder' is based on memory, experience and beliefs.

If the Computer is programmed with a response to a situation, then it will take over with automatic behaviours. [38] Therefore, the Computer can either give advice or take over.

Key Point

The Computer can only give advice that the Human and Chimp have put into it.

The Computer will need attention from time-to-time to remove unhelpful advice or false beliefs!

The two functions of the Computer

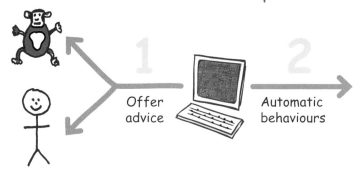

Reprogramming the Computer

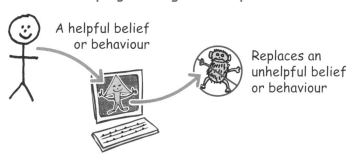

We often put Autopilots or Gremlins into our minds without knowing we have put them in. Usually we have to think through our beliefs in order to recognise how helpful or unhelpful they can be. Similarly, we need to recognise and replace destructive behaviours with constructive behaviours that can become habits.

Example of an Autopilot: *"What's the plan?"*

Here is an example of a practical Autopilot that produces an automatic helpful behaviour.

Whenever we meet a stressful situation, our Chimps naturally take over and often put us into an anxious state. The Chimp usually engages these anxious emotions and keeps focussing on the problem. This pattern can be broken, by programming your Computer with the automatic immediate response of asking, **"What's the plan?"** when faced with any stressful situation.

Example: *The Autopilot in action - Gordon and the burst tyre*

Gordon was driving to an interview for a job that he really wanted. As he drove along the road, he realised that he had driven across some broken glass. Two of his tyres had burst and the car was now immobile. Gordon's Chimp immediately went into meltdown and lost all perspective. The Chimp was focussing on the two tyres and the fact that the interview will be missed. The frustration, anger, anxiety or despondency that results could be overwhelming. However, Gordon had programmed his Computer to respond immediately to any setback with **"What's the plan?"**.

With this response, Gordon immediately accepted the situation, and by working with it, he formed a plan of action. He then engaged in a constructive pattern of working and kept his mind focused on the solution.

Finding and replacing Gremlins with Autopilots

Sometimes our unhelpful habits, behaviours or beliefs are obvious to us. At other times you have to stop and think about what you are doing or what your beliefs are, in order to find your Gremlins. Gremlins often work silently in our minds but have a great influence. For example, if we are regularly faced with a particular problem, we usually develop a learnt behaviour in response to this problem. If this behaviour goes unchallenged it becomes a habit. Habits can often be very poor coping strategies.

Example: *Leanne and her stressed Chimp*

Whenever Leanne's Chimp becomes stressed, it looks to her Computer for help. Her Computer has a Gremlin in it that states that: to relax you can have a drink of alcohol. Although this might relax Leanne, in the long-term it is a poor coping strategy for learning to relax. Leanne has to recognise the Gremlin and replace it with something more helpful, such as relaxation techniques or phoning a friend.

Finding Gremlins is not often so obvious or easy! The reason that it is often difficult to remove a Gremlin is because we don't usually recognise the unhelpful behaviour or belief. Also, they are often unique to us.

For example, if we had someone with low self-esteem, then they could have any number of beliefs that are contributing to this. These beliefs could include:

- I am not as good as other people
- Whenever I try something it doesn't usually work out
- I feel others are constantly judging me
- I believe that you have to have special talents to have good self-esteem
- I see myself as unattractive

This list is endless and could have some very unusual beliefs.

How do we find whether we are holding unhelpful beliefs?

The answer is to reflect on why you are getting the feelings that you are experiencing.

Once we recognise an unhelpful belief or behaviour, we need to replace it with a helpful truth or behaviour that resonates with you. In other words, ***Gremlins don't and can't go away unless you replace them with an Autopilot.***

Therefore, in the examples of the beliefs above we could find these beliefs by asking ourselves some searching questions and then replace the beliefs with some that are truthful but much more constructive.

"I am not as good as other people" could be found by asking what do you think is 'good' in another person? Your Chimp might answer, "If they are clever or look good". However, your Human might answer, "If they have sound morals and have a positive outlook". These are very different answers.

You now have the choice to decide which of the answers you want to work with: Chimp or Human? You might then find that, by choosing the Human answer, you do possess the qualities that you believe to be 'good'.

Key Point

A Gremlin can only be removed if it is replaced by an Autopilot.

Example: *Debra and her relationship*

Debra has had a run of unsuccessful relationships. She doesn't know if it is the men she is choosing or if it is something to do with her, or a bit of both. She now wants to explore what might be causing her relationships to fail. When we are trying to find Gremlins it can help to identify a pattern of unwanted behaviour. Her first step is to look back at her relationships and see if she can spot a pattern of behaviour that was unhelpful. She noticed that when her partner was late, she immediately started thinking that he was up to no good and deceiving her. She would then tell him this and explain that it was unsettling her. This might seem reasonable, but is it?

The Chimp puts Gremlins into the Computer

First, the partner might be someone who is not organised, and that is his prerogative. People offer themselves to us as they are, not to be criticised or moulded into how we want them to be. Debra can only help her partner to change if he feels it is a problem and he wants to change. Otherwise, she needs to accept him as he is or end the relationship.

This approach assumes that the problem is not within Debra herself. Here are two possible Gremlins. Her first Gremlin is in believing that if her partner is late then he could be deceiving her. Her second Gremlin is the coping strategy that she then uses to reassure herself: challenging him and explaining how she is getting upset. This challenging is unlikely to go down well with someone who might feel they are not being trusted or are being criticised. What could she do to break this unhelpful belief and behaviour? She can address the reason she feels vulnerable, which could be low self-esteem.

Alternatively, she might not be accepting the reality that many relationships do fail. However, starting with mistrust is very likely to lower the chances of success in any relationship. Maybe she has not accepted that for any relationship to succeed it must be based on trust.

Perhaps she has a hidden belief that she will never find the right partner or perhaps she believes that the right partner will have a role of always reassuring her. I am sure you can think of any number of reasons why Debra might be struggling and how she could turn this around.

Example: *Matthew and his search for a partner*

Matthew is unhappy because every time he meets a girl that he likes, he finds his Chimp is quite intense and worried about losing her. This often results in his Chimp sabotaging the relationship. He recognises his Chimp is intense but doesn't know how to stop it. Any unhelpful repeat pattern often points to a hidden Gremlin that is prodding the Chimp.

Matthew can ask himself, "Why do I become intense?" At first he might answer, "Because I am afraid of losing the relationship". That might be true, but he needs to search again because he hasn't answered **why he thinks he might lose the relationship**. This time he comes up with, "I think girls will only stay if I am constantly positive and fun to be with and this I can't do". Now he can check out his belief that unless he is always positive and fun to be with, then the girl will leave him. This is very unlikely to be true and is therefore a powerfully unsettling Gremlin. The Autopilot that could replace this Gremlin is: girls who love me will accept me as I am, and appreciate that some days I struggle to be positive in my outlook. This Autopilot might make Matthew feel more positive anyway!

The order of interpreting experiences

The Chimp is more powerful and faster to act than the Human. [40] [41] [17] Therefore, the Chimp always interprets information first and gets first chance to act. This explains why we often act before we think. The hallmark of our Chimp is its impulsive nature, without consideration of consequences. When our Chimps act impulsively, we often end up apologising sometime later. The same applies to emotional responses. When our Chimps act quickly, we can often overreact to a situation and become distressed. We usually gain some perspective after the event has passed.

The 'rules' of the mind

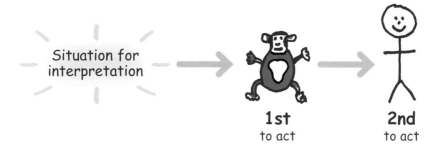

The order of action in the mind
- Scientific points

We have seen in Unit 1 that the thalami (T) send any received information to various areas within the brain, but most importantly, to the orbitofrontal cortex (OFC) first. The OFC then 'discusses' this information with different parts of the brain to gain some 'opinions' about what to do with it. The message therefore travels in various circuits and crucially passes through some significant areas before circling back to the OFC. During this circling through the brain, different areas contribute different 'opinions' and facts.

For example, the amygdala (A) helps to decide on whether we should choose fight, flight or freeze. The tract that joins the OFC and amygdala is called the uncinate fasciculus (UF). Somehow, this tract appears to give us a conscience, when we make decisions. It's no surprise that 'psychopathic' individuals have a small tract and appear to have no conscience about the actions they take.

The anterior cingulate gyrus (ACG) plays a big part in letting the OFC know whether something is good or bad and also the strength of how good or bad it is. The hippocampal formation (HF) stores factual memory and brings this into play and relays these facts back to the OFC.

At the same time as the limbic system is operating to make decisions, the message is also being sent, but more slowly, by the thalami to the dorsolateral prefrontal cortex (DLPFC). The DLPFC acts to bring rationality and logic to a situation, but because it is slow to act, impulsive decisions are often made before rationality has any influence. [5] [6] [42] [43] [44] [45] [46]

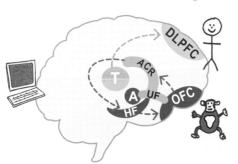

KEY

DLPFC - Dorsolateral prefrontal cortex
OFC - Orbitofrontal cortex
T - Thalamus
HF - Hippocampal formation
A - Amygdala
UF - Uncinate fasciculus
ACG - Anterior cingulate gyrus

To keep things simple, we will look at three different aspects that govern the way the mind functions:

1. **Power**
2. **Advice**
3. **Speed**

1. **Power**

There is a strict hierarchy when it comes to power. The Chimp is the strongest and is five times more powerful than the Human.[47][48] The Human has some power but not much! The Computer has no power.

Here is the power hierarchy:

Power

The **Chimp** can overpower the Human

Power is all about having the final say in decision-making. In the mind, power can be thought of as being the ability of one system to override the others using neurotransmitters or hormones. The Chimp can use these chemical messengers to block the action of the Human. This can account for why we want to think and behave in a certain way but often seem to do the opposite.

2. **Advise or influence**

Advice is in the hands of the Computer. The Computer can advise both the Human and the Chimp, and both of them must listen. They will find it extremely difficult to go against the advice.

Advice

The **Computer** can advise decision making for both the Human and the Chimp.

Advice is given in the form of beliefs, which either the Human or Chimp has previously put into the Computer.

The 'rules' of the mind

3. Speed

The speed at which each team acts goes in the order: Computer, Chimp then Human. The Computer is twenty times faster than the Human and four times faster than the Chimp. [49] [50]

Speed

The **Computer** will take over if it is programmed

Speed is determined mainly by two aspects. The first is the type of neurotransmitter used to pass on messages between neurones; these can be fast or slow transmitters. The second aspect is how myelinated (insulated) a tract is. This insulation will make the pathway faster to pass the message on. [51] [52]

Information entering the mind: *how this works in practice*

Please refer to the diagram below as you read this passage because the diagram will help you to visualise the order of the steps by which the mind operates.

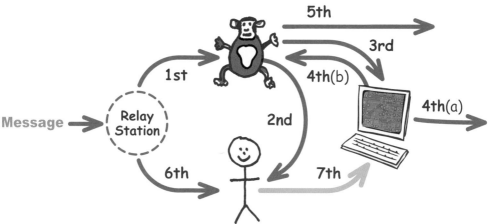

1st Step

When information enters the brain, the relay station first sends the information received to the Chimp. [5] [6]

2nd Step

The Chimp now does something remarkable. The first thing it does is to stop the Human from acting. It effectively knocks the Human out with a sledgehammer! [53] [54] [40] It does this, so that the Human does not interfere with what the Chimp needs to do. The Chimp needs to assess for danger and then act quickly, if it perceives any. [55] The Human, being slow to act, could hamper any emergency action that needs taking.

3rd Step

The second thing the Chimp does, having removed any Human interference, is to look into the Computer for any advice. At this point, we can see that the Chimp is in charge and is demonstrating its power.

4th Step

When the Computer is given the information by the Chimp, it has a choice to do one of three things:

1. If the Computer recognises the information from the Chimp and it is programmed to act automatically, then the Computer will take over and act immediately. [33]
   This is step 4a.
2. If the Computer has advice programmed into it, then it will feedback this advice to the Chimp. The advice usually takes the form of recalling previous experience, offering some context to the information or supplying truths and beliefs about the information. This advice could be helpful (an Autopilot) or unhelpful (a Gremlin) or it could come from the Stone of Life. **This is step 4b.**
3. If the Computer has no advice relating to the information it has received, it will *remain silent and hand back the information* to the Chimp.
   This is also step 4b – Handing back to the Chimp.

5th Step

The Chimp now acts, if the Computer has not taken over.
   All of these actions happen in less than a fifth of a second. [56] [57]
The outcome is that either the Computer acts or the Chimp acts. The only time the Human acts early, is if the Chimp doesn't feel the need to act. In this case, the Chimp goes silent and the 6th step comes into play.

6th Step

The Human has a chance to act, but only after the Chimp has decided not to act or has already acted. If the Chimp has already acted, this often leaves us apologising for what the Chimp has already said or done!

All of these arrows are fixed into place and is the way the mind will work. There is one final *optional* arrow. If we actively choose to programme the Computer with what we want to happen or what we want to believe, then arrow number 7 comes into play.

7th Step: *THE MOST IMPORTANT ONE!*

The 7th Step is *an optional step* and one that many people neglect. **This is the step that you can use to manage your mind.** It is all about programming the Computer.

If your Computer is programmed effectively, and cleared of Gremlins, then when the Chimp turns to the Computer, the Computer can take over. It can carry out your wishes and also settle the Chimp down. You can move directly to step 4a. This programming is something that you can do in your development time.

It might have struck you already, but there are no direct arrows going from Human to Chimp. The implication is crucial. It leads us to a most important key point.

Key Point

We CANNOT CONTROL the Chimp; we can only MANAGE it

As the diagram for managing the mind is quite complex, but very important, we will look at some scenarios to bring the process to life. These scenarios are all based on one simple example, but each scenario has a different programming of the Computer to demonstrate the steps in action.

Scenario 1: The Computer has no advice and isn't programmed
Scenario 2: The Computer has mixed advice from the Human and Chimp
Scenario 3: The Computer only has helpful advice or has been programmed to take over

How to operate within the rules

The example: *The failed lightbulb*

Terry has arrived home from work and noticed that one of the lightbulbs in his living room has failed. There is enough light from the other lightbulb fitting to light the room. Terry, from his Human, wants to change the lightbulb straightaway because it needs doing and there is no reason not to do it. Terry's Chimp knows changing the bulb will be a hassle. It would rather put off changing the lightbulb and sit down to watch the television. Terry's Computer is waiting to be woken up and then it will advise or act.

Scenario 1: *When the Computer has no advice or is not programmed*

The Chimp receives the information that a lightbulb needs changing: the 1st step. It prevents Terry's Human from thinking and then turns to the Computer: the 2nd and 3rd steps. The Computer has no advice to offer because neither Human nor Chimp has put any advice in. The Computer can't take over and act because it doesn't have a programme to do this. Therefore, the Chimp decides what to do, and sits down to watch television. There might be short-term gains from this, but in the long-term the task still needs doing.

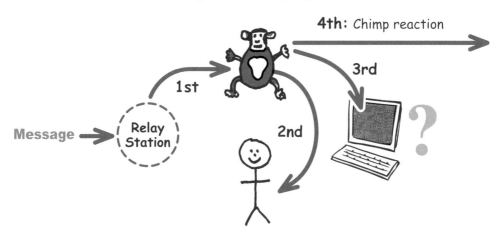

The Chimp can become unpredictable when the Computer is not programmed

Scenario 2: *When the Computer has mixed advice from the Chimp and Human*

If Terry has met this situation before or thought about it, then he will have put thoughts into the Computer.

In this case, after the Chimp has prevented Terry's Human from thinking, it will turn to the Computer to be given advice. This time the Computer has beliefs in it. The Human has placed some constructive and helpful beliefs (Autopilots) and the Chimp has put in some unhelpful beliefs (Gremlins).

In Terry's case, these are:

Autopilots
- When I act immediately I always feel better in myself
- When I act immediately, I don't have jobs hanging over me

Gremlins
- I can always put things off and usually I can get away with putting things off
- I need to feel in the mood to do things, and if I don't feel in the mood, then there is no point in doing anything

With these mixed messages being fed back to the Chimp, the Chimp chooses to listen to what will give it instant gratification. The Chimp chooses to put things off and sits down to watch television!

The 'rules' of the mind

The Chimp can still be unpredictable if the Computer is offering mixed advice

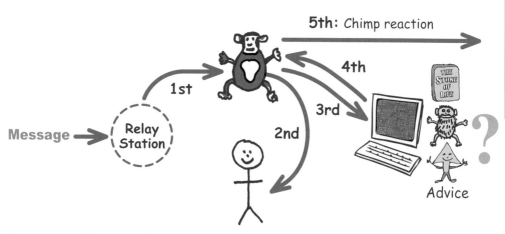

Scenario 3: *When the Computer only has helpful advice from the Human or has been programmed to take over*

In this scenario, the Chimp will still act but it will act in a very reasonable manner because the Computer will advise it appropriately.

The Chimp will become settled and has a reasonable reaction if the Computer has suitable Autopilots

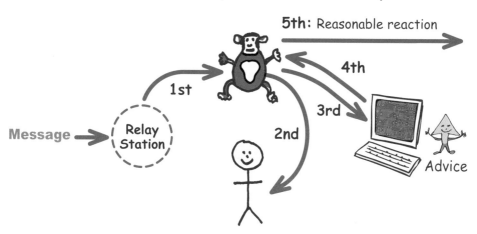

Terry has cleaned up his Computer by adding Autopilots and removing Gremlins:

Adding Autopilots that can apply *to all* unexpected events:
- I will only focus on the process for resolving any problem
- I am proud to be someone who is proactive and then relaxes
- I always feel better when jobs are out of the way

Removing Gremlins

Terry will do this by replacing them with Autopilots.

Gremlin: *I can still see, so I will change the lightbulb later.*

Autopilot: I see it as being lazy to put off what you can do now – I don't want to be lazy.

Terry's Chimp doesn't like the idea of being seen as lazy, so this drives his Chimp to act.

Gremlin: *I need to feel in the mood to do it and now I feel like I need to watch television.*

Autopilot: it's not about whether I feel in the mood, it's about doing what needs to be done.

The tidy Computer that is also programmed can take over. The Chimp would not then be involved. Therefore, whenever an unexpected event happens, the Computer will act immediately and automatically. This results in blocking the Chimp and Human from thinking.

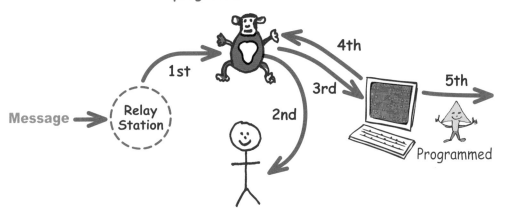

Does the Human ever act first?

Yes, but only by default!

The Chimp won't act if:
- The information is of no relevance to it
- The information poses no threat or concerns
- The Chimp has no agenda

If the Chimp has no interest in what it is being offered, then it will stop in its tracks. Therefore, the Chimp receives the information first but stops all actions. This means the message can now go directly to the Human. By default, the Human can now act first. The Human then consults with the Computer for any relevant advice. If the Computer knows what to do, it will take over. Otherwise, it will hand advice to the Human and the Human will then change the lightbulb, taking this advice into consideration.

The Human goes first if the Chimp is not interested or alerted

Implication of the rules of the mind

If you reflect on the way the mind receives information and the rules by which it works, then you can see just how important the Computer becomes in helping us to manage our behaviour, thinking and emotions.

The Computer will be considered in detail, as we progress through this mind management course, but first let's look at how we keep the Computer in good working order.

Tidying up the Computer

The Human and Chimp constantly put beliefs and behaviours into the Computer for storage. These will be used to advise both Human and Chimp. They also allow the Computer to take over, if it is programmed to act spontaneously, such as being able to drive the car to work without thinking about it.

How do we remove a Gremlin that the Chimp or Human has put into the Computer?:
- First, we need to recognise what unhelpful beliefs or behaviours are present
- Then we need to decide what beliefs or behaviours we want to replace them with

Finding true statements that challenge unhelpful beliefs is very effective at removing them.

Example: *David and the wedding speech*
David is the best man at a wedding and he is dreading giving the speech. If we look into his Computer, we might find the following two unhelpful beliefs:

1. I will make a mess of this
2. People will think I am incompetent

He can now tidy up the Computer and remove these Gremlins by using some truths. He can address his Gremlins and replace them with the following Autopilots:

- If I do make a mess of the speech, then at least I did my best. I am an adult and I can get over it
- Reasonable people don't care if I make a mess. They understand that most people are not gifted speakers. Unreasonable people are not worth worrying about

If David believes these statements to be true, then his fears will now subside, as he has replaced his Gremlins with the new Autopilots.

> **Key Point**
> *An important reminder: A rule of the mind is that a Gremlin can only be removed by replacing it with an Autopilot.*

Unit 2 Reminders

- The Chimp and Human can both:
  - Store beliefs in the Computer
  - Manage the Computer

- Gremlins can be removed by replacing them with Autopilots

- We cannot **CONTROL** the Chimp but we can **MANAGE** it
- We can manage the Chimp by programming the Computer

The 'rules' of the mind

Reminders Continued

STAGE 1

The two functions of the Computer

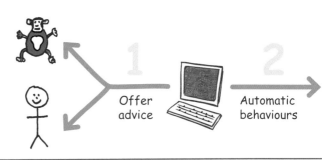

1 — Offer advice
2 — Automatic behaviours

The order of the steps by which the mind operates

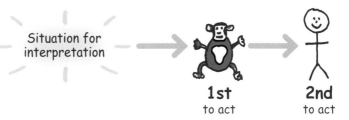

Situation for interpretation → Chimp (**1st** to act) → Human (**2nd** to act)

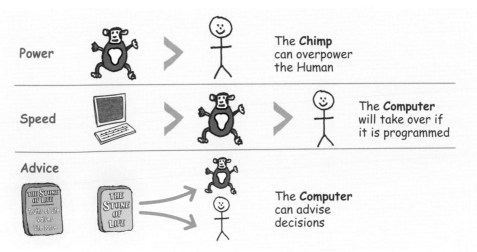

Power — The **Chimp** can overpower the Human

Speed — The **Computer** will take over if it is programmed

Advice — The **Computer** can advise decisions

The order of the steps by which the mind operates

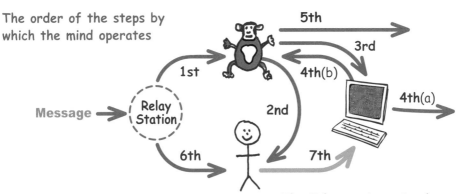

Message → Relay Station → 1st (Chimp) → 2nd (Human) → 3rd → 4th(a) → 4th(b) → 5th → 6th → 7th

The 7th step is optional, but is the most important step

Unit 2

Focus points and reflective exercises

1. *Finding and replacing Gremlins with Autopilots*
2. *Establishing Autopilots into the Computer*
3. *Appreciating how the system works*
4. *Application of the rules*

Focus 1: *Finding and replacing Gremlins with Autopilots*

Exercise: *Searching for and replacing Gremlins*

This exercise is for you to try to uncover beliefs or behaviours that are not helping you and then to look for alternative constructive beliefs and behaviours (Autopilots) to replace these Gremlins.

Look for times when you experience negative emotions and ask yourself:
- Is there a pattern of behaviour that I am repeating that is unhelpful?
  - What alternative behaviour could I try?
- What beliefs am I holding that are creating these emotions?
- Are my beliefs and expectations in a situation realistic?
- What beliefs and behaviours would I like to have that will make me feel better?

Focus 2: *Establishing Autopilots into the Computer*

Exercise: *Try out the "What's the plan?" Autopilot*

One rule of the mind is: The Chimp will react and take over if the Computer is not programmed for action or it doesn't have some sound advice to offer the Chimp. Try programming your Computer with the Autopilot of behaviour that asks, **"What's the plan?"** whenever you have a setback or are in a stressful situation. To do this, you need to recognise when something is troubling you or there is a problem. Once you have recognised this situation, you need to immediately ask yourself **"What's the plan?"**, and cease engaging with Chimp emotions.
By asking this question, you will be engaging your Human circuits to begin finding solutions to move forward with.

By practicing this change of behaviour, it will become the norm for you to become immediately proactive with a plan. If you always immediately ask, **"What's the plan?"**, you are very likely to find that you save a lot of time and energy, which would otherwise have been wasted on engaging with unhelpful emotions.

Focus 3: *Appreciating how the system works*

Below is a diagram of how the system would work in a specific situation. The example shows a scenario with some possible thoughts held by the Human, Chimp and Computer.

Example: *The Human wants to tidy up the house but it isn't happening!*

This simplified diagram is to demonstrate the principles involved. The Human and Chimp are on a different page. The Computer has both Gremlins and Autopilots in it. When the Chimp asks the Computer for advice, it will become confused because there are mixed messages. The Chimp has a ***choice*** of what it will listen to and it is likely to choose the Gremlins. Therefore, the house remains a mess.

We could challenge the Gremlins to see if they are true or helpful. For example, is it really true that we have to be in the right mood to tidy a house? The truth is much more likely to be that regardless of what we feel, once we get started, we will feel better and will get on with the job. This is a helpful and true Autopilot that can replace the Gremlin.

Exercise: *Completing the thinking*

Kieran wants to learn a new language, but he is struggling to get started.

Try to work out what each of his systems is likely to be thinking in the following scenario and then think of ways to replace the Gremlins.

Focus 4: *Application of the rules*

Exercise: *Applying the principles to yourself*

Reflect on your own situations, where you feel your Chimp has hijacked you. Try to elicit what each of your three systems is saying. This is likely to show you why you are being hijacked. It might help if you draw the systems out on paper. It is likely that each system will have several things to say.

The question to ask yourself is: "***What state is my Computer in?***"

- Is it lacking any advice and any programming?
- Does it have mixed advice from the Human and Chimp?
- Does it have only helpful advice and is it programmed to take over?

If you find any Gremlins, then replace them with Autopilots.

Unit 3
Developing yourself

Unit 3: *is about developing your Human. We will consider the advantages and great importance of choosing to operate in Human mode. We will look at how to do this by establishing your options. Finally, we will look at how the Human and Chimp systems can work effectively together.*

The advantages of choosing to operate in Human mode

Imagine the case of two hypothetical people:
- The first person is always emotionally stable, thinks rationally and is secure within themselves
- The second person is always emotionally unstable, thinks irrationally and is insecure within themselves

Emotionally unstable would include all aspects of being sabotaged by their own thinking, emotions or behaviours. Examples of this could include worrying unnecessarily, repeating destructive behaviours, always thinking negatively or becoming unhelpfully frustrated.

With these fixed states of mind, they present as two very different people.

Two states of mind

Emotionally stable
Rational thinking
Secure

Emotionally unstable
Irrational thinking
Insecure

Who is likely to succeed in life?

If we ask the question "which of these two people is likely to succeed in life?" most people would probably say, "Steve, the answer is obvious. The first person has a greater probability of success than the second person". Stop and think about what has been said. We have said that the person with a stable and secure mind is more likely to be successful and yet *we have not yet identified what it is they are trying to be successful at*. What is important here is that this holds true no matter how we define success or what we are trying to be successful in. Whether the goals are personal, professional, individual or in teamwork, it really doesn't matter. The first person is likely to do better than the second person.

It also holds true that the first person, being stable, is much more likely to be happy, content, confident and secure. Trying to succeed in business ventures, relationships, sport, healthy living, and anything else you would like to name, does not just rely on a good plan. It relies mainly on a stable person implementing the plan. [58] [59] [60] [61] [62] [63] [64] [65] [66] [67]

How do I get my mind to become stable and secure?

If someone asked me "How do I become successful, happy, confident or get a better quality of life?", I would suggest that they could ask a different question. A better question would be "How do I get my mind to become stable and secure?". This is because success, happiness, confidence and an improved quality of life are all more likely to happen when you have a stable and secure mind.

Occasionally, there are times when being unstable might help us to change, but it seems self-evident that the stable person is more likely to be productive and to deal with day-to-day situations more constructively. It is true that it is possible for unstable people to succeed, and some do. What we are looking at is *how to increase the probability of success.*

We can put in place the best processes and plans in the world and the most efficient systems but if the people who engage these processes or systems are inherently or intermittently unstable then the chance of success diminishes. Naturally, we want good processes and good systems in place, but we need the right people in the right state of mind to operate them. Even poor processes and systems can be made to work, when they are being operated by the right people in the right state of mind.

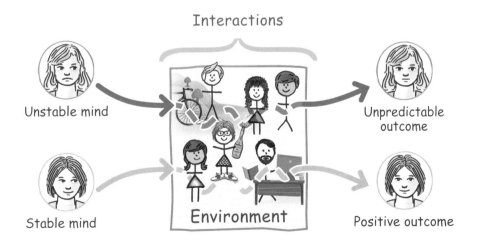

The ideal situation would be to have a stable person entering a great environment.

The self-evident factor, therefore, in raising the probability of success is the stability of the person themselves. This is so often overlooked.

> **Key Point**
>
> The critical factor in raising the probability of success is the stability of the individual person entering any process or system.

Get into a good place before you begin.

 MOVE

Unstable   Stable

But..... how do I make the move?

Earlier, we compared two hypothetical people, one was psychologically stable, and one was psychologically unstable. Obviously, they don't exist. Nobody is always stable, and nobody is always unstable. We move between these two states on a regular basis. These two hypothetical people actually represent the two thinking and analysing systems that operate within our minds: The Human and the Chimp. These two systems continually fluctuate, with nature dictating which system operates, *unless we intervene*. For example, nature will default our mind to the Chimp system if any 'threat' is perceived.[68]

Two fluctuating states of mind arising from two different systems

Unstable   Stable

If we learn to switch from the Chimp system into the Human system, we will have a way of shifting from an emotionally unstable mind to a stable one.

> **Key Point**
>
> Working in Human mode can put us at great advantage for being successful.

Example: *Planning some work*

Let's say that you are about to do some work and your Chimp is raring to go. If the Chimp leads, then it's likely that the process for doing the work won't be very well organised. It might not be in a logical order and your Chimp will not consider the long-term outcome. By stopping and resetting into Human mode, we can plan the best approach and consider what problems we might meet further down the line.

Example: *Henry and the difficult conversation*

Henry was about to confront his neighbour and knew that the conversation would be difficult. Henry wanted to let the neighbour know that the hedge between them needed cutting back, as it was growing onto Henry's drive.

If Henry isn't careful and approaches this in Chimp mode, then Henry's Chimp is likely to choose very emotional words, emphasise its annoyance and make sure it wins the conversation. Henry himself would just like to resolve the problem and be on good terms with his neighbour. If Henry stops and puts himself into Human mode before beginning the conversation then the outcome will most likely be a good one and Henry will still be on good terms with his neighbour. It only takes a few moments for Henry to stop and think things through and then be in the right mode before the conversation, to prevent his Chimp from hijacking him and causing unnecessary friction.

When you are about to have a difficult conversation, do you start it in Chimp mode without realising it? By reflecting before you begin, you can quickly switch to Human mode. You can then think about how you want to be during the conversation and consider how you want to approach it. If you remain in Chimp mode, then spontaneity might feel good at the time but have unhelpful consequences afterwards.

Stability is achieved by switching from Chimp to Human mode

Unstable → Stable

Brain activity
- Scientific points

Brain scanners can show us which part of the brain is working by 'lighting up'. This 'lighting up' (with bright colours on the scanner diagram) shows which part of the brain is using up oxygen from the blood supply. Using brain scanners, we can therefore see when the Chimp system is working. At that time, we are much more likely to experience feelings and thoughts of insecurity, have less rational thinking, have judgement based on feelings and be emotionally unstable; for example, we could overreact, have mood swings or feel anxious. [69]

Diagrammatic representation to illustrate how we can see the different parts of the brain working

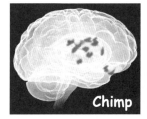
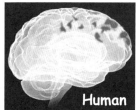

If we can change from operating with the Chimp system to operating with the Human system, this will show up on the scanner. When this happens and the Human is working, we show increased emotional stability and confidence, work with rational thinking and show sound judgement. [8]

Although working with the Chimp system can have its advantages, working with the Human system is much more likely to be beneficial. By learning to switch to the Human system and to operate with this system, we will be developing our Human.

Establishing your options

One way to switch systems from Chimp to Human is not only to recognise that we have a choice, but also to *know what that choice is*. We can find what our choice is by establishing *how we would like to operate*, as opposed to *how our Chimp might want to operate*, thus giving us that choice.

When we are in Human mode we can step back and look more objectively at how we would like to be. When we are in Chimp mode we simply engage with the emotions that situations evoke in us.

Recognising two options and making a choice will help to develop our Human system.

A fundamentally different approach

One way to significantly develop your Human is to recognise the option you have between 'reacting' and 'responding' in life. This point makes such an important distinction between the Human and the Chimp. It needs some reflection to fully appreciate it. At the end of this Unit there will be an exercise based on this principle. If you can grasp and recognise this principle in action, it can transform the way that you deal with life. [70] [71] [72] [73]

Chimps react – Humans respond

Impulsive emotional reaction

Acceptance with a plan

Example: *Cathy the runner*

Cathy is a runner who enjoys half marathon races. She was training for an important race, when her calf muscle suddenly tore. If she is in Chimp mode, she will *react* to the situation. She may fall into despair, become angry, be distressed or think catastrophically. All of these are a *reaction* to the situation, which is very natural but is unlikely to help.

If she is in Human mode, she might still be disappointed, but she will *respond* to the situation by first accepting the reality and facts, and then forming a plan of action to manage the injury. Once she has a plan in place, such as to rest, get the muscle assessed and re-adjust her training plan, her Chimp will settle down.

During the rehabilitation, if Cathy returns to Chimp mode, then she is again likely to experience some typical Chimp reactions. These could include: over doing things, becoming despondent or obsessively searching out different ways to get her leg to heal more quickly. The Chimp is likely to keep comparing where she is, to where she wants to be, and become impatient and frustrated.

This example demonstrates the very important principle that there is a world of difference between 'reacting' and 'responding' to events and situations. Responding means accepting the situation and then forming a plan to find a solution. *'Accepting' doesn't mean rolling over! It means establishing and working with the facts that are in front of you.*

Example: *Gerry and the MOT failure*

Gerry has booked his car in for its MOT. The following day he has an early start setting off in the car for a holiday with his family. Everyone is excited. At the end of the day he goes to collect his car and is told that it has failed the MOT and needs immediate urgent work. His wife, who is standing next to him, takes a deep breath in preparation for Gerry's Chimp's reaction.

Gerry's Chimp explodes. He expresses several irrelevant things: "Why is this piece of junk ruining my holiday? Why is this always happening to me?". The Chimp concludes: "I never wanted to buy this car in the first place. There will be no holiday this year".

Instead of this reaction, Gerry could have remained in Human mode and responded with dignity. He could have remained calm, accepted the situation and immediately put a plan in place to manage it. It isn't the end of the world.

There is a compromise: Gerry could exercise his Chimp by allowing it some controlled expression for a few minutes of managed expletives into the air, and then move into Human mode. Only Gerry knows the best way to manage his Chimp. An emotionally skilled individual can reach a point where the Human mode becomes automatic.

Developing yourself

Emotional situation → Exercise the Chimp for a few minutes → Allow Human to respond

Key Point

Chimps react: Humans respond.

Starting from where you are

The Chimp is by nature driven to protect us.

For most of us, this means we have an impulsive Chimp that wants to begin at the end point and not the start. Therefore, the Chimp will compare where we are, to where it wants to be, and compare what it wants to have to what it has actually got. It will therefore always perceive a lack. Our Chimps can then become despondent, frustrated or agitated.

The Human starts from where they are and what they have, and then makes plans to get to where they want to be. This is a constructive approach that usually enthuses us.

Key Point

Start from where you are and what you have got.

Example: *Jane and the hair salon*

Jane has recently started her own small hair salon. She eventually hopes to have regular customers with a turnover of around 12 people a day. She has currently built up to just three customers a day. If her Chimp looks at this situation, it will look to where it wants to be and only see how far off she is. The Chimp then becomes agitated by the current situation and feels worried that it might not get there.

If Jane switches into Human mode, then she will accept that where she is at the moment is fine in the scheme of things. In Human mode she will then make plans to increase her clientele, in order to reach her target.

Developing yourself as a person by choosing Human mode

Having established that it would be helpful to be in Human mode most of the time, and that we have a choice, how do we manage to do this?

We have seen that one way we can move into Human mode is by actively deciding on what we want to do and how we want to be. We first *create our options* and then *carry out our choice*. This helps our Human develop and helps us to move into Human mode.

'Developing your Human' effectively means being yourself, without any unhelpful interference from either Chimp or Computer.

> **Key Point**
>
> **The starting point for developing yourself, is to dissociate yourself from your machine, and work out how you want to feel, think and behave.**

As we have discovered, the Human within you begins with a blank piece of paper and writes down how they want to be. Therefore, to develop yourself, you need to make decisions on how you want to be, **BEFORE** you address the Chimp or the Computer's contributions.

Example: *Toni and her reflections*

Toni says that she reacts in life in a catastrophic way and presents herself to the world as someone who is overly anxious.

Toni's first step is to dissociate herself from her machine and stop blaming herself.

She does not overreact; it is her Chimp that is overreacting and being catastrophic. Her job is to manage the Chimp and prevent it from becoming distressed. After acknowledging that it is her Chimp, she can now work out how her Human wants *to feel and behave* in any situation. She could, for example, state that she wants to be calm and take things in her stride. *This defines what she is truly feeling and how she would be behaving, if her machine had not taken over.*

Toni has now established the two options to choose from:
1. **Her Chimp:** catastrophic and anxious
2. **Her Human:** calm and taking things in her stride

In order to go with her Human's choice she needs to manage her Chimp. She can do this by programming her Computer. This, we will cover in great detail, as we go through the course. The important point for now is that two options have been clearly established.

> **Key Point**
>
> **When developing yourself, it is important to recognise that there are always two distinct options open to you: what you want and what your Chimp wants.**

How the Human and Chimp systems can work effectively together

Although we can choose to work in either Human or Chimp mode, if we can get our Human and Chimp to work together, we are likely to obtain a better outcome.

Situation   Best solution

Can being in Chimp mode be a good thing?

Before we consider the Human and Chimp working together, it's worth asking a question: As our minds are frequently defaulting into Chimp mode, "*is there any time when being in Chimp mode is a good thing?*" The answer is definitely "yes". The Chimp can offer so much.

Here are two helpful aspects of our Chimp:

1. **Scanning for danger:** The Chimp constantly scans for danger, which is of great help to us. However, it often sees danger where there isn't any, to the point of becoming neurotic or paranoid and unhelpful!
2. **Intuition:** The Chimp can use 'intuitive' judgement [15] [74] [75], which can sometimes be just as accurate and helpful as 'rational' judgement. [31]

Example of intuition: *The new home*

When we first go to buy or rent a property, it is often our intuitive feelings that let us know whether this could be a future home or not. Although our impressions are based on our perceptions and some facts, it is hard to pinpoint why some dwellings feel warm and others not so friendly. Subtle factors, such as the way the sunlight falls into a room, creating an atmosphere, or the shape of each of the rooms, could make our Chimp feel more comfortable or less comfortable. The Chimp will intuitively notice such details. Often, we cannot quite put our finger on what it is, but we just know that there is something advising us from within.

This first impression can also be accurate when we meet new people. It all depends on whether or not your Chimp is good at using intuition and how correct its feelings are. Some Chimps are amazing at this, and others appear to be clueless! You have to know your Chimp.

Example of the Chimp being helpful: *The deadline and pressure*

Our Chimps, and often our Humans, like deadlines. Remus is about to take an exam. If he is in Human mode, he will commit to study on a steady basis and shouldn't have any problems, such as having to cram as the exam approaches. However, Remus's Chimp has been in charge for most of the course and the steady work ethic just hasn't happened. Now with just two weeks to go, the Chimp has gone into melt down. Its panic buttons are pressed, and it will now bring focus to the situation. It is Remus's Chimp that will effectively force him to study intensively (perhaps over doing it) until the exam arrives. Obviously, this is far from ideal, but it works for many of us! The Chimp both creates and solves the problem.

The Chimp both creates... and solves the problem.

When under pressure from any source, most of our Chimps will act and compel us to commit to tasks. This is because they will respect deadlines. Generally, Chimps hate to fail and characteristically attach great importance to any form of assessment or competition. The Chimp also works hard when it is made to be accountable. Therefore, being in Chimp mode can have its advantages and we will see a lot more of these advantages as we continue to understand our mind.

The Human and Chimp working together with decision-making and problem solving

There are two aspects to problem solving:

1. The recognition of the problem
2. The solution to the problem

The Chimp and Human take different roles when working together. The Chimp's role is to recognise that a problem is occurring and to keep focussing on that problem until the Human comes up with a solution and acts on this.

The Chimp will alert the Human to the problem by using emotions and feelings. These can be very uncomfortable to experience, but they alert the Human and prompt it to act. The Chimp can be persistent, but this is the Chimp doing a great job and therefore uncomfortable emotions when making decisions ought to be welcomed! The Human's job is to solve the problem with a solution by accepting the reality of a situation and then constructing a plan. This solution and plan will reassure and settle the Chimp.

If the Human remains inactive or doesn't come up with a solution, then the Chimp will continually focus on the problem. This means the Chimp will continue to produce uncomfortable emotions and thoughts, until the Human acts or the problem goes away by itself.

The mind creating awareness of a problem
- Scientific points

When the orbitofrontal cortex (OFC) receives information it communicates directly with many areas of the brain to search for warnings. Two specific areas help with this. The first is the ventromedial prefrontal cortex (VMPFC), which is a store for emotional experience. It is the home of intuition. When it receives information from the OFC, it checks to see if it recognises a pattern and then alerts the OFC to this pattern. The VMPFC also lets the OFC know whether this recognised pattern will result in a good outcome. At the same time, a second area, the anterior cingulate gyrus (ACG), detects conflict and anomalies and reports these back to the OFC.

The OFC functions on the basis that if something works then it will keep repeating it. When it doesn't work then it will change its reaction and try something else. The information the OFC receives from the VMPFC and the ACG, as well as other areas of the brain, can cause it to change how it will react to a situation; otherwise, the OFC just keeps repeating the same behaviour.

The VMPFC and the ACG are part of the Computer feeding back to the Chimp, the OFC. [43] [78] [79]

The VMPFC and the ACG are part of the Computer feeding back to the Chimp, the OFC.

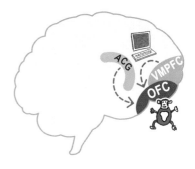

KEY

 - Ventromedial prefrontal cortex

OFC - Orbitofrontal cortex

 - Anterior cingulate gyrus

Key Point

Research shows that the best way to solve problems or make decisions is when the Human and Chimp work together. [47] [76] [77]

Example: *Rebecca and the study day*

Rebecca has to complete an online course on diversity for her work. She plans to get this done by working from home for a day. This seems to be a great plan. Rebecca knows the course has to be done and she will feel good when it's completed but her Chimp usually manages to sabotage her attempts to complete the course. Therefore, when she sets out her plans, she needs to have actions to manage her Chimp. These must resonate for her and keep her Chimp happy, so it's worth her consulting her Chimp to know what it will and won't do. Only she can determine how she does this because everyone responds differently to plans. Some things that might work for her could be:

- Rebecca could involve a friend and report back to her friend on her progress, as she does the course. Many Chimps don't like to be seen in a poor light by others. Therefore, by having to report back, the Chimp will take the lead and drive the work in order that it looks good.

- Rebecca could break the course down into small chunks, which her Chimp feels able to do. Asking our Chimps to do something that is overwhelming them will very likely lead to avoidance. Rebecca could for example, let her Chimp know that she will do just 20 minutes to get started then have a break for 10 minutes. With this plan of having a break after a short time, many people find that their Chimp will now engage with the task because their Chimp feels able to do just 20 minutes.
- Rebecca could find that she can do the task while listening to music that she likes. Some Chimps will tolerate doing something if there is pleasure attached to it. However, for some the pleasure becomes a distraction!

Example: *Getting fit*

Most of us want to be fit and do some form of exercise. The reality is that many of us don't carry out our plans. The reason is usually because we have not consulted our Chimps. We need to be realistic about what they will and won't do and make suitable plans to get the Chimp on board. Here are some ideas:

- Chimps like to work to routine. If we have an unstructured lifestyle then the idea that the Chimp will summon up the energy to exercise, when an opportunity arises, is unlikely to happen. Fixed routines can help to get the Chimp on board. If a fixed routine isn't possible then we have to think again.
- An alternative method is to keep a record of exercise sessions. For many of us, this will increase the chance of the exercise happening because Chimps like to see achievement.
- Joining a walking group or jogging group or a gym class is likely to work because nearly all Chimps are gregarious and love to socialise. Working out with a few friends is more likely to work than going it alone.

Don't react; respond: *The NEAT process*

There is a world of difference between reacting to a situation and responding to it. Chimps will naturally react to situations. However, remember that you do have a choice of which mode you want to operate with, so don't be fooled into thinking that you have to go with what comes naturally. Natural doesn't always equate to helpful.

Reacting is a Chimp in full flow. When we react to situations, usually on impulse, we haven't thought through the consequences and the outcomes become less predictable. Reacting to a situation doesn't usually result in the best outcome (there are of course exceptions). The Chimp will often continue to act on this initial reaction, which often results in the escalation of a problem.

Responding to a situation (accepting it and then having a plan to deal with it constructively) is the Human at their best. Most situations have better outcomes with this approach because it is based on fact-finding and rational thinking.

Assuming that you want to respond in future, rather than react, to challenging situations, then having a process to follow would help.

One process that you could try is the '**NEAT**' process.

N = Normal
E = Expect
A = Accept
T = Take care of it

When you recognise your Chimp is reacting or trying to react, begin by asking yourself "Is this normal *FOR A CHIMP*"?

The answer is always "yes". Anything is '**Normal**' and healthy for a Chimp. It might not be helpful, it might even be immoral, but whatever your Chimp does is normal for a Chimp! It doesn't help to judge the Chimp; only be concerned with managing it. This means that you can allow some managed expression of emotion, which will probably help your Chimp. Start by seeing the Chimp reaction as healthy and normal.

'**Expect**' your Chimp to react. It is doing its job. *The Chimp has no other way of engaging with a situation.* There are no surprises that our Chimps react impulsively, so expect to have hijacks and reactions from the Chimp.

'**Accept**' the way your Chimp is reacting. It is the nature of the Chimp. Acceptance means working with your Chimp's reaction. By accepting that it does react to situations, we can then work with this.

'**Take care of it**'; finally, make your plan to take care of the situation in the manner that you want to, rather than the way your Chimp might try to take care of things!

> **Key Point**
>
> *Don't try to stop the Chimp from reacting, but focus on preventing it from acting out its reaction.*

Example: *Responding not reacting*

Jay has a healthy Chimp. One of its features is to react very quickly in an aggressive manner. Jay has learnt that this is unhelpful but to be expected. An incident occurred one day, when Jay received a critical and unjust comment about his work from a colleague. Jay's Chimp immediately became poised for an attack. However, Jay's Computer had now been programmed to stop the Chimp in its tracks by immediately thinking: respond don't react. Jay had programmed in an automatic response to: take a deep breath, retain your dignity and calmly point out what has disappointed you, then to wait for a reply. Jay was able to manage his Chimp in this way.

Developing yourself

Example: *Tanya and perspective*

Tanya's Chimp demonstrates a characteristic of all of our Chimps: a lack of perspective. Whenever something unwelcome happens, her Chimp overreacts. Tanya recognised her Chimp overreacting and programmed her Computer to respond. This response was to stop, accept the situation and then look at the bigger picture. To look at the bigger picture, she would ask herself, "Will this matter in a year's time?". Tanya was late getting ready for work one morning and missed her bus, which would make her late for work. Her Chimp was about to react, when her Computer intervened. The Computer recognised the Chimp's distress but responded with "Accept the situation" and "Will this situation matter in a year's time?". The answer was "no". Having calmed the Chimp, the Chimp then began to challenge her reasoning, with, "But it does matter today and I might get the sack".

Tanya's response was simple, "I can't change the situation but I am an adult and I can deal with any consequences that might arise". As adults, this is true for all of us.

Tanya then formed her plan to prevent the Chimp from continuing with its worries. She distracted her Chimp by focussing on her journey to work and planning her weekend. This example shows how we need to get to know our Chimps and to know what truths to put into our Computer that will resonate specifically with us. It's all about programming the Computer!

Start from where you are and what you have

When we try to achieve anything, we naturally look to the outcome to see where we are going. Whatever you are trying to do, no matter how large or small a task, check to see if you are starting from where you are and what you have. This Human approach of accepting the situation and moving forward is much more likely to energise you than trying to start from the finished task.

Example: *Paying off a debt*

Jen had to borrow money when an emergency came up. She hadn't been managing her money well up to this point and already had a lot owing. The extra debt had pushed her Chimp into a panicked state. The Chimp could only look at the debt and see how much was owing. It wanted to be in a position of no debt and kept comparing Jen's present situation to what could be, if she didn't have any debt. This distressed her Chimp and brought feelings of dread and sleepless nights. If Jen remains in Chimp mode, she will always see the debt and how far she is from being free of it. This won't provide peace of mind.

By changing into Human mode and starting from where she is, Jen can accept that the debt is heavy but she is now starting to remove it. As long as she has a plan, she can stay in the moment of where she is and be happy with what she has paid off. By staying in the moment and starting from where she is, Jen can obtain peace of mind.

Example: *A small task but a necessary one*

For some of us, keeping our home tidy can be quite daunting, particularly if we have let things go a little or a lot! The reason many people then struggle to get started is because their Chimp imagines the finished tidy home and realises just how much work is needed to get there. This overwhelms the Chimp and it can't start.

By switching into Human mode, we still see the end outcome of a tidy home, but *we start from where we are*. We put our focus on doing just small amounts of tidying and see progress as we go. We don't become overwhelmed because we set our sights on just the next thing to do. This simple principle can be applied to all tasks. By breaking them down into steps and targeting the completion of a small step, the Chimp will usually be fine.

Example: *The minnows and the whale*

The above example demonstrates what I call the principle of 'the minnows and the whale'.

The Chimp will look for quick gains and avoid large tasks, even if these large tasks have large gains. Our Human will reason that important tasks go first, regardless of how large they are or the effort needed.

If we were asked to eat either a whale or a few minnows, it's obvious that the easy task is to eat the minnows. When we approach a daunting but essential task, our Chimp will tend to search for smaller tasks that we can do quickly in order to avoid the daunting task. For example, if we have to answer ten emails and one of them is essential to respond to but it is large, the Chimp will tend to reason that "I will just clear these nine smaller ones then I can concentrate on the larger one"… and the larger often never gets done. We are eating the nine minnows and avoiding the whale. It's never OK to eat a whale, so don't try!

How do we overcome this behaviour, which often leads to jobs not getting done? The secret is to break the whale down into edible pieces. In other words turn the whale into minnows and eat them one by one. We could break the daunting email into smaller steps that we can tackle one by one. By taking small steps, it encourages our Chimp to keep going.

How do we manage our Chimp if it doesn't cooperate?

What happens when we decide how we want to be, but the Chimp prevents us from carrying out our choice? When the Chimp is sabotaging our choices, there are two main ways of moving into Human mode.

Developing yourself

The first way

This is to manage unhelpful emotions, behaviours and thinking, *as they appear*. This is when our Chimps have hijacked us and taken over without our permission. We need to know how to quickly move back into Human mode. We usually get hijacked repeatedly throughout the day, so it is important to recognise when this is happening or has already happened. We can then deal with the situation quickly and effectively.

The second way

This is to **prevent** unhelpful emotions, behaviours and thinking from occurring in the first place. This approach means that we can ideally start and end our day in Human mode.

Two ways to move into Human mode

Recognising and stopping a hijack

Preventing a hijack from occurring

During this course, we will explore these two ways in depth, and will see just how critical a role the Computer plays. We will begin by looking at the Chimp in more detail and how to manage it when it has hijacked us. This will be covered in Stage 2.

Unit 3 Reminders

- Stability of mind is achieved by switching from Chimp to Human
- In Human mode, we can greatly increase our chances of being successful
- In any situation, establishing your Human and Chimp options will give you a choice
- Chimps react: Humans respond by accepting reality and forming a plan
- Humans start from where they are and what they have; Chimps start from where they want to be and what they are lacking
- When the Human and Chimp work together, problem solving and decision-making are at their best
- The Chimp's role is to quickly recognise a problem and let you know about it; the Human's role is to find a solution and reassure the Chimp

Unit 3

Focus points and reflective exercises

1. *Being in the right mode*
2. *Plans and making joint decisions*
3. *Don't react; respond*
4. *Start from where you are and what you have*

Focus 1: *Being in the right mode*

It isn't difficult to stop and take just a few moments to ask yourself if you are in the right mode before you engage in any situation. Sometimes, it's very easy to recognise that you are in Chimp mode. Being in Chimp mode can be with or without strong emotions. Once you have recognised that the Chimp is in control, you can then pause and often reset yourself quite quickly.

Exercise: *Am I in the right place before I begin?*

Which mode would be best?

This reflective exercise is about recognising when you are in Chimp mode and deciding if switching to Human mode would be helpful before you begin.

When the Chimp enters a conversation, it rarely thinks about what the other person would like from the conversation. The Human will consider what both the other person and you would want.

Before you interact with someone, consider which mode you are operating in, and if necessary, move into Human mode.

You could ask yourself, what do you want to gain from the conversation and what will they gain?

Developing yourself

Focus 2: *Plans and making joint decisions*

By reflecting on what we want to achieve, we can plan our actions appropriately. The problem is that our Chimps will usually modify the plans. The Chimp will only consider short-term outcomes or make sure that there is instant gratification. Therefore, when we plan, we need to consider how to manage our Chimps.

Exercise: *Do I have clear outcomes with a plan that my Chimp will work with?*

When you plan your next task or venture, stop and think about how to get your Chimp on board. In order to satisfy both, divide your plans into a rational Human approach and an emotional Chimp approach.

Consider asking yourself these questions, which could help:
- "What is my plan?"
- "What is in my plan to help my Chimp?"

Focus 3: *Don't react; respond*

Exercise: *Learning to respond rather than react*

Try spending a day **responding** to events and situations rather than **reacting** to them. As soon as you recognise your Chimp is about to react, try using the word '***respond***', to stop the Chimp momentarily, enough for your Human to take over.

Then, follow through with the **NEAT** principle and remind yourself that:
- This is '**Normal**' for a Chimp
- You ought to '**Expect**' the Chimp to do these things
- You can '**Accept**' the nature of your Chimp
- You can look for a solution to '**Take care of**' the situation
- Finally, with this plan, you can bring perspective and a long-term outlook

One way to establish the habit of responding is to count the number of times daily that you change your reacting to responding.

Focus 4: *Start from where you are and what you have*

Exercise: *Large tasks addressed*

If you have a challenge in the form of a large task, first decide on the final outcome, then form smaller steps to complete the task. Keep your eyes only on the smaller goals en route to competing the task. If you find your Chimp looking to the end outcome and becoming despondent or overwhelmed, refocus on the smaller goal. If the Chimp starts focussing on the final outcome of a daunting task, this can cause procrastination. Procrastination will be covered in detail later in the course.

Unit 4
Managing emotional reactions

STAGE 2: *will focus on the Chimp system and how to optimise working with it. We will begin the process of getting to know your Chimp and how to understand it.*

Unit 4: covers *various ways to manage your Chimp, including a three-step process.*

Our Chimps typically *react* to a situation, emotionally or impulsively, before our Humans have chance to think rationally. This Chimp reaction is normal and healthy but frequently unhelpful. In this unit, we will be looking at managing our emotions, thinking and behaviour, *as the Chimp is hijacking us*.

A natural reaction: *Exercising the Chimp*

Allowing the Chimp to express itself is the most natural way that most people's minds cope with difficult or stressful situations. They just release whatever emotion they feel. The Chimp is too impatient to wait or listen. Therefore, it struggles to pause, or be put on hold, and so has to be released for its exercise. Exercising the Chimp means allowing it to express some emotion. [80] [81] [82] [83] [84]

Impulsive expressing of emotion typically takes the Chimp a few minutes. As far as the Chimp is concerned, having a scream is the solution! [85] Allowing your emotions to be expressed constructively out aloud can be helpful, as long as the Chimp doesn't attack someone in the process, by aiming the emotion at them. Often the Chimp might need to express the same feelings more than once to settle down. However, the usual reason the Chimp keeps expressing emotion is because there is no solution to a problem it is facing. Emotion will keep on being expressed until the problem is solved. [86] [87] [88]

How long will the Chimp need to express emotion?

Addressing concerns and fears: *Boxing the Chimp*

Whatever problem you are facing, it is useful to make a list of the individual points that are causing concern. When you have the list, you can then address each concern individually. The solutions or answers to each concern will then 'box the Chimp'. Therefore, boxing the Chimp is about listening to the concerns it has and then answering each concern with facts, truths or solutions, so that the emotions can be managed or removed.

It is crucial that each answer to a concern must be accurate and reassuring. You can place these truths into the Computer in your mind, for use as a future reference by the Chimp. Otherwise, the Chimp will just keep on returning to the same emotions at some point later and it will find no reference when it looks into the Computer. Reminding yourself of truths, until they register in the Computer as Autopilots, takes time.

Grade A and Grade B hits: *A word about using truths to box the Chimp*

As we covered in unit 1, Autopilots are helpful beliefs and behaviours. However, although some beliefs might be true, they might not be that helpful or have a real impact on the way we feel or behave.

The truths that really resonate with us and cause us to change our behaviour, or our feelings, are the 'grade A hits'; and those truths that might be true but don't help us, are called the 'grade B hits'.

Grade A hit – *A truth that resonates with you and helps*

Grade B hit – *A truth that doesn't resonate with you and doesn't help*

The distinction is important because if we apply grade B hits to a situation, they won't help us. Whereas, when we apply a grade A hit, it will always help.

Example: *Terry and Sarah and the fear of flying*

Both Terry and Sarah fear flying. In order to help them, a psychologist has offered some truths:

- Flying is the safest mode of transport
- Some people spend their lives as aircrew, which shows how safe it is to fly

Terry had feared the lack of control when being on an aircraft. For Terry, the first truth truly resonates and he immediately feels better the more he thinks about this. It makes sense to him because he does a lot of travelling by car and doesn't fear this, even though he doesn't have control of other road users.

Sarah also fears the lack of control on an aircraft but the first truth just doesn't help her. However, the second truth does help her. She reasons that it can't be that unsafe if countless people travel every second of every day across the whole world and make a living by doing this.

For Terry, the first truth is a grade A hit. However, for Sarah, the first truth is a grade B hit and the second truth is a grade A hit. It is only the grade A hits that will box the Chimp.

Example: *Lucy and the car accident*

Lucy was driving to work and a car drove into the back of her vehicle. Before Lucy's Chimp could hijack her with any emotional reaction, she stated some grade A hits that resonated with her. These were:

- This will all get sorted out
- I will get to work eventually
- There is nothing I can do to change things
- I have a choice in how I want to deal with this situation: Human or Chimp
- I will be able to deal with any consequences
- I need to follow a procedure and hand over to my insurance company

It is critical that these truths resonate with Lucy otherwise the Chimp will not remain in its box.

Key Point

As we are all unique, we have to find our own grade A hits, which stop our Chimp in its tracks and help to stabilise our mind.

How to manage an emotional Chimp with a three-step structured process

We can use a three-step process for situations where the Chimp won't settle down.

The three steps:
1. Exercise the Chimp
2. Box the Chimp
3. Form a plan

Managing the Chimp in three steps

We will now look at two examples of how to manage the Chimp with the three-step process. The examples demonstrate common principles, which you can apply to your own particular circumstances.

Example 1: *The driving test*

Jason is about to take his driving test. The test is in one week's time and Jason's Chimp has become anxious and fears failing the test: all very natural and reasonable for a Chimp.

Step 1: *Exercise the Chimp*

In processing this situation, Jason first lets out his fears and exercises his Chimp.

He expresses all of his fears, which may include, for example:
- I think I could fail the test
- If I fail what will people think
- I can't face another test or take more driving lessons
- Failing will confirm that I am stupid
- I don't feel ready
- Failing means I have wasted my money on lessons

I am sure you can think of many more concerns or unsettling thoughts!

Step 2: *Box the Chimp*

By identifying and addressing each concern and fear, as they are expressed, Jason can address each one individually with some facts and truths.

If Jason doesn't uncover every fear that his Chimp has then he will find it difficult to settle his Chimp. Each response must resonate with Jason and put him at ease, otherwise his Chimp will continue to create a fuss. Here are some suggested answers that might work for him.

I think I could fail the test
- I have to agree with the Chimp because failing is always a possibility
- I am an adult not a child and I can deal with any setback, including a fail
- I will stop guessing at the result and have plans ready to deal with any outcome
- The test only assesses my driving on one particular occasion, it might not reflect the true picture

If I fail what will people think
- I can't change what people will think
- It's what I think that matters
- Those who care about me, care about me regardless of what I can and can't do

I can't face another test
- If it means so much to me, then I need to do the test again
- If I am not up to standard, then it would be wise for me not to pass
- I am not the only person to have to repeat a driving test

Failing will confirm that I am stupid
- Failing will only confirm that I have not reached a certain standard on a certain day with a certain test
- 'Stupid' is a derogatory term and I don't have to use this against myself
- Doing my best is all I can do

- Passing or failing a driving test doesn't define me
- I have to accept any limitations that I might have
- I am proud of myself for trying and for who I am

I don't feel ready
- It is likely that my Chimp will always feel that it is not ready
- The Chimp within me is trying to help me by stopping me from taking the test, in case I fail, but actually I want to take the test
- The test will tell me if I am ready or not, but I can't tell until I take the test

Failing means I have wasted my money on lessons
- I knew when I booked my lessons that failing the test was a possibility, but I wanted to take that risk
- I have learnt some driving skills and need to build on these
- I have to invest in lessons, and continue to do so, until I pass

As you can see, the answers to the fears and beliefs must be individual. None of us will have exactly the same answers. I am sure that you can come up with alternative facts or truths that may settle Jason's Chimp. The important thing is to challenge *your* fears and come up with answers that ring true *for you*.

When you have done this effectively, you can move into Human mode by engaging a plan.

Step 3: *Form a plan*

In Human mode we work with a proactive plan to go forward.

Jason could include the following in his plan:

- **Jason can decide on what he wants his mind to think about on the day of the test.** It would be unwise to set off for the test and allow his Chimp to take over his mind and allow it to decide on what he will think about. Instead he could take the initiative, be proactive, and tell his Chimp what he wants to think about and focus on. For example, he is going to focus on enjoying his day, regardless of what happens.
- **He could remind himself of the beliefs in his Computer**. It is always helpful to make sure the Computer is up and running with the Autopilots ready to answer the Chimp, should it try to hijack Jason. On the morning of the test, by going over the truths he has used to box the Chimp earlier, the Computer will go into action quickly if it needs to settle the Chimp.

- **He can prepare himself for life beyond the test.** He can do this by considering both of the possible outcomes; passing and failing; he can then be prepared for either outcome.

    A comment about 'failures': It is surprising how many people have told me about big disappointments in their life that caused them to take a different pathway. This pathway turned out to be better than their original plan and also gave them opportunities to discover something about themselves. 'Failures' can mean opportunities.

- **Making the most of every day.** Jason might choose to make the day a happy one because there is no reason why it shouldn't be and there are always learning points.

It does take time and effort to manage your mind by analysing your problems and finding solutions, but if you really want to acquire emotional skills then you have to do the work! [89] [90]

Example 2: *Mary and the executives*

Mary, who is an executive coach, has been helping to coach a team of four senior members of staff in a company. She has worked with each one to reflect on where they are at and has developed a plan with them. Three of the executives paid lip service to the exercise, and didn't do any work on themselves, offering lots of excuses. One executive worked well and fed back that she'd had significant improvement to her working practice.

The Chief Executive Officer (CEO or the boss!) of the company calls Mary into his office and tells her that the work she has been doing has had no effects and the executive group seem to be worse than when she first started with them. When Mary tries to discuss this with the CEO he refuses to discuss it and tells her that the contract will cease with immediate effect and asks her to leave.

This extreme example shows three common themes that occur in many scenarios:
1. **Unreasonable behaviour from a line manager:** there is *unreasonable behaviour* from the CEO in *refusing to discuss or give any explanation* to Mary.
2. **Undeserved blame:** there is evidence that Mary has done a great job with her plans, but the results are poor because three of the executives didn't act on the plans. Therefore, Mary has been *blamed for something that she could not control*.
3. **An injustice has occurred:** Mary is left feeling frustrated that she has been wronged and feels she can do nothing about it.

There might be many other themes in this scenario that you can think of and endless interpretations as to exactly what happened. The result is that Mary's Chimp is very likely to react and express emotion. This could range from anger to despondency. The Chimp's thinking could also range from thoughts of indignant hostility to self-reprisal or a crisis of confidence. This is because the Chimp has reacted to what it sees as a threat to its reputation and also a severe injustice, both of which have some foundation. *From our perspective, the actual reactions or emotions are not too important.* We are looking at the method of dealing with them.

Step 1: *Exercise the Chimp*

First we allow the Chimp to vent its feelings. This can be done alone or with a friend or colleague, as long as they know that this is just the Chimp exercising and therefore not to interact with it. It's important that Mary allows her Chimp to keep on going until the emotions begin to subside. By expressing her thoughts out loud and listening to them herself, Mary will better manage them. It is helpful *NOT* to try and give answers or solutions to your feelings at this point. It also helps if the person you chose to share your emotions with also knows that you are just letting off steam and that they don't try to engage with your thoughts at this point. Sometimes, when you have expressed your thoughts or emotions, just listening to them yourself will help to process them. If you prevent your Chimp from expressing its emotions, then there is a strong possibility that it won't allow you to move forward or to listen to what is being said. So, getting things off your chest is one of the most helpful things that you can do to move forward. [80] [81] [82] [83] [84] [86] [87] [88]

Step 2: *Box the Chimp*

Let's assume that the statements that follow are Mary's Chimp's reactions and concerns in this situation (and possibly her Human might agree):

- I am angry with the CEO because he didn't listen to me and as a CEO he should listen to me
- I have been judged on something that is not my fault and something that I could not control
- My reputation is damaged because of three executives
- I have lost some confidence in my own abilities because I should have made sure that the executives engaged

There are so many other ways that somebody's Chimp could react in this situation. If we can clarify all of the grievances or reactions, we can then address them individually.

We will look at some possible examples of how Mary's Human could address her particular concerns. Please note that they involve her Human reinterpreting how her Chimp has perceived the situation and also establishing some truths of life.

- The CEO was unreasonable but that is his prerogative. He might have had a bad day, he might be firing some arrows for someone else, or he might just be unpleasant. Whatever his reasoning, I have to accept that this is the way he chose to work on that day.
- Being judged on something I can't control or taking the hit for someone else is never easy emotionally (but it will almost certainly happen to all of us at some point). I can either accept this or I can try to correct it in some way.
- I have to work out a plan that will allow my Chimp to settle and put closure on the situation and only I know what will work for me.
- Sadly, it is true that others can ruin my reputation. What I can do is to remain true to my own integrity. At the end of the day, when I look in the mirror, it is only me that will stare back and not others. Therefore, if I can live with my own conscience, I will be in a strong place. Having friends to reinforce this, can also help me immensely but ***I do need to let them know*** if I want their help.
- Reputations are rarely destroyed because of the comments of others. Most people can see what could be happening and make their own minds up about someone.
- My Chimp is suggesting that I might not be up to standard. It is only natural for my Chimp to lose confidence because virtually all Chimps are over critical of themselves, if they are asked to assess how they are doing.

Step 3: *Form a plan*

Having expressed emotion and answered concerns, we now need to move forward with a constructive plan.

Mary's plan could include some of the following:
- Either, just let it go and accept that unpleasant interactions are part of life or write to the CEO and explain the situation and ask if he would like to meet again to discuss things.
- Write to the CEO with feedback only and not request a discussion. Explain that for future work to be effective it does need engagement from the participants and that only one did engage meaningfully. This might get this aspect of her grievance off her chest.
- Address the Chimp's fears that she has failed. By looking at the times that she has succeeded, and not to selectively look at the times she possibly hasn't, it might bring self-esteem back to the Chimp.
- Use the experience as a learning point and see what lessons can be learnt from this. If Mary really did think that she could have been better at ensuring the executives engaged, then this could be seen as a learning point rather than a failure.

None of us succeeds all of the time. It is unrealistic to think that we can.

Our Chimps don't think ahead and consider long-term consequences, therefore, each plan needs to be thought through and the possible outcomes considered. There is little point in having a plan that could make matters worse. [91] [92] [93]

Key Point

Setbacks and failures can always be seen as opportunities to learn from.

Five important points to consider when addressing concerns

1. **Express all concerns**
2. **Choose your listener carefully**
3. **Only cross real bridges**
4. **Exercise the Chimp before trying to listen**
5. **Seek help**

Point 1: *Express all concerns*

Express all concerns, no matter how irrational or catastrophic they might seem to be. When we speak our thoughts and concerns out aloud, the Human part of our mind listens and starts bringing perspective into the situation. Therefore, it is worth expressing any and all fears. Hidden fears, no matter how silly they might seem to be, can keep unsettling us in our unconscious minds because they have not been expressed and addressed.

The talking therapies
- Scientific points

When the emotional centres of the brain express our feelings by putting them outside of our mind, the dorsolateral prefrontal cortex (DLPFC) can then assess them and begin to process these emotions by bringing in logic and rationality. The DLPFC is the executive centre of the brain and will organise our thoughts, draw rational conclusions and add perspective. Talking things through by expressing them into the air is the basis for all of the talking therapies. This works by getting things out of our head and give words to describe our emotions. Once we can describe our emotions, we can then work with this to make sense of our feelings. In our model, the Chimp expresses feelings, the Human listens and adds perspective and logic, and the Computer then stores beliefs. [94] [95] [96] [46] [97] [98] [99]

Point 2: *Choose your listener carefully*

Some words of caution: If you are going to express any concerns, be careful whom you express them to. Often a partner or parent might not be the right person, if they can't handle your fears or concerns. Just make sure it is someone who will take things in their stride and is able to remain detached whilst they listen. This is why working with an independent person can help.

Point 3: *Only cross real bridges*

After expressing concerns, don't address those concerns that are not real but are based on assumptions or predictions.

Key Point
By not making assumptions and predictions and by going with the flow, we can manage each challenge, as it presents itself.

Point 4: *Exercise the Chimp before trying to listen*

Please remember to make sure that your Chimp has exercised enough, in order for it to be able to listen. We have to get things off our chest before we can listen well.

Point 5: *Seek help*

If you still feel you have no answers, then consider turning to someone else for help or advice because they might have the knowledge or experience that you don't have. It's great to be independent but struggling unnecessarily can create avoidable suffering.

A summary of the five points

Managing the Chimp with a pause button

In unit 2 we looked at the use of the Computer to manage the Chimp by using a pause button. We did this by programming the Computer to always immediately ask, "What's the plan?" whenever the Chimp begins to stress. Some people use the phrase "take a deep breath" whenever their emotions stir. It could be very helpful to have your own phrase to activate that will give you a moment to pause before the Chimp reacts.

How to manage the Chimp by giving it bananas

So far, we have looked at managing the Chimp by:

- **Exercising it** - allowing it expression
- **Boxing it** - settling it down by offering truths and answers to its concerns
- **Forming a plan** of action
- Using a **pause button**

There is an alternative way that can settle our Chimps down. This way doesn't offer solutions but calms the Chimp. We can offer the Chimp **a banana**!

Rewarding ourselves for doing something well or praising our efforts, even when we have not managed to achieve what we want to achieve, can settle the Chimp down.

Distraction is another method to remove the Chimp from being preoccupied with something that it distressing it, in the hope that with time the distress factor will naturally go away.

Managing the Chimp in a child
- Scientific points

Research shows that distraction is the best method for managing a very young child if they become distressed. Discussing and reassuring might help but distraction is far better. Our minds are built to follow a distraction, not to remain focussed. This is an obvious survival mechanism, which allows us to be vigilant in case danger suddenly appears.

The Chimp within the child will readily move on and quickly forget what was distressing it, if we distract them. As adults, our Chimps still operate by looking to distractions! Therefore, they can be a useful technique to give relief from unwelcome thoughts. [97] [98] [99]

Therefore, two different types of banana can be used to manage a Chimp:

- Reward
- Distraction

Example: *Byron and the interview*

Byron lost his job through redundancy. He has now attended several interviews. While he is waiting to be interviewed his Chimp becomes anxious and hijacks him. By the time he enters the interview room, he has become almost paralysed with fear and underperformed each time.

Byron has tried to box his Chimp; he has tried to go into Computer mode; he has tried to remain in Human mode; but nothing seems to work. He has a lot of insight into what is happening but he feels powerless to stop it.

He now tries the banana of distraction! First, Byron sets out in his mind the points that he hopes he will be able to get across during the interview. Once he has these few points established, he now sits in the waiting area and distracts himself by listening to some music. He knows that if he can distract himself, while he waits, then he can enter the room in a better frame of mind. Following the interview, he might also want to add a banana of reward, for his interview efforts, with a stop at his favourite coffee shop.

Unit 4 Reminders

- The key points of the three-step structured approach:
  1. **Exercise** - Allow your emotions to be expressed until the Chimp is exhausted
  2. **Box** - Identify the concerns that have evoked the emotion in you, along with any fears or interpretations of the situation. Then address each concern with logic, rational thinking and truths that resonate with you
  3. **Plan** - Formulate a plan that addresses each concern

- **Grade A hits** are truths that resonate and change our thinking and behaviour

- **Grade B hits** are truths that have little effect on us

- Having a **pause button** can give your Human chance to act

- Giving **bananas** of rewards or distractions is often a useful quick fix to help manage the Chimp.

Unit 4 Exercises

Focus points and reflective exercises

1. *Boxing the Chimp – Grade A and B hits*
2. *Giving the Chimp bananas*

Focus 1: *Boxing the Chimp – Grade A and B hits*

Exercise: *Boxing the Chimp*

Try to think of a situation that occurs regularly in your life that usually results in a Chimp hijack. For example, a person who always seems to say the wrong thing that winds your Chimp up or an unhelpful habit that you are trying to break but keep repeating, and this winds your Chimp up.

Work out the grade A hits that could stop your Chimp and box it. Imagine the situation occurring and practice using these truths. By repeating this exercise, you are *programming your Computer to respond* with the grade A hits, when the situation occurs for real.

Please remember that the truths you select to settle your Chimp must truly resonate with you and be grade A hits. They must not be simple truths that don't really resonate or do anything to settle your Chimp.

Focus 2: *Giving the Chimp bananas*

Exercise: *Giving a banana*

Dennis is an artist and painted some pictures. He has submitted them to a gallery for display and it would mean a lot to him if he were to be successful. He has been told he will hear in the next few days but his Chimp is stressing about it.

What distraction techniques could you suggest to Dennis in this scenario? Try to work some out before looking at the answer.

Answer

Obviously there could be an endless number of ways for Dennis to distract himself. The important point is that whatever he chooses will work for him.

He could try:

- Starting another painting, which would help whether his current work is accepted or not
- Making a list of things he has wanted to get done for a while and try and complete them
- Using the time to plan his future in more detail
- Getting out of the house and socialise with friends

The important point about distractions is making the effort to work out what will work for you.

Unit 5
Your mind in harmony
(The Chimp as a best friend)

Unit 5: covers *how to understand and accept the nature of your Chimp. We will look at how to recognise, accept and manage both negative and positive aspects of your Chimp. By the end of the unit, your Chimp will hopefully be your best friend!*

How to understand and accept the nature of your Chimp

An exercise: Pause for a moment and think of three characteristics of your Chimp that you regularly see being displayed in your day-to-day life. When you have the three characteristics, read on.

If you do this exercise spontaneously, or even try to describe your own Chimp, it's very likely that you will give three negative characteristics. [100] [101] Such as:
- It is too quick to jump to conclusions
- It worries and creates catastrophes out of nothing
- It reacts before it thinks.

There is no disputing that at times our Chimps can cause untold havoc to us and to others in our lives! However, if we keep seeing our Chimp with its emotions and agendas as a potential trouble, or even weakness, then it's not surprising that we don't have a good working relationship with it. The Chimp system can bring immense colour and enrichment to our lives. It can also be a major part of the basis for our success in life. [102] We need to appreciate this if we are going to have a good working relationship with it.

If you were describing a friend, you wouldn't just describe their negative aspects; you would give a balanced picture. For example, you might say that your friend can make errors. How do we get over errors that our friends make? We accept them for who they are and as they are and, if we can, we help them to overcome mistakes or weaknesses.

Acceptance of your Chimp, *as it is*, is critical if you are going to befriend it. If you do have a mainly negative view of your Chimp, it will not help your relationship with it. The aim of this unit is to reflect on the relationship that you have with your Chimp. It will help you to see your Chimp as your best friend and a power to harness in your life, by highlighting its positive aspects. [103]

Reflect on whether you are accepting the nature of your Chimp or fighting against it. Please remember the nature is genetically determined from birth. Only by accepting the nature can you begin to work with it. Try to see situations from your Chimp's point of view.

Common negative features of the Chimp

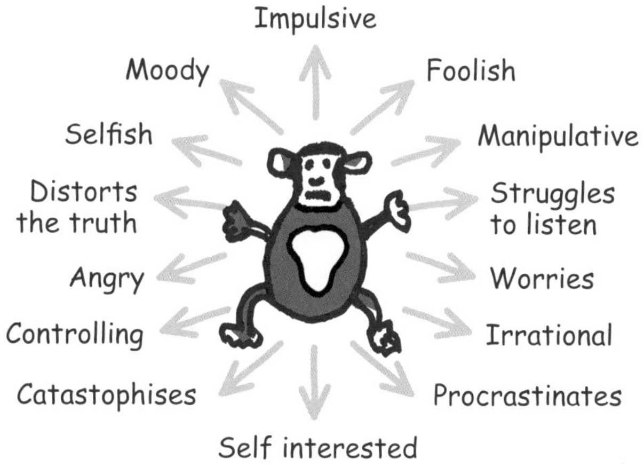

- Impulsive
- Moody
- Foolish
- Selfish
- Manipulative
- Distorts the truth
- Struggles to listen
- Angry
- Worries
- Controlling
- Irrational
- Catastophises
- Procrastinates
- Self interested

I accept you as you are.

I add colour to your life!

Example: *Duncan and the workload*

Duncan works on a switchboard and appreciates and accepts that the workload is heavy and relentless. Although Duncan understands this, his Chimp still overreacts whenever calls begin to stack up. In the past, Duncan has been disappointed with himself for becoming agitated and angry at the situation. Once he recognises that this agitation and anger is coming from his Chimp, and accepts the nature of his Chimp, he will see that his Chimp is acting in Duncan's best interests. What his Chimp is telling him is that the situation is untenable and needs a change in beliefs, approach or workload. It is now up to Duncan to act, and help his Chimp, by looking at his beliefs and also the situation in general. The nature of the Chimp must first be accepted and then its message understood.

The first step in managing our Chimp *is to accept it, as it is*. The Chimp is your best friend and is always trying to help you. It might not seem this way, but it actually is. Sadly, although it might be your best friend, it often appears to be a very inept best friend. It tends to overreact and become emotional. This doesn't mean it isn't trying to help.

So how do we get the more constructive side of our Chimp to appear?

Almost paradoxically, after what I have just said, it helps to first look at, and address, the negative aspects of our Chimp, and then to move on to the positive aspects. The reason for this is that until we accept the nature of the Chimp, we might always be puzzled by it and critical of it. This means we might be constantly asking why it keeps on repeating destructive behaviours, producing negative feelings or generating unhelpful thoughts that we don't want.

How to accept and manage the negative aspects of your Chimp

To understand the nature of your unique Chimp, try writing down some of your Chimp's unhelpful ways of working that you are going to have to learn to live with and accept. This doesn't mean that you have to put up with them! We can learn how we can reduce the likelihood of them appearing and how to manage them if they do appear. What we are considering is the nature of the machine that we have inherited in our genes and what we have to accept before we can start working with it. [104] Acceptance is recognising that something can't be changed so must be worked with or managed.

Key Point

Some in-built traits in our Chimps are normal and healthy but unhelpful.

I will describe some typical traits and characteristics that are very common in many Chimps. You might not relate to some, but if you can relate to them, try not to see these traits as a problem. If you do experience them, then it just means that you have a very healthy Chimp. You will have to live with these traits, but again I emphasise that you don't have to put up with your Chimp demonstrating them. When going through these examples, pause and see if you can recognise when they have occurred in your own experience.

Overreaction

Our Chimps are blind to the real world and only see a jungle. For example, Chimps initially see all sticks as snakes, until proved otherwise, and they therefore typically overreact, [105] [106] which in jungle terms would be appropriate. If the Chimp in a jungle decided to take a risk by assuming that the stick is not a snake, but got it wrong, then it might die from the snake attack. Therefore, the Chimp brain is built to fear and not to take risks. [48] [106] [107] In order to make sure it does this, it is programmed to 'overreact'. The Chimp brain works a lot with fear and has no way of countering it, other than reacting to it, and this it will do with gusto! It is our adult Human brain that can calm things down, usually by adding perspective or facts. [108] [109] Young children have no significant brain mechanism to deal with fear. [31] It is an adult that needs to reassure them, to counter their fears and prevent any inappropriate overreaction.

Insecurity is forever

Our Chimps by nature are insecure. By providing as much security as possible, we can limit this feeling. Putting in place such things as routines and having friends gives our Chimp a feeling of security. [110] [111] However, some things we can't be certain of, such as our health, and therefore we have to learn to manage any feelings of insecurity.

Our insecurity is based on our position in the food chain. We are not at the top of the food chain and can be preyed on, so it seems wise to be constantly vigilant. We are meant to be like this otherwise we wouldn't survive. Our Chimps perform this vigilance for us. [105] [106] However, this vigilance is rarely appropriate in modern society. Feelings of insecurity in any new setting or in any challenging situation are therefore normal and need accepting before we can manage them and hopefully remove them.

Our Chimps are always needy

Most Chimps crave companionship in its various forms. To deny this or to have the 'wrong' companionship has emotional consequences. [112] Similarly, the need for reassurance is perpetual in most Chimps. If your Chimp needs it, the simple answer is to ask for it. When you receive reassurance, the Chimp is likely to give out positive and pleasant emotions. Don't allow yourself to live with uncertainty when it isn't necessary. If you need to be reassured in any situation, for example about your value, whether you're doing a good job or whether you are loved, then simply ask. If the answer is not what you want, then at least you can deal with this fact, rather than leave yourself guessing.

> **Key Point**
>
> *Most Chimps need to be reassured repeatedly.*

Recognising this need in others around you can help if you are prepared to reassure them on a regular basis. For Chimps, it is only what is happening in the here and now that counts. Therefore, yesterday's reassurances are unlikely to be valid in its eyes. You might get quite frustrated if you think that constant reassurance is unnatural. Again, understanding and accepting that we all need regular reassurances in various areas of our lives will help us to appreciate that this is normal and also necessary. Clearly, if someone's Chimp needs constant *excessive* reassurance then this could be a problem to address.

Our Chimps rarely learn

Chimps can be trained. They also change their drives and needs with age. However, they rarely learn from a situation, therefore, don't expect them to. It's you, the Human, that learns and can put plans into the Computer to manage similar situations. [33] [34] It's not surprising that our Chimps can hijack us at any age with the same behaviours or feelings. I have had several people that I have worked with becoming dismayed because their Chimps repeat similar behaviours despite the person really making the effort to manage their Chimp. The usual comment I receive is "I haven't learnt anything!" It's important for them to appreciate that the Chimp hasn't learnt anything, but they have. There will always be moments when the Chimp escapes and repeats unwelcome behaviours. This is part of a healthy machine. This point is so important to accept. Otherwise, the person who is subjected to a Chimp hijack might go on to criticise themselves and perceive themselves as being stupid or out of control. This easily leads to low self-esteem and frustration. I frequently stop and ask them if these behaviours, emotions or thinking from the Chimp are normal – the answer is always yes!

Key Point

Any behaviour or thinking is normal for a Chimp!

We have mentioned previously that there is nothing that your Chimp can do that isn't normal for a Chimp to do. That doesn't mean any behaviour or expression of emotion is acceptable! It also doesn't mean that you are not fully accountable for your Chimp's actions. Hopefully these points that are being made will prevent anyone from beating themselves up when they suffer a Chimp hijack.

Our Chimps overeat

One of the most distressing misuses of the eating drive is when our Chimp overeats.

As our knowledge of eating behaviours increases, we are more aware of just how powerful the drive to eat is. Clearly, it needs to be a very strong drive, or we wouldn't survive. It is coupled with a release of some hormones that bring pleasure and other hormones that make us feel more secure. This paradoxically doesn't help when we live in a society with an abundance of food. This is because we might use eating to gain security and comfort, or use the eating drive as our main source of pleasure. [113] [114] Learning to manage the eating drive is a struggle and only you can find what works for you.

Key Point

It can be hard to appreciate the subtlety between being responsible for managing the drive and NOT being responsible for having the drive.

You should not feel guilty about having in-built drives, but you might experience disappointment if you fail to manage a drive, which is your responsibility.

Whichever way you look at it, the eating drive is here to stay and accepting it is the first step to managing it.

Helpful at all times

I have described some potentially negative traits of typical Chimps, but it is worth remembering that the Chimp is always trying to help us. It is not trying to be an enemy. It just operates with a different agenda and with different methods.

Key Point

The Chimp is trying to help us at all times.

Think about that key point. It is a real revelation to appreciate that your mind is always trying to help you. If you understand this, then you can work with your Chimp and not become distressed by it. The Chimp is always trying to be your best friend and ally.

The problem is that what the Chimp wants and the way that it wants to solve things aren't always what we want or agree with. This doesn't make the Chimp wrong. It is acting according to its nature. Battling with the Chimp isn't helpful; understanding it and working with it is constructive. It is worth reviewing how you see your Chimp and what you feel it can do for you. There is a true adage that: *we see what we want to see and we find what we look for*. If you always look for the bad points in your Chimp then it is very unlikely that you will have your Chimp as a friend.

How to recognise the positive aspects of your Chimp

So what are the good points?

Having accepted that the Chimp has traits we need to live with and learn to manage, what are its worthy points? The Chimp has a lot to offer and here are some of the good points:

Common positive features of the Chimp

- Engages others
- Streetwise
- Enthusiastic
- Protective
- Ambitious
- Sense of Humor
- Troop driven
- Intuitive
- Caring
- Motivated
- Energetic
- Gregarious
- Perfectionist
- Nurturing others

Streetwise

What we mean by streetwise is the ability to perceive what is happening around us, particularly any undercurrents, and to respond appropriately. Intuition is a major part of this. Decision-making with the Chimp's intuition can sometimes be better than logic because logic might lack facts or interpret the facts in a faulty way. [15] [74] [75] [31]

Streetwise
- Scientific points

The Iowa gambling experiment demonstrated the ability of the Chimp to be more streetwise than the Human. The experiment tested people's ability to detect some deceit. Subjects were asked to gamble by choosing one of four possible options. They were told that the four options would all randomly move their money up or down. In reality this wasn't true. Two of the options gave large rewards, but with repeat betting, would take their money down. The other two options gave small rewards, but with repeat betting, would take their money up. The DLPFC (Human) could verbally tell the experimenters what they thought was happening. The OFC (Chimp) is connected to our sympathetic nervous system. Therefore, if the Chimp detected something it would cause us to sweat a little on our hands. A galvanic skin responder was attached to a finger of the subjects, as they gambled, in order to detect any Chimp uneasiness.

After ten bets, the OFC (Chimp) caused sweating to occur if the two money-losing options were selected. The Chimp had worked it out. It then made the person move to select the two moneymaking options. When people were asked why they were avoiding the two money-losing options and selecting the two money-gaining options, they were not aware of this. Their DLPFC (Human) had not worked it out but the OFC (Chimp) had.

It took around fifty bets before the DLPFC realised that the options were biased. Our Chimps are streetwise and speak to us with that uneasy voice from within when there is something not quite right! [47] [115]

Perfectionism

It is common for someone to complain of being a perfectionist and this can result in them being unforgiving towards themselves or others.

It's worth explaining why perfectionism *is a very helpful trait for your Chimp to have!*

Problems only arise if we approach perfectionism via the Chimp and not add on the Human input. Here is the difference:
- The Chimp demands and expects perfection, even with uncontrollables
- The Human aims and hopes for perfection but accepts the reality that many things can't be controlled and we don't always achieve our full potential. [116]

Key Point

Perfectionism from the Chimp is a great drive to have, provided when we reach an outcome, we can switch to the Human system and rationally accept the reality that we rarely achieve perfect results.

Example: *Zak's geography test*

Zak wants to get 100 per cent on his geography test. Zak's Human and Chimp both agree that this would be perfection.

The Chimp's approach is to demand that Zak obtains 100 per cent. It believes that no matter what happens, if he tries hard enough, he will get 100 per cent. The Chimp cannot accept any errors or lack of knowledge on his part. It also cannot accept that the person who marks the script might not think Zak deserves full marks on every question. This might be a great motivational approach by the Chimp, but it sets the scene for possible emotional turmoil. There is an alternative!

The alternative is the Human approach. The Human is as driven as the Chimp in wanting to achieve perfection. However, the Human focuses on what it can control. Therefore, Zak's Human plans a study programme, plenty of revision and has a realistic mindset on the day of the test. All of these are achievable and under his control. He is committed to these plans and accepts that if he has done all that he can then he can accept the outcome, whether it is satisfactory or disappointing. If the drive for perfection from the Chimp is coupled with the reality of the Human, then Zak will be in a great place to get the best out of himself.

Example: *Betty and the 100 metres training sessions*

Betty is a very keen athlete. Both her Human and Chimp want the perfect training session every time she goes to the track. The Chimp is driven and determined to excel and will give everything towards achieving perfect training times. If Betty stays in Chimp mode during the session, she will harness the energy and enthusiasm of the Chimp, which is great. The Chimp's drive for perfection is helpful at this point.

If Betty stays in Chimp mode *during the session*, but changes to Human mode *as it ends*, she could achieve the best of all worlds. In Human mode at the end of the session, her Human can accept that Betty has done all that she can and will not become upset if perfection has not been achieved. In this example, we can see how the Chimp's drive for perfection is very helpful, provided we move into Human mode to manage the outcome.

Often the Human and Chimp want to achieve similar things

I EXPECT to run a new personal best in every session!

I HOPE to run a new personal best in every session.

Example: *Tommy and the sales*

Tommy is a businessperson who has a product to sell. He dreams of making huge profits from sales by converting every sales pitch into a successful sale. To achieve this, he works on some goals that will help to ensure each sales pitch will be successful:

- List the key points of the product
- Present the product well
- Prepare a timetable for reaching as many buyers as possible
- Set the right price

So far so good!

Here is the potential problem: Tommy's Chimp wants to set some uncontrollables as its goals. Tommy's Chimp adds the following to his list of goals:

- High uptake of product
- Customers being satisfied
- Customers agreeable to recommending the product

Chimp and Human goals are different

The problem is that this is a list of uncontrollables. To set these as goals is very risky because trying to achieve them is outside of Tommy's control. Tommy's Chimp has muddled up a dream with a goal. These are dreams because they are hoped for, but not under Tommy's control. If Tommy now rates himself on how he is doing with these uncontrollables, it could lead to a very negative state. He might experience stress, low self-esteem and feelings of failure. [117]

Key Point

If we are assessed on an outcome that we cannot control, then the usual result is stress.

Example: *Venus and the children's party*

Venus was organising a birthday party for her six-year-old son. She had planned the food and the games, and sent out the invitations. She was looking forward to the party but then her Chimp began to focus on the uncontrollables. Her Chimp focussed on whether the food was right, whether she had invited the 'right' people, whether the games would be entertaining enough and focused on just about anything that it couldn't control or guarantee.

What could Venus have focussed on? The things she could control are mainly inside her head! For example, she does have a choice on where she puts her focus and she does have a choice on how she will manage her Chimp. There are also many practical things she could focus on, such as making sure the food is ready, getting the games prepared and giving people a warm welcome. Effectively she could switch to Human mode.

Example: *Doug and the new girlfriend*

Doug believes he has fallen in love and feels sure that this is the right person for him. However, as is often the case, there are complications. Rhonda, the girl he has fallen for, is very keen on Doug but lets him know that she is bisexual and is also very keen on another girl. She cannot decide which is the right relationship for her. Doug's Chimp is pushing Rhonda for answers and is becoming more desperate to control the situation. Clearly none of us can control somebody else's feelings or who they are attracted to. To focus on trying to make someone love us is futile. What we can do is to focus on relaxing and just being ourselves. In situations like this, painful as they might be, all we can manage is our own behaviour and thinking. Managing our Chimp is the focus. This way we can present ourselves, as we are, giving the relationship the best chance of working out. What if it doesn't?

It is always worth thinking beyond any situation to see what would happen if things didn't go as we want them to. Often our Chimps worry about possible unwanted outcomes; that is their role! They focus on the problem until, you the Human, come up with a solution. The solution could be, to recognise that no matter what happens, life will go on and although there might be a period of grief, we do come to terms with setbacks and losses.

Harnessing the enthusiasm, energy and power of the Chimp

We can safely harness the power and energy of a perfectionist Chimp if we add perspective and realism from our Human and look to **controllables**. This is a very important concept in teamwork for business and leaders. If the leader puts demands on the team members to deliver something that is outside of their control then it is extremely likely that the team will suffer stress. [118] What we can expect from others is that they give their best (and this means maximum effort) and keep on trying to improve. Clearly, if someone can't reach a standard then there must inevitably be consequences, which may be unpleasant, but they don't have to be punitive.

Our combined strengths can forge a great team

I expect perfection.

I offer perspective and reality.

> **Key Point**
>
> *Perfectionism can be very helpful if it is connected to perspective and reality.*

One of the paradoxes of the Chimp is its ability to cause fear, yet also to overcome it. In times of distress or pressure, that might even be brought on by the Chimp itself, we can call on the Chimp for help! When people I have worked with in various situations use this approach, they have reported back how strange it is that their Chimp has suddenly come to their rescue with energy and enthusiasm. [119] [120] [121]

Example: *Demarcus and the 400 metres*

Demarcus runs the 400 metres. With a hundred metres to go, his Chimp would become fearful that it would not be able to finish and wouldn't respond to reassurances. However, if at that point he had called on his Chimp for help, his Chimp would very likely have come to his rescue with renewed energy and determination. This is because the Chimp is a troop animal and will naturally overcome its fear to help a troop member. The next time that Demarcus ran 400 metres, to his surprise, his Chimp overcame its fear and fought for him.

It might not work for everyone; but try to call on your Chimp for help the next time you feel overwhelmed.

Under pressure, paradoxically, the Chimp can step up if you ask it to.

Unit 5 Reminders

- Some in-built traits of our Chimps are **normal and healthy**, but unhelpful
- Acceptance of the nature of your Chimp is a must, in order to work with it
- Any behaviour or thinking is normal for a Chimp
- You are not responsible for the nature of your Chimp, but you are responsible for managing it
- The Chimp is trying to help us at all times
- Perfectionism is a good Chimp feature, provided it is connected to perspective and reality

Unit 5 Exercises

Focus points and reflective exercises

1. *The Chimp becoming your best friend*
2. *Focus on controllables*

Focus 1: *The Chimp becoming your best friend*

This reflective exercise is building on the unit 1 reflective exercise about understanding the nature of your Chimp. This time the exercise is not only about being realistic and accepting that our minds are pre-programmed but that the Chimp is our best friend.

Exercise: *The Chimp as a best friend*

Try and list the drives and traits of your Chimp that you need to accept and work with. Try imagining being your Chimp and experiencing the terrifying world that it lives in. Imagine not having a Human or Computer to befriend you and help you to see rationality and perspective! In this reflective exercise, you are trying to see your Chimp as a best friend, who needs support and understanding, rather than an enemy who is against you.

Focus 2: *Focus on controllables*

A practical exercise is to recognise how our Chimps focus on many uncontrollables. The Chimp demands guarantees in life. Whenever the Chimp cannot guarantee that something will happen, it is likely to become unsettled and worried. Therefore, by helping the Chimp to accept the uncontrollables and to focus on what can be controlled, we are much more likely to succeed in what we want to do.

Exercise: *The controllables*

As an exercise, see if you can recognise an area in your life where your Chimp is stressing about something that it cannot possibly control. Try to establish what can be controlled. Then, try to go beyond the difficulty, and work out what the consequences are if things don't go according to plan and how you will manage the consequences.

Unit 6
How to nurture your Chimp

Unit 6: *will look at how to nurture your Chimp by befriending, supporting and encouraging it. We will also consider the advantages of using commitment as opposed to motivation for getting things done.*

Befriending your Chimp by supporting it with your Computer

We have looked at how to manage the Chimp using the Computer. We can pre-empt the kind of things that might distress the Chimp and be prepared for them. This is a good way to befriend your Chimp. Here are two examples of how this works in practice and how we can help the Chimp to be less reactive by using the Computer.

Example: *Corey and the comment*

Corey works in a factory assembling washing machines. His line manager commented that Corey could be slowing down his assembly line, which was underperforming.

Corey's Chimp immediately reacted and tried to defend itself. If Corey could stop his Chimp from speaking, he would be able to approach this situation via his Human. In which case, he would first establish exactly what the line manager meant by his comment. Corey would feel uncomfortable if the line manager continues and says, "You're a great worker but you don't have the right equipment that others do have".

On the other hand, the line manager might be saying that Corey really isn't up to the job and Corey has to face that. Even in this situation, his Human would probably want to consider how he could improve his skills or might accept that the job is not for him. Whatever the comment meant, Corey's Chimp is left vulnerable because it does not have the Computer supporting it when a criticism occurs. To befriend his Chimp he can programme his Computer, for example, with the following beliefs:

- Take a deep breath and think before you speak
- Whenever a criticism comes, stop and clarify why it was made – gather the facts!
- Criticisms are often said hastily, then regretted and later reversed
- Listen to a criticism and see if there is some truth in it – don't defend the indefensible
- Criticism is only one person's opinion – it's not necessarily a fact

Defence from the Chimp can be detrimental

Example: *Janice and decision-making*

Janice regularly puts herself down because she feels she can't make decisions without becoming stressed. Janice appreciates that her Chimp is doing this but sees it as a negative aspect of her Chimp.

We can turn her impression on its head. The Chimp is actually trying to help and has become distressed because Janice is not understanding it and working with it. The Chimp is primed to protect us and make sure we are safe. It is still operating with a jungle outlook and to make a poor decision in the jungle could cost the Chimp its life. Therefore, any decision, no matter how trivial, will challenge many Chimps. Whenever the Chimp worries about making a decision, it is actually helping us. It is saying please pause, think about gathering any facts before deciding and also think about the consequences. This is always helpful. The role of the Human is to gather the facts and make a decision. [31] The Human also accepts that with any decision, there is a possibility that it might not turn out to be the best, but it can be lived with. The Chimp will always believe that it cannot live with a 'wrong' decision.

How do we turn the Chimp's worry and indecisiveness into a helpful situation? Here are some suggestions:
- First acknowledge that the Chimp is helping by asking our Human to take over
- Therefore, pause, go into Human mode and gather the facts
- Gain some perspective and let the Chimp know that most decisions are not critical
- Look ahead to possible consequences and outcomes from your choice of decisions
- Accept that any decision is a risk, BUT an adult Human can manage any outcome or consequence of the decision
- Remind the Chimp that not making a decision can be very stressful
- Make the decision and praise yourself for doing it

To be decisive when making decisions is a both an art and a habit.

Planning links to pleasure

Happy Chimps are easier to manage! It's stating the obvious, but looking after your emotional needs has to be given time and effort. One way to plan in pleasure is to sweeten any unpleasant tasks with a reward for when you complete them. Chimps usually like to work for rewards. Sweeteners can also be linked to activities. For example, some people find that listening to music whilst working or drinking coffee whilst in a long meeting helps to make the activity more pleasant.

Unpleasant tasks can be sweetened with rewards

Example: *Bill and the gym*

Bill runs a gym but membership has been declining. He understands that people want to become fit, but their Chimps feel the effort needed to do this is just too difficult to face each week. He decides to sweeten the gym workout by finding individual sweeteners for each member.

His solutions include:

- Treadmills with a screen showing short movies or educational programmes to fit with the time needed on the treadmill
- Time challenges or weightlifting challenges for those who enjoy trying to beat records or set personal bests
- A feedback group following the sessions for those who enjoy social aspects
- A team effort with small groups that encourage each other as they work out

Warning! Check that the things you think will bring happiness or pleasure in the short-term will also bring satisfaction in the long-term. Beware the short-term fix chosen by your Chimp! For example, there is nothing wrong with eating treats. However, if they only bring happiness whilst you're eating them and then unhappiness some time later, the short-term fix is unhelpful.

Don't let your Chimp compare itself to others!

Chimpanzees compare themselves to others within their troop in order to establish their standing. This is to ensure that they have proved their worth and can remain within the troop. The Chimp within us does the same and decides its worth by comparing itself to others and sees this as crucial to self-worth. [122] [123] [124] However, try to prevent your Chimp from comparing how you are doing in life to how others are doing, because it will usually distress your Chimp. Instead, the Chimp will settle if you compare how you are doing to living out your own values.

Remember that when we operate with the Chimp system, we necessarily look *outwardly* for our reference points. When we work with the Human system we look *inwardly* to our own values and with perspective. The problem is that when the Chimp compares itself to others, it usually plays down its own accomplishments and therefore sees itself in a poor light. Comparing how you are doing to others or to what they have, or to things that they can do or have achieved, will almost invariably lead to becoming unsettled and dissatisfied.

Humans check on where they stand by measuring how they are living up to their own values and standards. This is much healthier and will prevent some unnecessary distress. No matter how much we try, there will always be someone better than us, at whatever we do. However, there is no one better at being you than you.

Clearly an exception for comparing yourself to others is when the Chimp gains strength by doing this. For example, in sport the Chimp might find it useful to compare itself to its rivals in order to drive itself to train and compete.

Reassure your Chimp with values if it compares itself to others

Don't allow your Chimp to do a juggling act

I run a weekend conference each year for members of the public. It's an entertaining social weekend but has some serious themes. As a break from the discussion groups, I took in a lot of small apples that had fallen from some apple trees. I asked each person to see how many apples they could juggle. Most managed two or even three but no one got to four. The reason I did this fun exercise is to remind delegates that the mind is very similar when it comes to juggling. Most of us can manage to juggle two or three problems in our heads but once we have four or more, we can't manage. The mind doesn't do well at solving multiple problems simultaneously. What happens is that the Chimp takes over and moves from one problem to another repeatedly going in circles and solving nothing.

You can help your Chimp to escape from this juggling act by writing down each problem. You can then enter Human mode by addressing them one at a time and finding solutions.

Chimps also try to do a juggling act when writing out 'to do lists' that are too long. It is found that as a rule of thumb, a 'to do list' of six items is optimum, if you wish to get them all done in one day. Try not to overload or overwhelm the Chimp with jobs to do or problems to solve. Work out what works for you, but *the general rules are:*

- *Six priorities or less per day on a 'to do list'*
- *Just two problems before you write them down and solve them individually*

[125] [126]

Learning to say "no"

Many Chimps feel that they should always say "yes" when asked to do something. This can lead to cases of resentment or feelings that others should then also say "yes". What would be good to work out is why someone would feel obliged to keep saying, "yes" to every request.

Here are some of the common Gremlins of belief that can be lurking in someone's Computer:

Saying "no":

- Is proof that I am not a pleasant person
- Might upset someone
- Might make someone think badly of me
- Makes me feel bad about myself

Saying "yes":

- Makes me a better person
- Makes me more popular with others
- Will stop me from letting others down
- Will stop me from feeling guilty

There could be many reasons why some people allow these Gremlins to remain in their Computer and able to prod the Chimp into saying "yes".

These Gremlins need to be removed and replaced by suitable truthful Autopilots, such as:

Saying "no":

- Is a sign of respect to my Chimp and myself
- Is sometimes the right and appropriate answer
- Might upset someone, but "no" will be accepted by a reasonable person
- Is proof that I can use discernment

Saying "yes":

- Doesn't make me a better person
- Can be a sign of weakness in the eyes of others
- Could unsettle my Chimp and cause me stress
- Can be unhelpful and inappropriate

You are not obliged to say "yes"

The lesson to learn is, when asked to take on a new task, take your time and reflect. Be selective and decide whether "yes" or "no" is the most appropriate answer. You might need to start by checking the beliefs in your Computer and tidying them up. Then practising saying "no"! [127] [128]

Discussing your plans with your Chimp

Many plans are unsuccessful because emotional aspects are not taken into account and addressed. Concerns and fears that the Chimp might have are always better brought out into the open and discussed. Allowing your Chimp to exercise without judging it is helpful. The exercise itself can bring perspective as the Human listens to the Chimp. [81] [88] When I have encouraged people to speak their fears aloud, it's not unusual for them to come to the end of their dialogue only to smile. This is because they have now seen things differently just by listening to their own Chimp. As your Chimp expresses itself, your Human can listen. [81] Always consult your Chimp in any plans you make and allow it to express itself!

If you find talking uncomfortable, then you could try writing down your feelings. [129]

Example: *Courtney and the Wedding plans*

Courtney has completed her wedding plans but her Chimp is still unsettled, despite everything being ready. The problem is that the Chimp needs to express its fears, such as its concern that the weather might ruin the day. This is a real concern, but by talking things through Courtney can bring some perspective. She could redefine her perfect day, which doesn't rely on the weather. She could choose to redefine the perfect day as getting married, having friends and family with her, and enjoying the moment. Of course, if Courtney decides that the day will not be perfect unless the weather is good, then she must accept the consequences that the day could be ruined. This will inevitably generate negative emotions from her Chimp.

By discussing emotional aspects of any plans, we can avoid negative emotions.

Commitment and motivation

There is a big difference between commitment and motivation. To get things done, the Human uses commitment, whereas the Chimp uses motivation.

Commitment is a structured approach with a plan, which works ***regardless of how we feel***. Motivation relies on emotion to carry out a plan. Motivation can help to bring added enthusiasm, but it relies on feelings, which can change quickly. Motivation is like icing on the cake, where commitment is the cake. We can get things done without motivation because we can choose commitment. Motivation is in the hands of the Chimp and only you can determine what will motivate your Chimp. Small steps and fast rewards are the usual motivational drivers.

Example: *Violet and commitment*

Violet is a surgeon. She is performing a six-hour operation and is now three hours into the work. The operating theatre is hot, and she has become very tired. Imagine how the patient would feel if Violet went into Chimp mode and said: "you know what, I just don't feel motivated to keep going". Instead, Violet allows her Human to say to the Chimp, "it isn't important how you feel, it's about commitment regardless of feelings."

You always have a choice when taking on any task: you can either commit to it or go with how motivated you feel. It's not surprising that once we've chosen to use commitment and see the work being done, our Chimps often become motivated!

Start with commitment and motivation might follow

Commitment and motivation
- Scientific points
Various areas of the brain have been researched when looking at motivation. Examples include:
- The reward system: this involves dopamine release from the ventral tegmental area, which is aimed at the nucleus accumbens: [130]
- The orbitofrontal cortex: has a 'Go or No go' approach, so that if a reward ceases then the activity stops. For example, praise from others, without which an activity might cease [131]
- The anterior cingulate gyrus seeks rewards by removing inner conflict [132]

Motivation can be further divided into external and internal motivating factors.

An external motivating factor means that the reward is outside of the person, for example gaining money. Motivation will diminish without the reward. External rewards are predominantly based on the reward system, which then strengthens a specific behaviour by influencing the hippocampal formation. [133]

Internal rewards, such as self-satisfaction, self-praise or altruistic behaviour, are pleasing feelings within the mind and are mainly based on the dorsolateral prefrontal cortex.
Commitment is a rational choice formed in the dorsolateral prefrontal cortex. Here, facts and information are combined with logic to rationalise the benefits of doing something. Motivation can lack the use of rationality and be influenced purely by outcomes. Commitment is therefore based on a rational plan that does not depend on feelings.

The CORE Principle

Successful people usually have a structured approach to challenges. I have combined what I feel are the four major elements to a successful approach.

C = Commitment
- working out what it will take to succeed and being ready to deal with anything that might stop you. (See pages 176-178 for full details).

O = Ownership
- having and personally owning a plan of action. Therefore, both your Human and Chimp agree and endorse the plan.

R = Responsibility
- being responsible and accountable for putting the plan into action and measuring your progress and reporting back.

E = Excellence
- ensuring that you gave it everything, as this is your own personal level of excellence.

Efficient versus effective

The CORE principle might give a successful approach but the way that we work needs to be effective:
- **Effective** means that what we are doing will give us the outcome that we want

- **Efficient** means that while we are working, we are using our effort and time well, ***but*** what we are doing isn't necessarily going to get us the outcome that we want.

The Human chooses commitment and works with effective and efficient plans. The Chimp uses motivation and adds on efficiency.

When in Chimp mode, we can lose focus on what we are trying to achieve and put in a great deal of effort, while being efficient, but fail to do well. It is worth asking yourself are you working effectively or efficiently? In other words, are your actions going to achieve your aims? Sadly, we can often be very efficient but not very effective. Ideally, we would want effectiveness with efficiency.

Example: *The medical student*

Stephanie and Garry are medical students and have just eight weeks to learn all they can about paediatrics (childhood) before they take an exam.. Garry learns by summarising a huge text book and tries to cover all areas. His approach is very efficient because he has covered all of the subject. Stephanie learns by being effective. She first works out what is important to learn for the exam and saves time by focussing on the things she needs to know.

The result at the end of the eight weeks is that Stephanie is more likely to do better in the exam.

Ask yourself: Is what you are doing effective or is it just efficient?

> **Key Point**
>
> ***Effective working is better than efficient working.***

Unit 6 Reminders

- Nurturing your Chimp will help to bring the best out of it
- To get things done, the Human uses commitment, whereas the Chimp uses motivation.
- Commitment often motivates our Chimp

Unit 6 Exercises

Focus points and reflective exercises

1. *Planning links to pleasure*
2. *Managing multiple challenges*
3. *Commitment and motivation*

Focus 1: *Planning links to pleasure*

This is such a simple and obvious thing to do but we tend not to do it. Whenever you have a task to do that is not going to give you pleasure, try to be inventive and either link the activity to a pleasurable experience or try to add a reward at the end of the task for when you have completed it.

A laborious or tedious task can always be made easier if it is broken down and small rewards are added at each completed stage. The Chimp is the one who wants the reward; Humans just get on with the task. Therefore, if your Chimp is playing up, ensure the reward is in line with the Chimp's idea of a reward. For example, most Chimps like someone to appreciate what they have had to endure and once they feel they have had this understanding they settle down. Letting someone know what you have gone through, or achieved, is often a very good reward for the Chimp.

Exercise: *The 'when I get around to it' task*

Choose a task that you have to do but are struggling to get around to doing it: maybe something that has been waiting to be done for some time. Clearly, it is your Chimp that is stalling if you yourself want to get the task completed. Agree with your Chimp how you will break the task down and what reward you will give your Chimp when each phase of the task is completed. One of the secrets of keeping going is to refuse to engage with any Chimp chatter once you have begun. Although we cannot control our Chimps, we can be firm with them and learn to refuse to engage with them!

Focus 2: *Managing multiple challenges*

Most of us have multiple tasks to complete every day. Many people avoid the more difficult or time-consuming ones and instead apply themselves to the quick fix or easy tasks.

Exercise: *Forming and executing a priority list*

If you'd like to become more organised and efficient, try starting the day by writing out your six most important tasks and put them in order of importance. It's crucial to do this in a rational way and not allow your Chimp to bring in emotions and draw up an easy fix list. Remember that research indicates that a list of six priorities is likely to be executed and finished rather than a long list of priority jobs mixed with non-important jobs. [125] [126] Doing a six point ordered priority list on a daily basis can programme your Computer to work effectively. Mastering this habit could help you to become more organised and efficient.

Focus 3: *Commitment and motivation*

This focus point overlaps with the previous ones. It is really about actively removing your emotions from a situation and doing what you need to do, regardless of how your Chimp might feel. If you can establish, as a habit, using commitment rather than motivation, you will find that you will get things done in a timely manner and even your Chimp will feel better for this!

The secret to managing the Chimp in situations where motivation is an issue, is to learn to refuse to engage with the Chimp.

Exercise: *Managing the Chimp by actively blocking it*

Is there a task that you want to accomplish that you are struggling to complete? Try to begin with the word 'commitment'. In other words, I will do what needs to be done regardless of my Chimp's feelings. The Computer will take over with a plan, if you can maintain your focus on the process of what you need to do and not on how you feel about doing it.

This principle can be used in any area of life, and in any situation, where you want to achieve an outcome but your emotions could prevent you from performing your task.

Please remember: You can keep your focus on a process, by not looking at the whole process, but by only looking at the next step of the process. This will prevent your Chimp from focussing on how large the task is or on how long it might take to complete it. This can overwhelm the Chimp and paralyse you from starting.

Unit 7
Managing your drives

Unit 7: *will consider how we can manage powerful and compelling biological drives by drawing a line. The eating drive and the security drive will be used as examples to demonstrate how to recognise and manage your Chimp's input.*

Fulfilling the Chimp's drives

The biological drives that we possess are clearly there for a reason. They help us to survive and they also help us to thrive. Therefore, they are not unhelpful or destructive but rather they are there to be managed and utilised. [134] [135] [1] [136]

Key Point

All of our drives are helpful, if we manage them well.

Our Chimps have biological drives that need fulfilling. The starting point is to list the Chimp's drives that you will want to address.

The diagram shows some important drives:

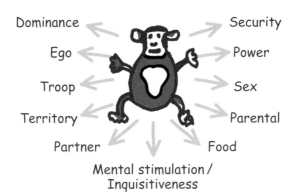

If the Chimp's drives cannot be met directly, then the Chimp will sublimate the drive to gain fulfilment. This means that it will find a way of fulfilling the drive, but not in the way it was originally intended. For example, the parental instinct can be sublimated onto a pet. The drive gains fulfilment and the Chimp is happy.

By going through the list of drives, you can select those that are strong in your Chimp and make sure that they are being fulfilled or sublimated **appropriately**.

If we don't fulfil the Chimp's drives, then this frequently results in discontent or the drive being applied inappropriately.

The Chimp could still gain fulfilment from sublimated drives

Example: *Valerie and the unrecognised drive*

Valerie feels that her life is not fulfilling but doesn't know why. She says she is happy and has lots of friends, a great family and a good job, which she enjoys. So why does she feel unfulfilled?

One of the drives that nature has instilled within us is to search for mental stimulation, which can present as inquisitiveness or curiosity. By finding excitement, taking on challenges or experiencing novel situations, we are constantly learning. This would improve our chances of survival. This drive is not as obvious as most of the others, but it is nevertheless a very important one. Nature drives us to learn.

Valerie has to first recognise the drive and then decide how she would best fulfil this. It could be by exploring new places, by taking on a charity challenge, by joining a new group or by learning a new skill. When she has engaged in an activity that fulfils her drive then she might find the contentment that she is missing. What she will have done is brought purpose and some direction into her life.

Managing your drives by drawing a line

Fulfilling drives directly or by sublimation is healthy but there is a slight problem! The Human knows where and when to draw the line and also when to be satisfied when fulfilling any drive. The Chimp on the other hand doesn't know where or when to draw the line and is only temporarily satisfied with what it has achieved. [137]

The Human needs to recognise and manage the Chimp's drives.

The Chimp just keeps on redrawing the line. If we don't help the Chimp to stick to the line that has been drawn, this can result in unhelpful emotions and destructive outcomes.

Example: *Marcus and the drive to achieve*

Marcus has a Chimp with low self-esteem. Marcus thinks that if he can pass his driving test on his first attempt then he will feel so much better about himself and his self-esteem will be fine. Marcus takes the test and passes first time. He is elated but this elation doesn't last. Before long, the Chimp reminds Marcus that there are lots of people who pass first time and it isn't that great an achievement. Therefore, the Chimp decides that if he can manage to be selected for the local football team then he will have truly achieved something special. He works hard on this and is finally selected. The elation is again short-lived. Once again the Chimp dismisses the achievement as not being that great and searches again. The Chimp has no idea where to draw the line. If Marcus does not intervene this can continue indefinitely with no improvement in self-esteem and, even worse, a lowering of self-esteem if Marcus should fail to reach his goals.

The Chimp can downplay its achievements

Managing your drives

Example: *Sheila and improving her status*

Sheila has an office management role that is well paid. She has always had dreams of owning her own home and having a good car. By hard work she has achieved this. Her problem is that she now wants a better car and a bigger house. She always seems to spend more than she earns, and whatever she buys doesn't seem to satisfy her. To the Chimp, owning possessions will offer security. The problem is that the Chimp cannot draw the line and feels that what it has is not enough. It is healthy to have a drive for status but only if it is contained. Her Human has to come to the rescue and draw a line. The line is to let the Chimp know that when it has achieved what it set out to do, it will enjoy it and not try to keep on improving at the expense of its peace of mind and happiness.

In-built drives

Our in-built drives cover all of the survival and thriving requirements for the species or an individual but vary from person to person in their strengths. Drives also change in strength and importance during our lifetime. What we will do now is to focus on two of the main drives as examples: eating and security.

Individuals differ in the strength of their drives

Key Point
All drives are based on survival and are helpful but need the Human to manage them.

Managing the eating drive

The eating drive is clearly critical to survival. In order to make us eat, the body experiences a craving for food. The body supports this craving by releasing pleasurable hormones and transmitters when we do eat. [113] [138] [139] Therefore, eating has a craving aspect to it and also a pleasurable aspect to it. This could almost be seen as a food addiction that we all have! It is understandable that many people will struggle to contain the drive. [140]

In order to manage the drive practical measures need to be taken! It's no use telling the Chimp that "instead of eating crisps all night, we are going to go for a walk", if the Chimp then puts a couple of packets of crisps in your coat pocket as you set off. Being realistic, we have to put in place elements that will make it virtually impossible for the Chimp to act out its drives. The Chimp's approach is to overeat for survival and pleasure but doesn't know where to draw the line because it doesn't appreciate that overeating can cause serious health problems. Sadly, there is no set of instructions to follow when it comes to managing Chimp drives because everyone is unique, and you have to work out what will work for you.

How can we help the Chimp to draw the line? This all depends on why the Chimp is being driven to eat. If we know why it is doing this, then we might be able to change the Chimp's behaviour. It might be eating because:

- It is trying to feel full because it thinks feeling full is normal
- It has developed poor habits
- It makes poor food choices
- It is using food to manage stress

Here are four more examples illustrating common reasons why a Chimp might keep eating.

Example 1: *Joseph and low self-esteem*

Joseph has low self-esteem and worries about social interactions. In order to make himself feel better he has learnt to use food as a comfort. [141] This has caused a weight gain and his Chimp is now berating Joseph even more. The answer is to recognise what the real problem is and to address this. While Joseph learns to build his self-esteem he can consider different ways to gain comfort. For example, he could talk through his feelings and be understood. [142] [143] Alternatively, he could build his self-esteem by getting involved with a charity or join an interest group. Joseph no doubt recognises that his comfort eating just adds to his problem.

Example 2: *Tina and her beliefs*

Tina appears to have the belief that she is a food bin. Her Chimp is listening to a Gremlin in her Computer that says that no food should ever be wasted. Therefore, if any food is left over she must eat it. It's a useful exercise to search for any beliefs you might be holding that need challenging and removing.

Example 3: *Jason and pleasure*

Jason is concerned about his health because he just can't resist certain foods! The release of dopamine in his brain gives great pleasure. Food can activate the release of dopamine but so can lots of other things. [113] [138] [139] Jason has developed a habit of using food to gain pleasure, without considering if he could substitute something else for the food. The important point here is that he needs to replace the food with something he finds more pleasurable but healthier. For example, this could be jogging, reading a book, watching a movie or meeting up with friends.

There are alternative sources of pleasure

A change of mind and a change of food
- Scientific points

Nature has cleverly programmed us to vary our diet. The hypothalamus gives us a hunger drive. [140] [141] Our orbitofrontal cortex then searches out our favourite food. It compels us to eat our favourite food and to find the food attractive. [142] [143] [138] [113] [139]

Once we have eaten so much of our favourite food, the OFC then changes its mind and perceives our favourite food as being unpleasant. This makes the OFC move on to other foods. This is nature's way for making us eat a varied diet with different foods. This is why if you stick to a diet that only allows you to eat a few chosen foods it tends to work. What these restricted food diets are doing, is helping your OFC, the Chimp, to become repelled at the thought of eating too much of the same food. Imagine being allowed to eat only bananas and carrots for the day. After eating a few, even though you might feel hungry, you're unlikely to keep eating more of them.

This inability to eat too much of the same food also accounts for why our Chimps overeat at buffets by varying what they eat and also why after eating a heavy meal and feeling full, the Chimp still manages to eat some dessert.

If you are allowed to eat what you like, then you can keep varying the food until the Chimp has overeaten and feels uncomfortable. To prevent overeating, feeling full is a last resort: an inability to vary the food is a much better way.

Example 4: *Rachel and her grazing habits*

Rachel doesn't eat large meals but continually grazes on small amounts of food throughout the day. Even after a reasonable size meal she continues to pick at various foods.

Rachel's experience is that she can keep eating provided she changes the type of food that she is eating. She loves pizza and can happily eat her way through a large pizza. However, when she feels full from the pizza, she finds the idea of eating more pizza unpleasant. However, if she is then offered some cheesecake, somehow she can find room for this. Her Chimp has changed its perception of the pizza and that makes the Chimp look for another food that might be appealing.

The solution is to decide on just a few healthy food substances that can be eaten all day long and to draw the line there for the Chimp.

Substitute two healthy foods you can eat all day

Managing the security drive

The drive for security includes such things as: having defined territory, shelter, support from others, familiarity and routines. The security drive can overlap with many other drives, such as territorial, eating or partner finding. This drive has been adapted to fit into the society that we have created and this causes some difficulties. For example, we have previously considered how the opinions of others are a strong feature to provide us with security. Even within a chimpanzee troop, the 'opinions' of other members of the troop are important. Therefore, we see chimpanzees trying to impress or curry favour with other chimpanzees. [144] With the arrival of social media into our society, we now have a mass of opinions to contend with, and some will not be favourable. It's impossible to please everyone and this causes our inner Chimp some stress!

The need for approval by others
- Scientific points

One of the functions of the Orbitofrontal cortex (OFC) is to alert us as to what is accepted socially. [145] The OFC therefore helps us to look to others to ensure that we are popular, have approval and stay within social norms. People who have some damage to the OFC can fail to do this by not recognising when they have made a faux pas. [146] Typically, we are careful not to say something that would be offensive to others and this effectively keeps us from being rejected. Rejection or perceived rejection can have devastating effects on our psychological health. The OFC is the centre for ensuring that others will accept us. [147]

The meaning of security to the Chimp and Human

Security for the Chimp and Human is different. Both of them look for security by employing internal and external factors. However, the Human puts a great emphasis on the internal factors and the Chimp puts a great emphasis on the external factors.

Security for the Chimp is not the same as for the Human

- Possessions
- Familiarity
- Friends (Troop)
- Achievements
- Territory

(Chimp thought: External factors mean more to me.)

- High self-worth
- High self-esteem
- Living by values
- Acceptance

(Human thought: Internal factors are more important.)

The Chimp is basically looking to build a secure feeling from what it can achieve, what possessions it has, what people think about it and what territory it owns. Whereas, the Human looks to implement values, self-worth, self-esteem and inner peace to find security. Both Chimp and Human use self-image but in different ways. The Chimp wants a good self-image to impress others, which gives it a sense of security. The Human wants a good self-image for self-approval by living out and fulfilling their values.

Practical management

Clearly, it helps to make both Human and Chimp feel secure. In order to do this, we need to obtain a balance when it comes to the Chimp. The Human can achieve a sense of internal worth and security not just by looking to their values but also by working with the reality of life. Therefore, the Human accepts that we have to live with some measure of insecurity because we cannot control every factor in our lives. The Chimp cannot accept any sense of insecurity. Even when it obtains a level of security, it will often find something to be worried about and draw a new required level of security. This means the Human must intervene and bring the Chimp back into reality with some home truths. These grade A hits will settle the Chimp down.

Here are some possible grade A hits:

- Security can never be absolute
- Every day things will change and I cannot stop this
- Some days will be good and some not so good
- I am an adult and can manage whatever comes my way
- There are always people who will help me
- To stop my Chimp from worrying, I need to draw a line

Example 1: *Holly the artist*

Holly is an artist who has an accomplished record of producing many pieces of excellent artwork. Her Human and Chimp both wanted to do this and some years ago set out to achieve it. They might have had very different reasons for wanting to produce excellent artwork, and for wanting to achieve, but at the time of setting out they were in agreement.

When Holly achieved her objectives, her Human was proud and could reflect on what she had achieved and hoped to do more. However, her Chimp's celebration was short lived. It soon dismissed what she had achieved and wanted to raise the level again by re-drawing the line of what it takes to feel a sense of achievement. The drive to prove itself and to achieve is strong.

In this example, the Chimp has demonstrated several ways in which it is predisposed to act:

- Whatever the Chimp achieves it will soon belittle
- The Chimp fails to recognise and celebrate the value of what it has achieved
- Whatever standard is reached, the Chimp will need to re-set the level in order to reassure itself that is it good enough
- The Chimp stresses about not being able to achieve the new higher standard or even the same standard again

Holly's potential problem is not recognising what is happening and stopping the Chimp from re-drawing the line.

The Chimp can dismiss any achievements and redefine what makes it happy

Example 2: *Dan as a provider*

Dan is the father of three young children and wants to provide them with the best quality of life that he can. He works hard and has a mortgage on a house that he was proud to buy for his family. They have a car that gets them around and they have a reasonable life style.

Here comes the dilemma; Dan's Human can see that the family are doing fine and that his quality time spent with them is good. He is fulfilling his parental and security drives.

Dan's Chimp has decided to redraw the line and is looking to keep on improving things. It will drive him to make more money and might well look for a better house and a better car. There is nothing wrong with this. However, it is up to Dan to decide whether to allow the Chimp to re-draw the line. If it redraws the line, there are potential consequences that could diminish the quality of his family life. If he stays with the status quo or makes small improvements, he won't compromise the family quality time. Dan has to decide where he wants to draw the line.

I have lost count of the number of parents who have been referred to me, who describe the same problem of not getting the balance right for their family. They have tried to improve things, which is good, but ended up sacrificing what quality of life they had. Only you can decide what is right for you. What I would like to draw your attention to is that our Chimps will continually re-draw the line because it is in their nature to do so, but they often do not consider the consequences of their actions.
[148]

> **Key Point**
>
> *Always consider the consequences of re-drawing the line.*

Example 3: *Gemma the linguist*

Gemma is learning a language and has not found it easy. However, she has managed to be able to communicate at a basic level. Her Chimp is now dismissing this achievement and is pushing her to learn much more. This is distressing Gemma because the Chimp is also fearful that it won't be able to improve. This is very typical of our Chimps; they push us to do things, then have a meltdown when it comes to doing them!

This scenario demonstrates a very different aspect to re-drawing the line. This time her Chimp is probably doing the right thing in pushing her to re-draw the line. The Chimp's problem is not celebrating what has been achieved and trying for the new line without fear. Sometimes there are advantages to re-drawing the line and our Chimps are right to encourage us to do so. Her Chimp just needs support to commit to trying.

The examples demonstrate a principle that the Chimp operates by. There is nothing wrong with re-drawing the line that we originally drew, unless we miss out on celebrating what we have achieved and also don't consider the consequences of re-drawing the line.

Key Point

It's good to celebrate when you have reached any line that you have drawn.

In all of the examples above there are no rights or wrongs, there are just thought-through decisions. Dan may well take on a more stressful job or move house to better his family and this could be the right decision for him at that time. He may have thought through the consequences and be prepared for them. Gemma may have decided to stop at the level that she has reached and enjoy what she has achieved.

Key Points

- **The drive to find security is different for Humans and Chimps.**
- **The Human puts an emphasis on its search for internal security; the Chimp puts an emphasis on its search for external security.**
- **The Human needs to help the Chimp to draw the line in its search for external security.**

Redrawing the line needs managing

The drive to achieve can become a millstone around our necks and paradoxically result in us becoming unhappy, dissatisfied and often feeling like we are failing.
[149] [137] [150]

Example: *The elite athlete*

I once worked with an elite athlete at the highest level within their sport. Their target was to win a gold medal in a major championship. As we approached the championship, they were completely convinced that if they achieved the gold they would be satisfied for life. The day of competition came and they excelled. They got their gold and also achieved a world record. Their elation was off the scale. Some weeks passed and we met again. Their Chimp had completely redrawn the line of what constitutes success and what is good. Their Chimp then offered me one of my favourite Chimp quotes: "I know I set a world record but I didn't set a good world record".

Already the Chimp realised the world record they had set could possibly be broken and it wasn't happy. Sometimes we have to pull our Chimps into line and remind them to enjoy and celebrate what we have achieved, no matter how big or small. Thankfully, the athlete saw the funny side of this and went on to enjoy their gold medal.

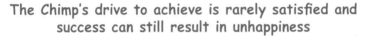

The Chimp's drive to achieve is rarely satisfied and success can still result in unhappiness

Can we avoid the pitfall of inappropriately re-drawing the line?

Yes, there are things we can do that can help us not to fall into potential pitfalls when considering whether to re-set our goals or standards:

- Humans know where to draw the line and whether it's helpful or not to re-draw the line
- Start with a blank page and decide where you want to draw the line that marks success for any particular project or drive fulfilment. It is likely that the Chimp will agree at the start of the project on what constitutes success and an acceptable standard
- Each person is different, and you must decide what works for you
  For example:
  - How much do you want to earn that you believe will keep you happy?
  - What possessions do you want to have, that your Chimp will agree to?
  - What is the lifestyle that will fulfil your basic drives, such as food, sex and territory, while not compromising your values? Remember that the Chimp will agree at the time, but it is quite possibly going to try to change the line once it reaches that level
- When you have reached your line, or as close as you can get to it, celebrate your success and don't allow your Chimp to take this success away
- If you are deciding to re-draw the line, then look at the consequences or impact that changing to a new level will bring
- Consider the rewards that working to a new line would bring and make sure they are worth it
- If it is a drive that is going too far, then try and work out an alternative way to satisfy the drive
- Check that a drive isn't being used to satisfy an area where it doesn't belong

With the demands for wanting more, are our Chimps drives insatiable?

Not necessarily. They can be, if you don't manage them. You have to work out what your own drives are and recognise the needs you have. Some people find they do reach satisfaction in life, and have a very healthy status quo and lifestyle, without constantly wanting to better themselves or have more. It's a bit like having a thermostat in your head for achievement or possessions. The problem is, that at times, some of us appear to have no top end to our thermostats! We have to recognise when to be content and when to push on.

Feelings of missing out

Some people experience a feeling that no matter what they do or where they are, they feel as if they are missing out: a learnt behaviour. [151] Sometimes, in the moment, all seems well but before long the same feelings return. It is often described as a feeling of something missing inside. Being dissatisfied or always wanting to be in a better place can be a Gremlin: a learnt behaviour. A more constructive behaviour is to accept that life could naturally have moments of inner peace, which you can enjoy, without having to constantly search or strive to create or find those moments.

Can the line change over time?

Yes! As life changes, so do we, and it might be that you need to adjust the line up or down. It might be hazardous staying with the same line throughout life without challenging where you have reached and what is appropriate now.

Very different drives behind choosing an activity

Lottie is a Rugby coach and she has a large squad of players. She wondered why each player had chosen to play rugby. By discussing this with the players, she found that the players had very different reasons. Here are some of the answers they gave:
- Just to enjoy the game
- For health and fitness
- For team bonding and common purpose
- To prove to others that I can do this
- To prove to myself that I can be good at something
- To define myself
- To have a feeling of power and success
- To be part of the best team around

We can see that many of the reasons given are based on primitive survival drives that have been transferred into sport. This can therefore be a very healthy way to fulfil the Chimp's drives. Lottie can now develop aspects of the game that will help to strengthen each player's commitment to the team, and to the game, by promoting the drives each individual has. Provided the players can recognise what is behind their desire to play, the Chimp can appropriately keep re-drawing the line to improve performance or it can be helped to manage a drive that is becoming destructive. In some cases, the drive being used to play the sport might not be appropriate and it is worth changing this.

How people fulfil drives is unique to them

Unit 7 Reminders

- Our biological drives are helpful, if we manage them appropriately
- Drives can be constructively used in different ways
- Always consider the consequences of re-drawing the line
- It helps to celebrate success on reaching the line you first drew
- Humans use internal values for security, whereas Chimps use external assets and achievements

Unit 7 Exercises

Focus points and reflective exercises

1. **Reflection on drawing the line**
2. **Reflecting on the security drive**
3. **Reviewing your drives**

Focus 1: *Reflection on drawing the line*

This focus point is to consider the drives that you are experiencing in your life and deciding which ones are worth investing energy into in order to satisfy them, and which ones need to have a line drawn under them.

Example: *Colin and satisfaction*

Colin and his Chimp strives to better themselves. However, the Chimp is never satisfied and instead of reflecting on what it has got and what it has achieved, the Chimp only sees what it could have and what it lacks. This can be helpful to drive Colin further, but it has to be managed. Colin recognises the need to stop and reflect on what he has and be grateful for this, however, he struggles to convince his Chimp. The way that Colin manages the situation is to agree with the Chimp to strive for better but to stop and have some days where striving ceases and relaxation, gratefulness and celebration take over.

Exercise: *Reflection on managing your drives*

Reflect on how much time you spend striving to fulfil drives and how much time you spend relaxing, being grateful and celebrating successes.

Focus 2: *Reflecting on the security drive*

This focus point is to reflect on one aspect of security: feeling at peace with yourself.

The security drive we have is often seen as just a physical one; securing a home, a job, a partner and so on. The aspect of feeling secure within yourself is often missed, and yet this is arguably more important than any of the physical or other psychological security needs.

Respecting yourself, by living out your values, can help immensely by giving you a ***sense of security and peace within yourself.*** Clearly, we need to provide physical security but an inner sense of security can often make up for any lack of physical or other psychological deficits. Effectively what we are doing is going into Human mode ***before addressing our physical or psychological Chimp needs***, instead of entering into Chimp mode, which usually loses perspective when it comes to security.

Exercise: *Addressing your security drive*

Try to practice entering Human mode and create inner security before addressing any outer security issues. This practice can be done by imagining security issues that could arise during your day, such as suddenly feeling vulnerable by someone making a criticism directed towards you. Turn to your values and be at peace with yourself first before returning to address the problem that has presented.

Focus 3: *Reviewing your drives*

As we possess some very strong and compelling drives, it can be a useful exercise to review these drives to see how many we are fulfilling in a constructive way. The drives can be fulfilled either directly or we could sublimate them in a healthy way

Exercise: *Reviewing your main drives*

Complete the table to review where you are with each of the drives: then take action as needed! Some of the drives will be covered in later units; therefore at this point it is a measure of where you think you need to put your attention. Explanations of each drive are given to guide you.

Drive	I am happy	I ought to rethink	This needs addressing
Security			
Territorial			
Shelter			
Eating			
Dominance			
Sex			
Partner			
Troop			
Ego			
Role in life			
Stimulation			

Here are some explanations for each drive:

The **security drive** is fulfilled when the Chimp is feeling relaxed and content. Evidence that the Chimp is not feeling secure can either be by feelings of apprehension or worry or it can be by behaviours, such as trying to control people, events or surroundings inappropriately.

The **territorial drive** can be thought of as the extent to which your Chimp feels content with the boundaries it sets to protect what it believes belongs to it. This can include physical boundaries, possessions and your role in life. How relaxed is your Chimp about these boundaries being well defined? How relaxed is it about managing any of these being invaded or threatened?

The **shelter drive** is easier to define because it just means; do you have the right safe and private space that your Chimp needs? For some people this is really important, whilst for others their Chimps can be happy sharing space.

The **eating drive** is self-explanatory. Is it being managed well for the level of health, fitness and weight that you want to be?

The **dominance drive** is healthy if used appropriately. The Chimp will use this to show its superiority over others and to suppress others in order to elevate itself. The Human can use this drive by sublimating it into helping others and supporting them to improve their self-esteem or sharing and helping them to overcome adversity or problem-solve.

The **sex drive** is complex. It is arguably the strongest and most difficult drive that we have to manage. Everyone differs in the way that they experience this drive and how it presents but here are some common reasons why the drive can be challenging:
- It is meant to be strong in order to perpetuate the species
- It is compelling because sexual tension builds an appetite for sex
- The drive can be displaced inappropriately towards others
- It is influenced by moral values that can conflict within individuals
- It is influenced by society's values
- It can be inappropriately linked to other drives, such as dominance and security
- It is frequently perceived as a source of embarrassment or denial by some individuals

The question that you can ask yourself is: "Do I feel at ease with my sex drive and how it is being fulfilled?". If the answer is "no", then it is wise to address the areas that you think are causing the unease. These could be about behaviours, beliefs you hold or situations that you are in that need addressing.

The **partner drive** is another complex drive. We are driven to find a partner in order to secure the species; therefore this drive is usually strongly linked to the sex drive but not always. Reflect on how you want to employ your partner drive and making it practical and in line with your values. This will help you with input into your relationship or to finding a partner.

The **Troop drive** is such a critical drive to get right. This will be covered in greater detail later in the course. It is added here for completeness. It is sufficient at this point to measure whether you feel that you have a group of unconditionally supportive people in your life. If the answer is not a definite "yes" then you can address when we reach this later in the course.

The **ego drive** is about feeling good about yourself. There are many interpretations when it comes to the word 'ego'. We will consider it as having a positive image of yourself and a great relationship with yourself. This is paramount to developing good self-esteem. The question that will help you to improve your self-esteem is to ask yourself at this point, "How good am I at separating myself from my machine and celebrating the real me, not the me that the world sees after interference from Chimp and Computer?"

The **role or purpose drive** defines how we see our place and usefulness to others, society and self. Be wary about how you measure this. Your Chimp will always look at what you are achieving in terms of tangible measures, such as money, prestige or accomplishments. There is nothing wrong with this but more importantly is this in line with your values? Your Human will look at measures such as, how much energy and happiness that you bring to others. This drive is strongly linked to a sense of satisfaction.

The **stimulation or pleasure drive** is about your innate need for physical and mental stimulation. It encompasses curiosity and discovery and new experiences. How much are you employing this drive, which will bring richness to your life? Do you have plans in place that work for you and keep this drive satisfied? How much are you looking after your own happiness? What efforts are you making and most importantly, are they working?

There are of course other very important drives that you might want to add, such as parental drives and the drive for independence. Many of the drives overlap and can be worked on under different titles. The point of the exercise is to consider how you are managing the powerful driving forces that will compel you to act.

Unit 8
Emotion – learning an internal language

STAGE 3: *focusses specifically on understanding and managing our emotions.*

Unit 8: *explains emotional messages from the mind and how to work with these. By the end of the unit, you will be able to speak to the Chimp in its own language.*

How the Chimp constantly sends us messages using emotion

Our body communicates with us to let us know what it needs us to do. It can do this physically by making things uncomfortable for us, or alternatively, by making things feel better. For example, when we need to drink we become thirsty and this automatically drives us to find fluid. If we get too hot, we search out shade or try to cool ourselves down. These examples demonstrate how our physical bodies send us messages and it is up to us to interpret this message and act appropriately.

Likewise the Chimp constantly sends us messages. Whatever experiences we have in life, the Chimp will always interpret the experience first and then send a message in the form of an emotion to the Human. The Human needs to interpret this emotional message correctly before acting.

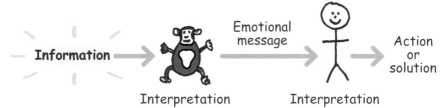

The difficulty with the Chimp is that we often don't understand what it is trying to say to us and therefore we can misunderstand or misinterpret emotional messages.

For example, if your Chimp becomes worried, it could express anger instead of anxiety. If you don't recognise this, then the anger could be misunderstood and mismanaged, which could make matters worse. Learning to understand these emotional messages relies on your understanding of how differently the Chimp and Human communicate. The Human speaks with a rational language based on facts and logic. [8] The Chimp speaks with a language based on intuition and emotion. [7] [3] [152] [153] They can both speak a little of the other's language but in order to be effective, our Human needs to learn to understand and speak to the Chimp in its own language.

Emotion - learning an internal language

Learning the emotional language of the Chimp is a skill because although the emotions offered by the Chimp might be helpful, they can also be over-emphasised, inappropriate or even misleading. [154] [155] [156] [157] [158]

> **Key Point**
>
> *The Chimp is constantly communicating with us and it helps to learn its language.*

Two different languages

Example: *Two messages*

I once worked with an Olympic athlete that used to complain of becoming very anxious before competitions. This is a fairly predictable response from a healthy Chimp system. We looked at this anxiety from the Chimp's perspective. She learnt that her Chimp was merely sending a message to warn her. The Chimp had interpreted the situation as the athlete putting herself in danger by taking on unknown Chimps that could be stronger than her. When the Human interpreted the Chimp's message and added some rationality and perspective, she learned to reassure her Chimp and then focus on the process of competing.

All was going well until the Chimp sent another message to her. This time the message was a feeling of 'anxiety', despite her apparently being in a good place.

Her body had released adrenaline, in readiness to compete. [159] The adrenaline had produced uncomfortable physical symptoms, such as a dry mouth and rapid heart rate. This was a good sign, showing that her body was preparing for the competition. However, her Chimp had misinterpreted the adrenaline as being anxiety. There was no anxiety just adrenaline. Anxiety occurs when we add negative thoughts to these symptoms of adrenaline. [160] [161] As the Chimp had interpreted the symptoms as being anxiety it had added beliefs such as; "I am anxious", "I am nervous", "I am not ready", "My body feels heavy", "I don't think I can do this". All of these thoughts are very unhelpful and will use up lots of energy. Once the Human realised what the Chimp was saying, and interpreted the symptoms correctly, the Chimp settled again, even though the symptoms of adrenaline continued. The symptoms were even welcomed as a positive sign that she was ready to compete.

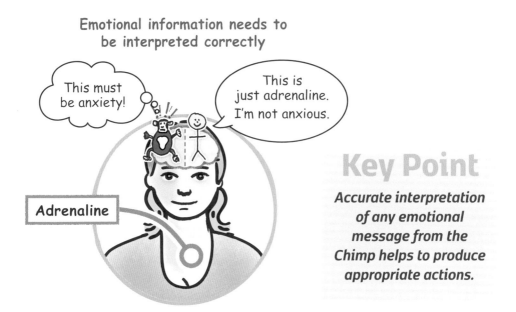

Emotional information needs to be interpreted correctly

Key Point

Accurate interpretation of any emotional message from the Chimp helps to produce appropriate actions.

The type of emotion within a message might not matter

Example: *The urgent call*

A relative had left a message on Kerry's phone, asking Kerry to call her urgently. This will evoke an emotional reaction from Kerry's Chimp. Depending on the relative, the emotional message sent from Kerry's Chimp could range from fear to annoyance. What the Chimp is doing by giving an emotional reaction, of any kind, is sending a message to Kerry's Human asking it to find out the facts and then make a plan. Notice that her Chimp could give any type of emotion: agitation, worry, anger or feelings of indifference. The important point here is that there is an emotional message that needs to be acted upon. The type of emotion that the Chimp uses is not necessarily significant.

Therefore, the correct interpretation of *ANY* emotion that Kerry experiences is that the Chimp is asking Kerry's Human to establish the facts and make a plan. [162]
It's good to recognise that often the type of emotion contained in the Chimp's message is not important, but it is what the Human does with the message that is important.

Emotion - learning an internal language

The Chimp doing its job

Example: *Interpreting a message from the Chimp when serious news arrives*

Imagine that you are told some serious news regarding your own health. In this situation there is likely to be a naturally strong emotional reaction from your Chimp. Everyone might react differently, but regardless of the emotion sent, the message is the same. The Chimp is saying, "*please can you, the Human, reassure me by finding out the appropriate information from the relevant person. This information must satisfy me and might need to be told to me on several occasions*".

Messages that keep on being sent

If we keep getting the same emotional message from the Chimp, it is usually because we are not acting on this or not acting effectively. [163] [164] [165] For example, when somebody upsets you, your Chimp will have an emotional reaction. Whatever emotion is experienced, it merely represents a message from your Chimp. The Chimp then waits for you to decide what to do with the message. It is your responsibility to find a solution to stop the message from being re-sent. If you don't change something, or act, then the message will keep on being repeated until you do act. The Chimp might even change the emotion being sent because it is trying to get through to you, but it is still the same message. Therefore, constant or repeated emotions, such as worry, anger, irritation or emptiness will persist or keep being repeated until you act on this emotional message and sort things out.

For example, a message might begin with you feeling hurt and wounded about a comment made by a friend. If you don't resolve the issue, then the message might change from feelings of hurt to feelings of anger or even despair. Try not to engage with the actual emotion but rather appreciate the emotion *as a message to act on*.

If we do not respond to important messages, they are eventually put into an 'unresolved issue box' within the mind and will emerge some time later, or just keep unsettling us because they are effectively unfinished business.

The Chimp's messages to us

The Chimp can use emotions to communicate with us, in order to ask us to do different things.

> **Key Point**
>
> *Try not to engage with any particular negative emotion, but see it as a message that needs to be understood and acted upon.*

Here are some examples:

- **To ask us to find a solution:** probably the commonest communication. For example, the Chimp becomes frustrated or worried and is asking the Human to find solutions to remove the frustration or worry.
- **To ask for reassurance:** for example, in a new situation, such as entering a room full of people that we don't know, the Chimp becomes vigilant and on edge.
- **To ask us to interpret something:** for example, if the Chimp has had a bad experience and needs us to check that it has interpreted this correctly.
- **To ask us to fulfil a need:** the Chimp will use an emotional message to ask us to fulfil a need. For example, feeling lonely is the Chimp's way of telling us to search out a partner, friend or company.

Remember that the Chimp doesn't look for solutions; that is not its role. The Chimp's role is to either take immediate action to escape danger or to send an emotional message to the Human to take over and work rationally. [152] [153] [166]

Example: *Tammy and irritability*

Tammy works hard and enjoys her work but finds that her Chimp is very quick to become irritable. Tammy recognises this emotional message of irritability keeps appearing but doesn't know how to stop it.

What Tammy can do now is to ask herself what her Chimp is trying to convey as a message. There are numerous possibilities and Tammy's role is to find the right one and the right solution. Here are some possible interpretations of what the irritability message is saying and corresponding solutions from her Human:

1. I think the message is that you are being unreasonable in your expectations. We need to have realistic expectations of what is likely to happen at work and to work with the reality, even if we don't like it
2. I think you are telling me that you are tired and need to have more down time and better sleep. We need to plan in some relaxation and recuperation and develop a regular sleep pattern
3. I think you are saying that you don't feel you are given recognition for what you are doing. We need to ask for feedback and encouragement
4. I think you feel overwhelmed with work and it is worrying you. We need to be assertive and to let this be known to our line manager
5. I think you are saying that you are worrying about things outside of work and you are bringing them to work. We need to address personal issues and then work will be fine

How to deal with uncomfortable or negative emotional messages

We can all experience strong negative or unpleasant emotions, such as anger, frustration, annoyance, guilt, anxiety, fear, sadness, and dejection. Most have a major impact on the quality of our lives. Therefore, it would seem wise to create an Autopilot that automatically sees any negative emotion as a catalyst for change. The Autopilot could be to refuse to engage with the emotion. Then, to ask the question: "How can I act constructively to remove these emotions?"

Negative emotions are not there to be endured but are there to prompt us to act.

> **Key Point**
> *A negative emotional message needs converting into a constructive action.*

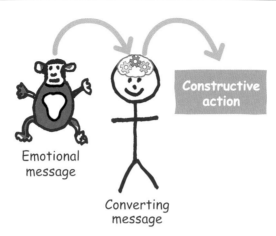

What else can you do with a negative emotion?

A crucial point is to recognise *what you can do* with a specific negative emotion. Clearly, some emotions won't just disappear because you want them to. However, engaging with them continually isn't helpful either. What options have we got? So far, we have used the emotion to form a plan and act. Alternatively, if this isn't working, then we can:

- Reject the emotion
- Replace the emotion with a more beneficial one
- Accept the emotion and work with it
- Process the emotion

It is a skill to discern which option will be most effective when dealing with unwelcome emotions. Sometimes, we can use more than one method. Here are some examples of each:

Reject the emotion
There are many times where we can choose to reject a negative emotion by just stopping and thinking about how useless or unhelpful it is. We don't have to engage with an emotion just because our Chimp has offered it. There are lots of times when the Chimp could produce nonsensical negative emotions that we can confront and reject by choice: a cancelled concert, getting a speeding ticket, getting caught in a traffic jam, making an error, being delayed; the list is endless.

Replace the emotion with a more beneficial one
Emotions from the Chimp are just a message and they can be interchanged. Therefore, a negative emotion can be swapped for a positive emotion. For example, mishaps of all kinds frequently lead to unhelpful emotions. However, we can sometimes choose to laugh at situations or ourselves, and replace unhelpful emotions with mirth or amusement. Negative feelings from self-criticism can be turned into positive feelings by praising effort and offering self-encouragement.

Accept the emotion and work with it
If we are grieving for a significant loss in our lives, such as a person, a job, or a friendship, then it's pointless to try and change or remove this particular emotion. We will be going into grief reaction in great detail later in this course but for now we need to understand that this is an emotion that *we have to accept and then work with* in order to manage it.

Process the emotion
An example of an emotion that might need processing would be our Chimp's reaction to failing an important exam. Processing the emotion means coming to terms with it by expressing our feelings and then using rational thinking. It usually involves talking out loud or writing down our feelings. We then need to have a plan to move on. Again, we will go into great detail on how to process life events later in the course.

Key Point

If your negative emotions continue then you will need to come up with a different plan.

How to speak to the Chimp in its own language

The Chimp speaks and understands feelings and emotions, not logic and facts.

Therefore, it would help to try and speak to the Chimp in its own language of feelings and emotions. Here is an example of a different way to get the Chimp to diet with you!

Example: *The diet for weight loss*

The Chimp goes after food because of the feelings that it gets from eating. Therefore, if we can speak to the Chimp in terms of feelings, it is much more likely that we will get our message across.

Before eating, ask your Chimp what it is *feeling* about the food it is about to eat. Then ask it how it *would feel* during the action of eating: both the positive and negative feelings. Finally, ask it to imagine the meal has ended and ask it how *it would then feel,* and again get it to imagine how *it would feel* thirty minutes after eating.

Go further by discussing with your Chimp how it *will feel*, with positive and negative emotions, the next time you step on to the weighing scales.

By doing an exercise in *honest feelings*, the Chimp might well be the one who helps you to eat the way that you want to eat. This is because the Chimp is likely to be in conflict with two different feelings. Remember that the Chimp doesn't work with facts, so don't add factual details; only speak with feelings. [69]

Speak to the Chimp in its own language

Example: *Jim and his job loss*

Jim is aware that logic and rationality are not helping to settle his Chimp.

He appreciates that facts and common sense are a foreign language to the Chimp but that it can communicate fluently in feelings and emotions. He now applies this principle to his situation.

Jim has worked for a company for nearly twenty years. Out of the blue the company announced that it was going to make redundancies and Jim was informed that he was one of those selected. Jim's Chimp was angry and upset because it felt that others who were being kept on were not as dedicated or skilled as he was. This might or might not have been the case but continuing to state all the facts and logic in the world is unlikely to placate his Chimp. Jim might fully understand and accept that life is often unfair, or it could be that he really does understand and accept the redundancy had to happen, but this still won't help his Chimp.

Jim has decided to speak to his Chimp with emotions. He asked his Chimp how long it wanted to be upset for? Would it like to get angrier or less angry? He asked his Chimp if it would prefer a different emotion? None of these questions are truly about facts or logic; they are just about emotion. He asked his Chimp how it wanted to feel and will it achieve this by using the emotion that it has? Sometimes, the Chimp will change its stance purely by discussing the emotion it feels.

Unit 8 Reminders

- Emotions are usually messages sent from the Chimp system
- The type of emotion sent is not as critical as the message it aims to convey
- Understanding the message and using it to effect change is a constructive way forward
- Unwelcome emotions are not given to be endured but rather to be seen as messages to prompt appropriate action

Key Point

Chimps use emotions to dwell on: Humans use emotions to act on.

Unit 8 Exercises

Focus points and reflective exercises

1. *Choosing the best emotion*
2. *Managing emotional messages from the Chimp*
3. *Speaking to your Chimp in its own language*

Focus 1: *Choosing the best emotion*

This focus point overlaps with focus point 3. It is about talking to the Chimp in its own language by asking it to choose an alternative emotion. The Chimp often chooses to express an emotion that is unhelpful. Some questions that you could ask the Chimp are:

- What are the alternative emotions that you could choose from?
- How will a different emotion make you feel?
- What effect will a different emotion have on others around you?
- Will an alternative emotion help the situation?

Example: *Anger to disappointment*

A common emotion to replace 'anger' is 'disappointment'. If your Chimp can change an emotion to express how it feels, the new emotion can still be as effective but have a much more positive impact on both your own health and those around you.

Exercise: *Changing to a helpful emotion*

The next time you are experiencing a negative emotion, try asking yourself if there is an alternative more constructive emotion that would do just as well or even better than the one you are experiencing. Try not to engage with a negative emotion, which might just give it more energy. Prompt the Chimp to offer you a new and more constructive emotion to start working with, so that you can use this new emotion to form a plan for moving forward.

It is a skill to change emotion but with practice it can be achieved. If you do find it difficult, try to express out loud the negative emotion and exercise your Chimp first, before shifting across to a more beneficial emotion.

Focus 2: *Managing emotional messages from the Chimp*

This exercise is about managing unwelcome emotional messages, interpreting these messages and then coming up with some plans from the Human.

Exercise: *Interpreting and acting*

Think of possible interpretations of an unwelcome feeling or emotion from your Chimp or Computer and then decide on what action to take. If you struggle, speak the emotion out with a friend, as they can often see the interpretation and solution.

Forming a table might help to structure how you convert emotions into actions. You might want to personalise it by adding additional rows – for example, who could I call on to help me to get this right?

Practical table to convert emotional messages into actions	
What is the emotion being expressed?	
Why did the Chimp send this message?	
Is this my Chimp or a belief from my Computer?	
If it is my Chimp – What constructive action do I convert the emotion into?	
If it is my Computer – What belief do I have to challenge and replace?	
If I can't act immediately, when do I put my plan into action?	
How will I celebrate when I have completed the task?	

Focus 3: *Speaking to your Chimp in its own language*

Exercise: *Speaking in emotional language*

This is an exercise in working with your Chimp rather than trying to box it with truths. If your Chimp has strong emotions, and logic doesn't appear to be working, try to allow your emotions expression. Then, talk to your Chimp about how long it wants to have the feelings it has and how it might feel in an hour's time or a week's time or even longer. By asking the Chimp about how quickly its emotions and mood can change, it can help the Chimp to recognise that its own emotions can be fickle and change rapidly.

Unit 9
Disguised emotional messages explained

Unit 9: *will unravel the disguised and potentially confusing emotional messages sent by the mind. Managing emotional scars and ghost emotions is also covered with panic attacks used as an example.*

Types of 'disguised' emotional messages

When the Chimp sends us a message in the form of an emotion, it isn't always appropriate or straightforward. The Chimp's message can often be confusing or misleading, for example, inappropriate laughter instead of sadness or suppressed fear presenting as numbness. [163] [167] [168] Making sense of emotional messages will help with understanding and managing emotions in general. We will consider seven specific ways by which the Chimp communicates with emotion:

1. Straightforward appropriate emotion
2. Replaced or altered emotion
3. Over-emphasised emotion
4. Displaced emotion
5. Mixed emotions
6. Suppressed and repressed emotions
7. Projected emotion

There are so many possible interpretations to every emotion experienced that only you can decipher what is going on in your unique situation. To demonstrate these different types of messages, we will look at several examples to offer some insights into the way the Chimp works.

1. *Straightforward appropriate emotion*

Straightforward emotions need little explanation. They are appropriate to the setting and can be easily understood. Examples would include: getting annoyed with someone who is deliberately being obstructive, or feeling sad when someone tells you gloomy news. These are both appropriate reactions to the situation and easy to interpret and act on.

2. Replaced or altered emotion

Sometimes the Chimp will operate by replacing an expected emotion with a different one. [167] [168] [83] [169] This is usually because some emotions are easier to express than others.

For example, anger is easier to express than grief. It is well recognised that after a significant loss, people can experience intense anger and might direct this at someone close by. [170] [171] [172] [173] Remaining angry can prevent the grieving process from progressing, so it is worth recognising when an emotion has been replaced. When the anger has been recognised as replacing grief, grief can be addressed and this will cause the anger to subside.

Frustration that hasn't been addressed can alter its presentation and turn into anger. The Chimp has turned the feelings of internal frustration into anger so that it can direct this outwardly. Again, the answer is to go back and address the original emotion of frustration. When frustration is addressed and removed, the anger will subside. The commonest cause of frustration is approaching a situation with unrealistic expectations. Therefore, changing expectations to more realistic ones can remove the frustration.

> **Key Point**
>
> **Many of the Chimp's preferred emotions are easier to express than the more appropriate emotions.**

When a replaced emotion occurs it is worth trying to work out what would be the more appropriate emotion. When you choose the appropriate emotion the replacement or substitute emotion usually subsides.

For example, if we encounter setbacks, frustration is easier to express than disappointment. Both might be a reasonable reaction to a situation, but the Chimp will substitute and alter the emotion to its preferred way of reacting. [174] [175] By recognising this, we can opt to go with disappointment and work with this to resolve any setbacks. When we say we are frustrated, it sends a message back to our own mind that could result in us having further negative feelings. Whereas, if we say we are disappointed, then it sends a message back to our mind that could calm it down with a feeling of acceptance.

By opting to use a different emotion the Chimp will express itself in a different way.

> **Key Point**
>
> **It is always worth considering what alternative emotion you could use if the one you are experiencing is not constructive.**

Example: *A relationship loss*

It's interesting to note that when people fall in love they usually face the world and all of its challenges with a new sense of confidence. However, when a relationship ends the opposite can occur. This complete loss of confidence in most areas of their life can be devastating. If this disproportionate reaction can be recognised and accepted, then it could be restricted to just one area of life: the lost relationship. The loss of confidence can also be changed into the more appropriate sense of uncertainty.

3. *Over emphasised emotion*

Some reactions and expressed emotions from the Chimp are over emphasised and appear to be out of proportion to the situation. [69] Whenever this happens, it is usual to find that there are beliefs underpinning the reaction. In other words, the Computer system has a Gremlin fuelling the Chimp to react. [176]

Gremlin of belief

Key Point

If there are unhelpful beliefs then they need to be found and replaced with helpful beliefs.

Here are two examples of underpinning beliefs: one showing an undermining influence and one showing an intensifying influence.

Example: *The critical remark – an undermining influence from beliefs*

Tony has a belief that his line manager does not value him. When Tony hands some work across to his line manager, the line manager comments that the details would have been better set out in a table. Tony's Chimp then reacts by feeling terrible and uses this single comment to confirm to himself that he is not valued. This theme is a common one.

The resolution: Tony first checks out the truth behind his belief. By talking directly to the line manager, he can ask what the line manager feels about Tony's work. If he needs help then this can be sought; if he is doing well and he is valued then this can be established. It would also help to ask the line manager how he operates. If he makes comments that he doesn't feel strongly about, then they are only suggestions not criticisms. Frequently people in positions of authority might make remarks that are not criticisms but just throwaway comments. It's important for Tony to know how the line manager operates. By establishing these facts Tony will have a better basis to work from. The over-emphasised emotion in reaction to the comment can then be prevented.

Example: *The sales result – an intensifying influence from beliefs*

Arlene works for a small company that sells novelty gifts. Her belief is that the gifts will soon become very popular and that the company will therefore prosper. A small contract has arrived in the post and Arlene has reacted with great enthusiasm. Her colleagues can't understand why she is so excited by such a small contract. To her Chimp, it represents an indication that her beliefs are right and hence the apparently over-emphasised emotion. ***Our beliefs will give energy to our emotions***, so it is important to check what beliefs we hold about situations that can provoke over-emphasised emotional responses.

4. Displaced emotion

Displaced emotion occurs when an appropriate emotion is removed from its context and placed somewhere else. [169] [175] A simple example would be, when someone has a bad day at work and comes home and shouts at the cat. The cat didn't do anything; it was just the recipient of the person's frustration at work. The emotion could have been appropriately expressed to the right person at work rather than brought home and displaced onto the cat. It is not unusual for us to displace emotion onto people who are close to us, rather than express emotion in its rightful setting.

Not all displaced emotion is negative. A particular displaced emotion can sometimes be seen when someone cares for others rather than looking after themselves. Looking after others is good, but not if you are neglecting yourself. If you are in need of some care and attention yourself, then it's not wise to displace your own needs.

Example: *It's not about the goldfish*

Some years ago during a hospital clinic, I was presented with a common scenario. I think you will find this example easy to follow and see what was happening, before we even reach the end of the story! A new patient, (we will call him George) had arrived along with a friend at the outpatient clinic.

The friend insisted on seeing me alone first and was very keen to explain what had happened to George. George was a young man who had a good job and lived alone. He said that George had lost his wife about two years earlier and had not really engaged socially since then. He told me that George loved his wife; but when she unexpectedly died, he had shown no sign of emotion at all. All of his friends were surprised and wondered why he hadn't grieved.

Time passed and after two years George's pet goldfish died. When the goldfish died George fell apart and cried his heart out. The friend was amazed and said that it seemed like George loved the goldfish more than his wife. After the goldfish died, George couldn't hold his life together and there were serious concerns for his psychological health, and he needed assessment. The friend asked, "Why did George love the goldfish so much?".

Once I had the full picture, it was quite obvious what was happening. George had not been able to grieve for his wife and had suppressed his emotions for two years. The death of the goldfish gave an opportunity to displace and express his bottled up emotion. It turned out that several years earlier George and his wife had been to a funfair and won the goldfish on the day of their wedding anniversary. The goldfish became symbolic to George and he felt that through the goldfish he still had a link to his wife. The death of the fish symbolised the death of his marriage and his wife, and hence the grief began. Displaced emotion is very common. I explained this to George's friend and told him, it's not about the goldfish. His friend looked at me puzzled. He then said, "So, why did George love the goldfish so much?".

Our Chimp's emotional circuits operate by expressing emotion but very frequently displace it where it doesn't belong. [69] This can lead to all kinds of problems, especially to those who are on the receiving end.

Displaced emotion and suppressed emotion can both work together and cause problems in your current life, but actually relate to past experiences that have not been dealt with. [177] [169] Recognising displaced emotion is not always easy.

Key Point

Try to ask yourself if the emotion that you are experiencing is in line with the situation and if not, search for any unresolved matters that you may be holding on to and not addressing.

5. *Mixed emotion*

Mixed emotions are common and are just the Chimp looking at a situation or problem from different angles and then reacting to each one. [178] A few simple examples will demonstrate how they occur.

Mixed emotions need separating in order to deal with them

Example: *The lost child*

Teresa had gone to the shopping precinct with her four-year-old child Bruce. While she was distracted, Bruce had wandered off and got lost. On realising what had happened, Teresa's Chimp had gone into a panic and began frantically searching for the child. After twenty minutes, which seemed like an eternity to her, the child was found. Teresa's Chimp could now see the situation from two different angles and therefore produced a mixed emotional picture. She could present as annoyed or she could present as relieved. It would not be surprising to see her both hugging the child and chastising him at the same time. This mixed 'attack' and 'protection' of the child is not confusing if we can see why it is occurring. Complex mixed emotions need insight and understanding if they are to make sense.

Example: *Colin and the death of a suffering relative*

Colin had a great relationship with his father. Unfortunately, Colin's father had been slowly deteriorating with Alzheimer's disease for several years. His father had suffered and been increasingly distressed by the confused state that he had developed. Colin struggled with this for years and now with the death of his father, Colin has mixed feelings.

It is always sad and often painful to lose a relative that we love. However, if the relative is ill and suffering, you can reasonably expect to have mixed emotions. On the one hand, you can view the situation as the loss of a loved person, which hurts, and you wish they were still here. On the other hand, you see a person who is suffering or is no longer recognisable as themselves, and it is a relief to see them released from suffering. This relief can present as inappropriate guilt.

Dealing with these two different emotions might require two different approaches. It is better to express each emotion separately then address it, before tackling the next emotion.

Colin can first discuss the reasonable sense of relief and inappropriate guilt before allowing the grief to be expressed.

Example: *Eli and the separation*

Eli has recently found his marriage in difficulty and, at the request of his wife, they are undergoing a trial separation. He has found his emotions are jumping all over the place. He wrote down the several emotions that he was experiencing.

These included:
- Fear
- Loss of interest in everything
- Feelings of low self-esteem
- Anger
- Self-loathing
- Yearning

As he addressed each one, they made sense and the reasons underpinning them could then be addressed. For example, his self-loathing made sense when he realised that he was blaming himself for the break-up and was searching for all the things he felt he could have done better. It might be very helpful to reflect on what happened and how he could change things in the future, but to allow destructive self-loathing is unhelpful in the extreme.

6. *Suppressed and repressed emotion*

Previously in this course, we considered why it was important not to stifle emotion. Here, we return to look at this in more detail.

Suppressed emotion is when we consciously push an emotional reaction down within our minds, ignore it, and don't allow it to be given any expression or attention. [83] [179] [180] This is different to rejecting a Chimp's offer of emotion. Rejecting an offer means we have addressed it, made a decision, and let the Chimp know.

Repressed emotion is when we unconsciously push an emotion down within our minds, so that we might not even be aware we are pushing it down and ignoring it. [181]

Whenever we bury an emotion within our minds, it will try to find expression. [182] This expression might not be recognised as belonging to the suppressed or repressed emotion. [183] It is understandable that as we go through life some events and experiences might be too painful to deal with at the time, and become lost in our unconscious mind.

Suppressed and repressed emotions don't usually stay buried!

From time to time we might recall these events, but avoid addressing them again. The effect this has will vary tremendously from person to person, depending on a host of factors. However, the result of suppression or repression is usually a feeling within, that all is not well. Suppressed or repressed emotion can change and present in any form, such as being depressed, anxious, angry, unsettled, or with low self-esteem. [184] [185] These emotions usually result in unwelcome behaviour. For example, if someone was abused as a child and has not come to terms with this, it can cause maladaptive behaviours to develop. [183]

Example: *Olive and confidence issues*

Olive works in a small shop and has a woman as her line manager. There have been several incidents involving Olive and her line manager feels that things are not going to work out. The three incidents were as follows:

- Olive reacted with verbal aggression towards the line manager when the line manager asked Olive to re-count the till monies, as they didn't seem to add up.
- The second incident was when Olive was arranging a display in the shop window. She suddenly broke down and had to be given time out because she felt that the display was not good.
- The final incident was when Olive had tried to help a customer by packing her bag, but the customer told Olive she might be better employed serving the next customer. Olive then shouted at the customer and told the customer she was only trying to be helpful. The line manager had to intervene and apologise to the customer.

The line manager concluded that Olive had an unsuitable temperament to work in the shop.

The three incidents are not the problem. The problem was that Olive's mother had psychologically abused Olive during her childhood. Olive could recall many occasions where her mother would control her and criticise her. This led Olive to have low self-esteem. As the issues from childhood had not been addressed, in her case, they had continued into Olive's adult life. Olive herself would probably not be aware that her unconscious mind was reliving some of the abuse. It isn't surprising that any female line manager might be perceived by Olive's Chimp to be disempowering her and being critical of her. At the same time, Olive will be self-critical because her Chimp will be searching for evidence to prove that she is incompetent. Even a friendly word from a customer could be taken the wrong way. With help, Olive can turn all of this around, address her past, and not allow her past to influence how she behaves today.

Many of us carry some buried emotions and when we are ready, it would be good to address them. It can be helpful for anyone to deal with past issues in a constructive way with a trained therapist who can guide the person through.

7. Projected emotion

Occasionally we project our feelings and beliefs onto other people. [186] This means we are not comfortable with the feelings or beliefs we have, so we get rid of these feelings or beliefs by imagining that they belong to someone else. We literally believe that someone else is experiencing the beliefs and feelings. When we confront them about 'their' feelings and beliefs, it is not surprising that they might be baffled or disgruntled. The simplest example can be seen in children in a playground.

Example: *Rick, Paul and Lucy*

Rick quite likes Lucy, but he feels embarrassed to say so. His friend Paul is standing next to him when Lucy approaches. Rick's Chimp says to Lucy, "My friend fancies you". Rick has projected his feelings onto Paul. Rick actually believes that Paul has these feelings and will become upset if Paul denies it.

Example: *The patient and doctor*

Jake is a patient in hospital being visited by his friend Mahmud. Jake is happy with his treatment, but Mahmud doesn't think the treatment is good enough.

The doctor comes to the bed and asks Jake how he is doing. Jake is about to say, "I feel fine" because he actually does feel fine, but before he can speak Mahmud's Chimp jumps in. "Well doctor, Jake does not feel fine because he doesn't think you are doing enough for him". Jake is alarmed and disputes this and Mahmud gets upset. Projecting emotions is common in teenagers. For example, if a teenager doesn't like someone they often accuse that person of not liking them. It's often a cry for reassurance; the teenager wants to be liked by the person.

Managing projected emotions

If you can recognise a projected emotion it is worth trying to address the real concern behind it, without uncovering the projection and challenging it. The patient in the example above could have tactfully explained that perhaps they were giving the wrong impression, but then go on to clarify that they were really happy with the treatment. For the teenager example, simply expressing reassurances that they are liked, and giving examples of what is likeable about them, could resolve the problem. It usually doesn't help to confront someone who is projecting emotions with the fact that they are doing this. This will usually bring denial in the person and further problems.

Real emotions and 'Ghost emotions'

The Human and Chimp work differently when it comes to awareness of time. [187] [188] When we are operating in Human mode, we are able to recognise that emotions are specific to a fixed time. We can re-live memories in our head and experience the emotions again but still recognise that they don't belong to today. [189] When we are operating in Chimp mode, we will not be able to put a time period to emotions. The Chimp does not have an awareness of time. Emotional experiences happen outside of time for the Chimp. It can suddenly experience an emotion from years ago and feel it just as strong and real today. [14] Our Chimp re-lives the experience, as if it is happening now. [190]

The real emotions: We will call the emotions that we experience in response to everyday situations that are happening to us now, the 'real emotions'. They are in the 'here and now'.

The ghost emotions: These are the negative emotions that emerge from our past experiences and suddenly appear to haunt us! These ghost emotions appear from 'emotional scars'; experiences that have significantly affected us and memories that we have to learn to live with, that we have already processed. Here are some examples of emotional scars that can bring back these ghost emotions:

- Losing someone very close to you
- Failing in something that was important to you
- Making a mistake that had severe consequences
- Being let down by someone who meant a lot to you
- Being unjustly treated or misrepresented

These are the type of 'emotional scars' that we can obtain as we go through life. They are painful events and we often visit them from time to time, almost trying to change what happened. We have processed the events but somehow they seem to reappear because they have become scars that we have to learn to live with. It is these scars that produce the ghost emotions that haunt us. These are emotions that belong to the past and really don't help us in the here and now. They are redundant messages.

Why is it important to distinguish between real emotions and ghost emotions?

Real emotions can be dealt with as they are happening. This means, for example, we could re-think about why we are getting the emotion and then work to remove or replace it. We can treat this as an emotional message and act on this.

Ghost emotions are much more difficult to work with because they are fixed on an event, which we can't change. The event has become a painful memory, so it will keep presenting itself. We are going to look at how to deal with emotionally painful past events in due course, but here we are going to look at how we manage the emotion that presents. Of course, it's best to deal with the memory of the event if we can, but the reality is that it will often remain and we have to manage it, as it presents.

Managing Ghost emotions

Ghost emotions are best dealt with by recognising they are not in real time and if the event has been processed, refusing to engage with the emotions. Ghosts are effectively a learnt emotional reaction from the past, rather than a message. [14]

Example: *Panic attacks*

Panic attacks are a very common experience. Panic attacks turn into ghost emotions, so it is worth looking at them, as a way of managing this particular example of a ghost emotion.

Panic attacks are a cluster of symptoms that can appear unexpectedly and are extremely severe in nature. [191] They have *two aspects* to them.

1. The first aspect is a group of physical symptoms, such as a very rapid pounding heart, sweating, feeling nauseous or faint, or shaking.
2. The second aspect is a group of beliefs that accompany the symptoms, such as a belief that the person is dying (possibly from a heart attack), about to faint, vomit, or lose control.

A panic attack starts with the Chimp

The sense of panic is overwhelming and a terrifying experience to have. Repeat panic attacks can be seen as an emotional experience rather than an emotional message. There is nothing that your Chimp is asking you to do with these symptoms and therefore they represent an experience rather than a message to be acted upon. Therefore, our Human role is to learn to manage the symptoms. The initial panic attack could arguably be seen as a message from the Chimp to the Human to sort out whatever is causing the attack, but after that, if the cause has been sorted, repeat attacks no longer act as a message. [191]

Once addressed, the panic attacks subside slowly with less and less occurring but they still do occur, and therefore present as ghosts. These ghost attacks can take a few weeks to several years to vanish and occasionally they might even appear to gain strength, but in reality they will fade with time.

The main point here, is that in order to manage these ghosts they need to be recognised for what they are: inappropriate symptoms that need to be discounted from being important. They are just repeating a learnt pattern of behaviour and appear unpredictably. [192] So the question is: if these are just ghosts from the past and have no relevance to today, how do I stop them from happening?

The starting point is to recognise that when these symptoms occur, your Chimp is engaging with them and then panicking even more. You have to decide that enough is enough and take the lead. ***Only you can do this.***

To be clear:

- The Chimp experiences the symptoms and engages with them
- The Human rationalises the symptoms and takes control of the situation

Taking the lead won't immediately stop the panic attacks, but with time they will subside and finally disappear.

The method of removing the panic attacks is three fold:

- Manage the physical symptoms
- Manage the beliefs
- Recognise them as redundant messages

I have worked with countless sufferers of panic attacks, as they are a very common experience and extremely unpleasant. Once the sufferer comes to accept that these attacks ***are not dangerous just very unpleasant and inconvenient***, they begin to manage them and then finally to ignore them. The anticipation of an attack is usually worse than the attack itself and again this anticipation needs to be addressed.

There are lots of excellent websites online to explain how to manage panic attacks. Here are the fundamentals of the management:

Managing the physical symptoms

Whatever symptoms you are experiencing:

- Slow your breathing down
- Try to find somewhere quiet
- Tighten then relax your muscles
- Do some activity to take your thoughts off the symptoms
- Try to focus on something pleasant that will distract you

In other words, take control and have a plan.

Managing the beliefs

Check your beliefs before and during the attack:

- You will not come to harm from a panic attack
- The attack will usually be short and you will fully recover

- You might feel you can't breathe but you will be fine
- Panic attacks don't kill people and don't do any long-term damage

It's important to make sure that you manage your thoughts during an attack. Otherwise your mind will run wild and cause the symptoms to feel worse.

Recognising the panic attacks as redundant messages

If you have resolved any causes of the attacks, then:

- Recognise that these are **Ghosts**: a habit
- They stem from a memory, not a current situation
- You need to disengage and give them no time or energy

This isn't easy but will work if you persist. If things don't improve then clearly get professional help. Clinical psychologists and other therapists are experts in this area.

Ghost emotions can become a problem in their own right!

Ghost emotions typically come from emotional scars. To avoid engaging with these ghost emotions it is important to recognise they are not current. You have the choice to disengage and refuse to give them energy. For example, when a relationship ends there will be some time needed to process the pain and give yourself some compassion. However, when you do get back onto your feet, it is important to recognise when a ghost emotion has appeared, and not to give it any of your time.

If you do engage with a ghost emotion, then you will keep re-enforcing it, so that it will keep coming back. [193] [194] The ghost emotion then becomes a separate problem in its own right. Being trapped for years by ghosts is not helpful. Decide when you want to move on and reject the ghosts.

Ghost emotions can be ignored

Unit 9 Reminders

- Emotional messages are often disguised and need to be recognised and understood if they are to be managed correctly
- Some emotions are not responses in real time, but responses to past experiences
- Once an emotion has been processed, beware the ghosts from emotional scars!

Unit 9 Exercises

Focus points and reflective exercises

1. *Mixed emotions untangled*
2. *Real messages and ghosts*

Focus 1: *Mixed emotions untangled*

Whenever we have an emotional response to a situation that involves conflict or uncertainty, there is usually a mixture of emotions.

Exercise: *Untangling emotions*

This exercise is about working with emotions and not just expressing them. When you next experience conflict or uncertainty, untangle the different emotions and then address each one by searching for underpinning beliefs. When you have worked out the beliefs that are causing a particular emotion, check that these beliefs are rational and truthful. If they are, then try and work with them to make plans. If the beliefs are unrealistic then challenge them and change them into truthful realistic beliefs, before working with them.

The usual reason that emotions don't pass is that they result from a situation that is being fought against instead of been worked with.

Focus 2: *Real messages and ghosts*

An 'emotional scar' is a term I use to describe the unpleasant memory of an event in our lives. It remains with us and comes to life from time-to-time or even feels like it is there all of the time. It seems like whatever we do, we cannot process the event. These negative emotions are 'ghosts' of the event. We don't have to engage an emotion just because it comes into our head. If we can't stop our Chimps from engaging, then we can revisit the event and try and replace the negative emotions with different more constructive and appropriate emotions.

Example: *Carlos and school bullying*

Carlos was doing well at school and was quite happy until a new boy arrived. This boy began to bully Carlos. Carlos did all the right things in standing up for himself and reporting the bullying as it occurred. Despite doing all the right things, the bullying continued and Carlos felt unsupported and dejected. After leaving school,

Carlos managed to obtain a good job and the support of his company. The bullying had gone from his life. However, Carlos kept on getting unwelcome emotions when the past events entered his head. If Carlos just can't gain some perspective and put to bed his past experiences, then he can make a decision. He can decide to simply refuse to allow his Chimp to engage with the emotions, whenever he is reminded of the bullying he had endured. Ideally he needs to process the event, but occasionally in life we are stuck with an emotional scar.

Exercise: *Managing the ghosts of an emotional scar*

Some events that you have experienced might have left some emotional scars. If you have tried to process these and they do not seem to be going away, then try to apply the principles involved here. We can learn to recognise the emotional ghosts as not being applicable to the current day, and try and exchange the emotions for more appropriate ones or simply refuse to allow our Chimp from engaging with them. Try to recognise when emotions are ghosts and bring yourself back to the current time and live in the moment.

Unit 10
Expressing emotion with insight and change

Unit 10: *will cover how to express emotion effectively and constructively. We will look at exercising the Chimp with understanding and change, and how to address emotions that are bubbling under the surface.*

How do we express emotion effectively and constructively?

Expressing emotion is something we do automatically every day. There are ways of making this expression effective and constructive, as opposed to ineffective and destructive. Emotion channelled in a positive way can bring about beneficial changes. [195] [196]

The following examples demonstrate some of the effective ways in which to express emotion and also some of the pitfalls of not constructively expressing emotion.

It's not what you say; It's often what you don't say

When people interact with each other, one of the problems is not what they say but often what they don't say. We share our lives with partners, family and friends and so many of us forget to let them know what they mean to us. It isn't that we take them for granted; it is usually that we just assume that they know what they mean to us.

The Human part of the brain might very well appreciate this point but the Chimp characteristically doesn't. Our Chimp needs to be continually reassured. [197] It will vary from person to person but generally there is a need for appreciation and being loved. [198] [199] There is also a need to feel welcomed and wanted. If you feel this way towards any individual it might be worth letting them know. For many, it will be a boost to their self-esteem and a reassurance to their Chimp. [199] [200]

Example: *Silence can be costly*

Rachel and her partner Jane have been together as a couple for many years. However, they don't communicate their feelings towards each other very easily. Rachel has always had the belief that Jane knows how much she feels about her and this is said by her actions. The potential pitfall here is that Jane may well understand this, as a Human, but her Chimp might not. The Chimp usually needs constant verbal reassurances and if it doesn't get this, then it starts to reinterpret intentions. [197] It would help if Rachel gave just five minutes a day to express what she feels towards Jane, and also express her appreciation that Jane makes a choice, every day, to stay with her. Reassurances can prevent Chimp uneasiness.

Expressing positive feelings

If you try to check on how you feel throughout the day, you might be surprised by how many positive feelings you experience but do not express. If we don't express feelings such as happiness, gratitude, sense of well-being, and so on, they can go unrecognised. Having asked people to specifically look for positive emotions throughout the day and then to express these verbally, they report back that when they do this, life seems better. Expressing positive feelings and demonstrating these seems to make the feelings come to life.
[195] [196]

It helps to recognise the good emotions we have regularly

The moaning habit removed
- From Gremlin to Autopilot

Our Chimps can learn to act in certain ways and this behaviour can then be stored in the Computer as a Gremlin or an Autopilot. [33] [110] The Chimp is not only scanning for danger but also for situations that it is not happy with. [55] The problem is that if we leave our Chimps unchecked then they can develop a habit of moaning and complaining: the moaning Gremlin! [201] [202] The Gremlin and Chimp then take it in turns to just moan or complain. Here, we have a Gremlin mimicking the Chimp. We are all susceptible to this Gremlin because every day there will be times when things don't go the way that we want them to. When the Gremlin takes over and acts, the Chimp isn't actually reacting at all. The Computer has become programmed to moan, every time something happens that we don't like or want. This is so fast, that we almost do it without realising that we are moaning habitually.[201] If you can recognise this, then try and develop an Autopilot of acceptance and solution finding, so that every time something unwelcome happens, you learn to respond positively, even with a smile.

A constructive response to unwelcome situations

Try to develop an Autopilot that instantly sees a negative outcome as something to overcome or learn from.

Example: *The traffic lights*

Jim is travelling by car and is gradually moving into Chimp mode because it seems that every time he approaches a traffic light it turns red. His Chimp is stirring with irritability and impatience. It has personalised the red lights as being aimed at him. His Chimp expresses that the whole system is ridiculous and needs changing. His Chimp has now decided to engage in a battle with the traffic lights. It races the car towards each light but still fails to win, which brings more anger and expletives. Before he had reached this absurd place, he could have set up an Autopilot that asks him to consider why he is experiencing this agitation and where else it might be appearing in his life? These unhealthy emotions could be a sign of stress, learnt impatience, feelings of low self-esteem, loss of perspective, being over-tired, unresolved problems or a whole host of other reasons. [203] In order to stop the same impatience presenting in other areas of his life, it could be very useful for Jim to stop and reflect on what he needs to address. If he feels it is just a straightforward reaction to the traffic lights then he can deal with this, but some reflection might be revealing.

We tend to repeat behaviours across all aspects of our life, not just limit them to one situation. [193] [194] An alternative response to the red traffic lights, which might be more rational, is saying to his Chimp that life doesn't always go the way we want it to and the traffic lights are not personally trying to wind me up! Therefore, I could use the red traffic lights to relax for a few minutes and reduce the tension I seem to be under. With some inventiveness, you can change most irritants into a point of learning or a trigger to stop and reflect.

Key Point

We can learn about ourselves from our Chimp's reaction.

Provoking the Chimp in others

Expressing your emotion towards another person can be a productive or destructive thing to do. Before expressing your emotion, it is helpful to consider what affect this will have on the other person, and is this what you want to achieve.

Example: *The untidy person*

Suppose you have a colleague or flatmate who is very untidy and this is irritating you. Your intention might be to try to get them to see how their untidiness is affecting you and also to get them to try and be tidier in the future. If you know these are your intended outcomes, then consider how you want to express your emotion and what approach you need to make, in order to achieve your outcomes. Therefore, it is important not to allow your Chimp to attack the person by directing negative emotion at them. You might feel that it will achieve what you want from the person and you might well be right, but there are usually much better ways of engaging.

Negative emotion directed at anyone is very likely to provoke his or her Chimp to defend or attack. These are the most natural Chimp reactions. [41] Allowing yourself to get into Human mode will bring in reasoning and also an awareness of the possible consequences of your actions. Reasoning can be an alternative and more helpful way to work than expressing emotion. However, if your Chimp also wants or needs to express itself, then ask yourself if this can be done constructively. So, you could try explaining how it makes you feel, when you see untidiness. You could also express how you feel **when things are tidy**.

Expressing emotion with insight and change

The problem, with a situation like this, is that you might find you have very different values or acceptable limits from your flatmate or partner, for things such as tidiness. If this happens, then your choice is to accept that it is likely to always be this way or to move yourself out of the situation, if possible. It can also be the case that there are solutions but that your Chimp just doesn't want to accept any of them. It's always helpful to ask yourself if the problem lies within yourself and what you could do to help. For example, think of the obvious: could your Chimp be more tolerant or less critical?

Example: *Fay and her untidy husband*

Some years ago, I had a patient who was struggling with depression. He had responded to treatment and was back for a review and was now doing well. Despite all efforts by his wife, Fay, to encourage him, he was still very untidy and this was distressing her. Fay explained that he had always been untidy but she couldn't accept it and was aware that she was nagging him. It was clear that this man would never be as tidy as Fay wanted him to be. Some people are better than others with organisation skills! Her husband did not feel he had a problem with untidiness.

Years of disgruntlement and no change were now taking their toll. Fay had a choice, she could either accept that this aspect of her husband wouldn't change or she could walk away and see if someone else will accept him, as he is. After thinking about this, Fay said that she realised that she would rather have an untidy husband than no husband at all. Sometimes we have to decide to change our stance and accept a situation. Only the Human can do this; a Chimp can't.

> **Key Point**
>
> **Sometimes, we have to accept that life doesn't always run the way we want it to: A useful Autopilot is to accept this and change our stance.**

Exercising the Chimp without criticising others

Situations involving conflict or frustration are never easy to manage. Our Chimps usually express emotions that are unhelpful. [7] [3] A constructive way to exercise your Chimp, rather than express emotion, is to allow your Chimp to ask non-critical and non-judgemental questions. Asking questions to someone can bring about changes in their behaviour because they will be able to reflect on what you are asking and what they offer as answers. [204] [205]

Example: *The critical parent*

Tom's father constantly criticises him, which he finds irritating. Tom is a twenty-eight year-old man and he resents the approach his father is taking with him. If Tom unhelpfully expresses the emotion he feels towards his father then it is likely to create further problems. Even if Tom feels better for doing this, will it achieve what Tom wants in the long run?

Expression of emotion aimed at people

Tom says he wants an adult relationship with his father and would like his father to respect his feelings. One approach that Tom can use is to ask questions that might help his father to go into Human mode and to reflect on what is being said. For example, Tom could ask what relationship his father would like to have with him, now that he is twenty-eight. Clearly this must be done with the appropriate tone of voice! Tom could also let his father know what he hopes for their relationship and ask if his father wants the same. Tom can ask what effect the father thinks his criticism is having on Tom. If Tom can remain non-emotional it is likely that the father will eventually stop expressing emotion and be able to have a calm discussion. Most people when approached in this way, will at first be in Chimp mode and have an emotional reaction to the questions. However, they will then calm down and move into Human mode if they appreciate that the questions are being posed in a non-judgemental way and as a means of obtaining a constructive outcome.

Exercising the Chimp with insight and change

This section is quite a critical one because it is important to understand that exercising the Chimp isn't just about letting off steam.

Letting off steam can be useful in its own right; however, exercising the Chimp doesn't need to end with just expressing emotion. To exercise the Chimp effectively we need to have some insight into why it reacted in the first place and then do something about this to prevent it from occurring again. Many people that I work with initially fall short of this and think that exercising their Chimp is only about an emotional outpouring. Understanding why your Chimp needs to be exercised in the first place, and addressing this, can improve the quality of your life.

Therefore, effective exercising of the Chimp really has three parts to it:

1. Expressing emotion constructively
2. Finding out what caused the emotion
3. Making changes to see it doesn't happen again

Example: *The supervisor*

Pearl is a supervisor in an office and her Chimp has become agitated because her employees have not delivered the work expected of them. She has let out her Chimp and expressed into the air her frustrations with the situation and finally calmed down. Now with a smile, she begins again to solve the problem. However, she hasn't really addressed the true reason for the Chimp's agitation because she has not considered what provoked it in the first place. It might be disappointing that the work wasn't delivered but it could be, for example, that her expectations of what can be delivered are unrealistic and this is the real reason for the agitation. Alternatively, it could be that the colleagues need support and training in order to do the work. Therefore, Pearl will only keep repeating the scenario of agitation without moving forward, until she understands that the solution could lie within her or be achieved by giving support to her colleagues.

> **Key Point**
>
> *Whenever you exercise your Chimp try to understand why it needed to exercise because this might bring about change in your own beliefs or approach.*

> **Key Point**
>
> *Reacting to an outcome without looking at the cause usually leads to a repeat of the situation and a repeat of the emotions.*

Example: *Adele and the repeated worry*

Adele has a repeated pattern of allowing her Chimp to worry about what people think about her. Whenever she reads an unfavourable comment on social media, her Chimp worries about who is reading it and what they will think.

When she experiences this, she talks to her close friends who reassure her Chimp and after a few days Adele feels better. It's great that Adele has taken the first step of exercising her Chimp by talking to her friends. However, this isn't really effective exercise because it is a repeat pattern and the cause of her Chimp's worry is not being addressed.

How can Adele address the cause? We first need to find the cause. If we ask Adele to tell us what she thinks, it's likely that she will eventually be able to express the situation something like this:

- "I find unfavourable comments destructive because I don't have a great relationship with myself and I have no answer to the comments"
- "My Chimp worries about what everybody thinks about me"

These would be common thoughts for many people in this situation. If we then ask: "How would you like things to be?", we might get a different answer.

Now Adele's Human will answer:

- "I want to have a great relationship with myself"
- "I want to be able to accept that comments are opinions and not facts"
- "I want to dismiss opinions that do not matter to me"

These are the statements that need to be implemented in order to prevent her Chimp from reacting to any unfavourable comments made.

Each statement needs to be worked on to establish them within the Computer. Her Chimp also needs reassurance that it is never alone with these things and her friends are there with her.

Following up after expressing emotion

Whenever we express emotion towards others, we usually need the people who are receiving it ***to understand*** why we are reacting and ***also acknowledge*** what we are experiencing. One of the reasons we express emotion is to communicate our feelings to others so that they will understand. [86] [87] [88] After you have expressed emotion, it will help you immensely to follow it through with the other person and check to see if they have understood why you have expressed emotion. If they do understand then usually it is followed by an acknowledgement of how you feel.

It is quite straightforward to ask them if they have understood and appreciate why you feel the emotions that you do. Being understood is a very helpful way to process emotion.

Expressing emotion with insight and change

Example: *The doctor's error*

When I was setting off in my career as a young doctor in psychiatry, I had a patient whose brother had acted very poorly towards her. She was unable to get over this because she had a belief that he owed her an apology and unless this apology was forthcoming she would continue to be angry and upset. It seemed clear to me that she needed to change her view and accept that sometimes life is not fair and he might never give her an apology. By not accepting this, she was only hurting herself. It all seemed very logical.

I saw her a number of times and compassionately explained (at least I hope I was compassionate!) that unless she moved on, she would be stuck with anger. We seemed to be going nowhere until I sat down and reflected on what I was doing. I then got inside her head to see her world from her point of view. It all still seemed so irrational but it then dawned on me that of course it was irrational because she was working with an irrational part of her brain (her Chimp - I didn't have the Chimp Model then). Once I took her Chimp's viewpoint, I suddenly really did understand and empathised with her irrationality of believing that everyone should and would act appropriately.

The next time she came in, I told her that I agreed with her and said that I really did understand why her brother owed her an apology and I also understood her anger and upset would remain until he apologised. I explained to her that it disturbed me thinking about it because I could see the pain that she was going through. Also, that I appreciated how she had suffered and how she had tried to accept it but couldn't; *but she had tried.* She burst into tears and told me that it was such a relief that someone understood her and appreciated what she was going through.

Almost like a miracle she switched into a different mode of thinking (her Human). She sighed and said she also understood that sometimes things don't happen that ought to and then accepted the situation. She was able to move on without getting the apology from her brother.

It would be some time before I could truly appreciate that I was dealing with two very different systems in her mind. The two systems being, the Chimp that might not be looking for solutions, but rather understanding, and the Human that would implement solutions; if only the Chimp could move on and allow the Human to do its job! We will return to this crucial point later in this course.

My role is to express my unhappiness

My role is to listen and understand before finding solutions

Key Point
Our Chimps need understanding and recognition of what they feel they are going through, no matter how irrational it might seem. Only then can they deal with certain emotions and allow the Human to move us on.

Colluding with the Chimp

The last example might almost seem like colluding with the Chimp. When some Chimps exercise ***unreasonably*** it would be inappropriate to accept their outbursts or the lack of responsibility by the person allowing this expression of emotion. Emotion that is damaging to others might be understandable but cannot be condoned. Unreasonable expression of emotion is sometimes very obvious in its presentation but it can be subtle. The earlier example of Pearl becoming agitated by the lack of work output by her staff might be a good example of unrecognised damage. Her unrealistic expectation of what can be achieved and the emotion expressed by her Chimp could be damaging to the staff. It would not be acceptable to allow this emotion to be repeated unchecked.

Key Point
Understanding the Chimp doesn't mean colluding with it.

Emotions that are bubbling under the surface

There are times when we hold emotion in and we do ourselves no favour. It is a matter of judgement of when and how you decide to express emotion. Selecting the right time, place and way of expressing emotion is helpful to be effective and productive. This is something that we can all learn and get better at.

There is a potential danger when we don't exercise the Chimp. If we suppress or don't acknowledge our emotions, then they might turn inward and become very destructive toward us. [192] At worst, they can lead to anxiety or depressive illness and at best they will surface in the wrong place and at the wrong time. [206] It always helps to express how we feel to an understanding person.

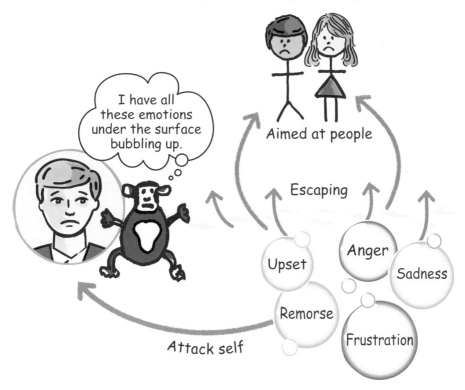

Emotions can escape and be displaced anywhere

Key Point

It can be very destructive to hold emotions and feelings in.

Example: *Bobbie and her exam results*

When Bobbie was at school, she didn't do well in her final exams and felt that she had underachieved. These results haunted her for years. Her Chimp could not accept what had happened and it constantly wanted to go back and do the

exams again. These negative and unhelpful emotions would be just under the surface in her day-to-day life and kept breaking through over many years. Clearly this has left an emotional scar in her life. This doesn't mean that it can't be settled down by revisiting it. Some likely beliefs she could hold are:
- I know I could have done better
- People weigh others up on their exam results
- Everyone thinks I am not very clever

The beliefs that Bobbie has are all negative. This is a good indication that they are Chimp driven. Her Human can review her beliefs and then manage the Chimp and get it on to the same page as the Human.

Some facts are as follows:
- There are many times at important moments that we underperform
- Sometimes, we don't know why things happen
- There are different forms of intelligence
- Intelligence develops continually during life with the brain being fully matured at around the age of thirty
- School exams are a snapshot of where we are at that time in life
- Is it really important what others think?

Having established these facts, Bobbie can now let these truths sink in. Sorting out emotions that are under the surface is really about revisiting events and seeing them differently. By seeing them differently, we can process the emotions that we feel.

Unit 10 Reminders

- It is a skill to express emotion constructively
- Underlying reasons why a Chimp needs to exercise ought to be explored
- Exercising the Chimp in an effective way, needs to be followed up with insight and change
- Chimps like to know that they have been understood, once they have finished exercising

Unit 10

Focus points and reflective exercises

1. *Exercising the Chimp, but not in a supermarket!*
2. *Addressing emotions that are under the surface*

Focus 1: *Exercising the Chimp, but not in a supermarket!*

A reminder: To exercise the Chimp effectively means that after we have expressed emotions or feelings, we then ask ourselves "How do I prevent these feelings from recurring?". In other words, we look to see what caused us to have a need to exercise the Chimp, and then make sure it doesn't happen again.

This focus point is therefore about recognising and addressing why the Chimp needed exercise to prevent it from happening again.

Exercising the Chimp doesn't have to be a loud rant or scream. Exercising the Chimp is about expressing feelings and emotions, so that we are able to make sense of them and employ them beneficially. Therefore, just talking through feelings, without intense emotion, is a good exercise.

Don't exercise your Chimp in a supermarket; take it to a safe compound. This means express your feelings in the right place and with the right people present. These are people who know what you are doing and won't interact or take the Chimp too seriously, until it settles down! Alternatively, exercise your Chimp alone.

Exercise: *Putting in place a regular structured exercise for the Chimp*

Fix a time of the day when you can think through your feelings and emotions and express these verbally or in writing. [142] [129] If you can't manage every day then try to allocate a weekly slot that is dedicated to releasing your emotions in order to reflect on them. Whenever you release your emotions try to move forward by addressing any causes underpinning them.

A structured way to approach this would be to write down a list of the areas of your life that are unsettled or uncertain at that time. Then talk through each, remembering that the object of doing this is to express emotion and then make plans for change. Most of the change will be in your own head and your approach to situations. Is it expecting too much? Is it being unreasonable? What beliefs are in your Computer that could be prodding the Chimp? For example, do you have a Gremlin in the Computer that is stating that life should always go according to plan?

When you have found a Gremlin replace it with an Autopilot that is a helpful and realistic belief.

Focus 2: *Addressing emotions that are under the surface*

This focus point is specifically looking at emotions that have been bubbling under the surface for some time. They often break through and influence our day-to-day life.

Exercise: *Removing hidden emotions*

This exercise is to search for and deal with recent events in your life that have left an emotional mess behind them; in other words, unfinished business.

To manage these bubbling emotions could mean that you have to work out how to address each irritant because they will each need their own expression.

For example, some methods could include:

- Expressing feelings to a friend to get things out of your system
- In Human mode, confronting the person or event that has left unpleasant emotions
- Writing a letter detailing your feelings, which can be sent or shredded.
  The exercise of writing things down can be very cathartic
- Revisiting the event and checking what beliefs you are holding that are causing the emotions to persist. When you have found the beliefs and replaced these Gremlins with Autopilots, it's important to dwell on the Autopilots so that they are firmly embedded into your Computer. This might take a few sessions to accomplish.

Unit 11
Changing habits, beliefs and behaviours

STAGE 4: *focusses on the Computer system.*

Unit 11: *covers how habits are formed by looking at the reward system in the brain. Changing habits is never easy. The Triangle of Change will be introduced to help you to change and form new habits.*

How habits are formed

Habits are formed when we carry out a pattern of behaviour or thinking that we deem to be successful. Therefore, if we try something out and it works, we will keep repeating it. The more we repeat an action, the more it will enforce pathways in our Computer system. This leads to an automatic programmed action that is carried out without being challenged. [207]

Habits being formed
- Scientific points

The science behind habit formation is very complex involving many areas of the brain but we can simplify things to get the principles involved. Warning: you might still find this heavy going!

When we perform an action that we find successful, we set in motion a chain of reactions within the brain. This chain of reactions will result in habit formation.

One significant pathway is called the reward pathway. This reward pathway is involved in all kinds of habit formation, including some drug addictions.

The pathway has three separate areas connected together in a chain: the locus coeruleus; the ventral tegmental area and the nucleus accumbens.

The first area, the locus coeruleus is the main centre in the brain for producing noradrenaline. Noradrenaline is an alerting neurotransmitter and keeps us vigilant. When the locus coeruleus becomes stimulated it releases noradrenaline, which travels up the brain to stimulate the second area in the chain, the ventral tegmental area. The ventral tegmental area then releases dopamine, another neurotransmitter, which stimulates the third area, the nucleus accumbens. The nucleus accumbens when stimulated brings pleasure. This therefore encourages us to repeat whatever we are doing. [208] [209]

Habits being formed *(Continued)*
- Scientific points

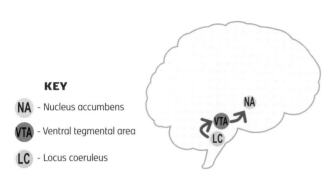

The reward pathway

First step:
LC releases noradrenaline to stimulate VTA

Second step:
VTA releases dopamine to stimulate NA

Third Step:
NA receives dopamine: anticipates and experiences pleasure

KEY
- NA - Nucleus accumbens
- VTA - Ventral tegmental area
- LC - Locus coeruleus

Habits being programmed
- Scientific points

In order to repeat our action, and form a habit, the nucleus accumbens does two things:

1. The inner part or core of the nucleus accumbens connects to the motor cortex, which controls our physical movements
2. The outer part or shell of the nucleus accumbens connects to the amygdala to involve emotional input. Some authorities feel that the shell is an extension of the amygdala

The nucleus accumbens has now combined actions and emotions to repeat our experience and form the habit. [210] [211]

Habits being formed by the Nucleus Accumbens

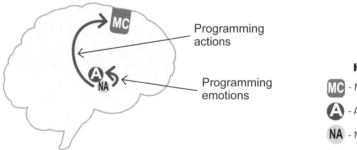

Programming actions

Programming emotions

KEY
- MC - Motor cortex
- A - Amygdala
- NA - Nucleus accumbens

If we carry out habits because they are successful, why is it that we can form unhelpful or destructive habits? [193] [207] [212] The answer lies in the difference between the way the Human and the Chimp define success.

Key Point
The Human and Chimp define success differently, which can lead to helpful and unhelpful habit formation.

Habits are formed depending on who defines success

The Chimp works impulsively and wants immediate gratification. It doesn't think about the longer-term consequences of its actions. [213] [214] [215] The two key ways in which the Chimp defines a successful habit are:

1. Immediate gratification
2. Pleasure or avoidance of pain

Therefore, the Chimp will reinforce any behaviour or belief that results in either or both of these.

Example: *Jodi and his unfaithfulness*

Jodi's Human wants to have a loving and caring monogamous relationship with his partner. He defines this as ideal and successful. However, his Chimp defines success as gaining short-term pleasure at the expense of sacrificing Jodi's morals. The Chimp will choose short-term pleasure rather than long-term peace of mind.

The Chimp won't think about the consequences of its actions unless it gets caught out. The covering lies that might be needed are also used to block his conscience. The Chimp has now established a habit that appears to work for it.

The habit will be established if the Chimp puts pleasure above monogamy. Clearly, this example has lots of avenues to explore and won't be quite so simple.

The Orbitofrontal Cortex (OFC) definition of success
- Scientific points

The OFC defines a successful outcome, as not being stressed and achieving immediate gratification. Therefore, behaviour that achieves immediate gratification or relief will be repeated. The OFC functions by reacting immediately to any stimulus it receives. The problem with the OFC definition of success is that, it might not only be temporary but also have long-term negative consequences.

The marshmallow experiment

The principle of this experiment has been repeated in various ways. It demonstrates the difference between the OFC definition of success and the Dorso-lateral prefrontal cortex (DLPFC) definition of success.

An adult offers a young child, who is around four years of age, a marshmallow. The adult must leave the room and tells the child that before they return the child can eat the marshmallow. However, they can wait until the adult returns, and if they do wait, they will get two marshmallows. The adult then leaves the room. The experiment shows that those children who eat the marshmallow immediately or after a short time appear to use this strategy throughout life and do not do well in later years. Those children who wait to get two marshmallows are using the DLPFC and thinking about the longer-term gains. These children apparently continue this pattern of thinking throughout life and are much more likely to be successful. Effectively, what is being said is that using the OFC system is not as helpful in the long-term as using the DLPFC, when making some decisions. Alternatively, it could be put as: operating with an impulsive OFC (Chimp) and defining success with this, is not as helpful as operating and defining success with the DLPFC (Human).

My concern with this experiment's possible conclusion is that it might be assumed that children and adults cannot learn how to recognise and switch systems. Learning to switch systems and manage the OFC is the basis for the Chimp model. I think we can all gain insight and operate how we want to and not be hijacked! We can all be successful. Those children who had dominant Chimps can learn the skill of managing them. [216] [217] [218] [219]

As the Human and the Chimp define success differently, it is important to establish what you see as success and not let your Chimp define what success means to you.

Example: *Jill and the habit of drinking*

Jill has always enjoyed socialising with her friends in the local pub drinking beer. She goes there each night, drinks two pints and plays darts. She sees this as being a 'successful' habit because it brings her lots of pleasure and says the drink stops her from experiencing low moods. The problem is that Jill has recently been diagnosed with pancreatitis and blood results show that her liver is struggling to deal with the levels of alcohol that she is drinking. Her Chimp has 'rationalised' that the drink brings her pleasure and plays a necessary part in helping her to avoid low moods. Therefore, the habit will continue because the Chimp only sees success with her drinking.

Jill's Human clearly sees that if the drinking is being used to prevent low mood it is a poor coping strategy. Her Chimp is using excuses to continue the pleasure of drinking.

Success for the Human would be to enjoy the evening but to limit or stop the drinking, and to use a healthy alternative for preventing low moods. Jill's choice becomes: The Chimp's short-term gratification with health consequences or the Human's long-term gratification with health benefits. The first step to solving any problem is to recognise it.

Habits being formed

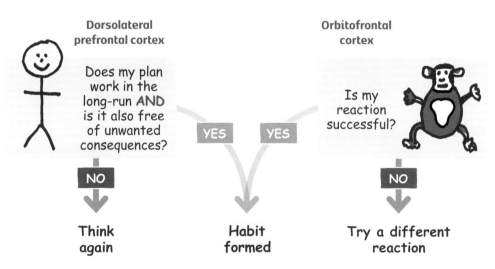

Habits can define the norm

One of the difficulties with the Chimp, having establishing a habit, is that it becomes the expected and the reference point for what is normal. Anything else can feel wrong, uncomfortable or perceived as unusual, even if the habit is not what our Human wants.

Example: *Lester and his eating habits*

Lester struggles with weight and his Chimp has established an eating habit at meal times. His Human would like to change his eating pattern but the Chimp has already defined the norm.

Lester eats a sizeable meal that leaves him feeling full and satisfied. After his main meal, he eats a dessert and following this, he eats a biscuit with his coffee. For Lester, this has become the established 'normal' eating pattern. If he tries to change this, it is likely to feel wrong. Even cutting out the biscuit won't be easy. However, if he perseveres with a smaller main meal and refuses a dessert and his biscuit, after a few weeks he will have established a new eating habit. He will now have this as his normal eating pattern and to go back to his previous habit will feel 'wrong'.

A habit seen as normal behaviour

It surprised me when I initially worked with repeat offenders that if they hadn't committed a crime for a while, some of them perceived this as doing really well. The rest of the population see behaving morally as normal behaviour. If an offender remains with the idea that not committing a crime is doing well, then they will see offending as being an acceptable normal behaviour. Establishing a healthy or helpful habit needs to be seen as normal and anything other than this as being unacceptable and not normal. If we view our habits in this way, then we are programming the Computer to advise the Human and Chimp what is and what isn't acceptable. This can help immensely for establishing the behaviours we want.

Habits are influenced by self-image

Most habits are formed and maintained by underpinning beliefs. These beliefs can come from either Human or Chimp. [220] A strong influencing force on how we behave is the way that we perceive ourselves. If we perceive ourselves as a hard-working individual, then we are more likely to work hard to demonstrate that feature. Habits can be enforced or changed by the way that you perceive yourself.

Example: *Eric and his self-image*

Eric says he struggles to get things done and often procrastinates and puts things off. He doesn't like this and it makes him frustrated that he can't seem to get his act together. The reality is that his Chimp is being prodded by a Gremlin in the Computer.

It's a big Gremlin of false self-image. The Gremlin that has now taken over is a belief that Eric has of himself. He perceives himself to be a person who cannot organise himself or get things done and is easily distracted. This self-image is false. Eric wants to get things done and he is not lazy nor is he a disorganised person. Eric's Chimp might have those characteristics. Eric is an organised and hard-working individual, but the Computer Gremlin keeps telling Eric that it is him that is lazy and disorganised.

Eric could remove this Gremlin and replace it with an Autopilot, a true self-image, which is someone who gets on with things immediately, as and when they need doing. He is likely to act on this belief. His habit of procrastination or avoidance can be replaced with a habit of immediately acting and of being hard-working.

It can be very productive to check on how you see yourself, and we will do this as an exercise at the end of this unit.

> **Key Point**
> *Your self-image often dictates your behaviour.*

How do I change unhelpful habits?

The Triangle of Change

If we look at what will increase the probability of someone altering their behaviour, habits or thinking, three factors appear to be necessary. I have called these factors the Triangle of Change. We can therefore use the Triangle of Change to assess whether we think that change is likely to occur.

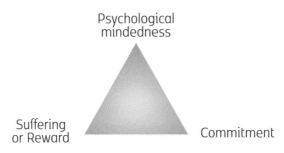

The Triangle of Change

Suffering or reward

A crucial point in dealing with any change is to appreciate that if we are not suffering enough or if the reward is not big enough, then we are unlikely to change our ways. [221] [222]

Example: *Jasmine and her car payments*

Jasmine is buying a car by paying monthly instalments. She is often late with payments and has been warned by the car company that they will repossess the car if she is not on time in future. Jasmine believes that these warnings are idle threats. Therefore, she does not perceive any suffering will occur because she will have her car. She also thinks that if they do carry out their threats she will just find another car dealer to work with. However, when she fails to make the next payment on time, the car company repossess the car. After suffering a lot of stress because of wasting lots of time pleading to get the car back, she finds her life without the car becomes very difficult. After further hassle of reorganising a new car sale, with a different company, she now owns a car again. As she now realises just how much suffering will occur when she doesn't keep up with the payments, she is much less likely to be late with her payments.

Changing your perception of suffering or reward

The perception of both suffering and reward are not necessarily fixed. We can increase either one by thinking through our situation and the consequences of not changing a habit.

Example: *Max and his marriage*

Max has been married for ten years to Julie. He loves Julie but has stopped paying attention to details in their marriage. He believes that the marriage is sound, and although not perfect, it will be fine. Max isn't likely to change his habits because he sees no suffering or reward. However, if he stops and thinks about the rewards of his relationship and doesn't take them for granted, he can increase his perception of these rewards. This in turn will instigate changes in his behaviour. He could also stop and realise that most marriages that break up are not necessarily in a bad way, just a bit neglected. He can see the reasoning that if he doesn't pay the attention that his wife deserves, then somebody else might give her that attention. The thought of the loss of his wife to someone else, might lead him to become aware of what suffering he might feel and help him to pay more attention to their relationship.

Reflecting can increase neurotransmitter release
- Scientific points

In the reward pathway, importantly, the ventral tegmental area (VTA) releases **variable amounts** of dopamine depending on **how pleasurable or advantageous a habit** is to us. This is very important because we can use this fact to help us form or remove habits. By thinking through the rewards that could happen and making positive changes, the VTA will release more dopamine, which will bring us anticipated pleasure. This higher level of dopamine in turn will instil the new habit being formed in the nucleus accumbens. Therefore, anticipation of reward will enforce new habits, and this can be achieved by reflection, particularly focussing on the benefits and rewards of change. [223] [224]

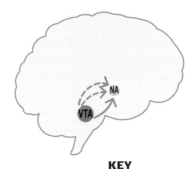

Variable amounts of dopamine released depending on your beliefs

KEY
VTA - Ventral tegmental area
NA - Nucleus accumbens

Psychological mindedness

Psychological mindedness can be defined in slightly different ways. Generally, we think of someone as being psychologically minded when they have an awareness of how the mind works and the effect that emotions and beliefs can have on self and others. [225]

> **Key Point**
>
> *If someone is psychologically minded, they will be open to challenging their own behaviours and beliefs.*

Therefore, psychologically minded individuals would be able to see a different point of view and reconsider how they perceive and interpret events. This ability to challenge their own position and beliefs, and to take responsibility for their own actions, and not blame circumstances or outside influences, is a major catalyst for change.

Example: *Vince and the habit of undermining others*

Vince often finds he is in conflict with others and feels he has to justify critical remarks he has made about people. If Vince is not psychologically minded he will always look outside of himself and find evidence as to why he has commented on someone. He will justify this comment and he will not see it as undermining someone. The habit will continue and lead to further conflict.

If Vince is psychologically minded, he will begin by looking to himself and his approach. He might well see faults in others but he will ask if his habit of commenting is actually helpful. He will think of the reason why he is doing this. For example, could it be that he has low self-esteem and uses criticism of others to elevate his own standing? He will also think about what he is trying to achieve by doing this and what effects it will have on others and ultimately on him. What Vince is doing is looking at the impact of his own actions and the reasons for them and how his mind is working. In our model he will be identifying what is likely to be his Chimp establishing an unhealthy pattern of working and then rethinking with his Human what he really wants to be doing.

Commitment

Commitment is a key factor for change. Commitment doesn't just mean that you make a statement of action and fully intend to carry it out. This is really just an emotional Chimp stance that is working with feelings. [226]

Commitment means forming a plan that will work for both your Human and Chimp. It will work on three counts:

1. Your Chimp, as well as your Human, is in agreement with the plan
2. Your plan involves establishing what it will take to make it work
3. Your plan has considered the potential pitfalls that might stop you and has a strategy to deal with these

Therefore, commitment means sitting down and making some effort to prepare your plan for the habit change.

Example: *Robert and his daily exercise*

Robert believes that daily exercise is good for him. He has tried repeatedly to either jog every day or to go to a gym. This habit never gets established and he continues to remain unfit. Why can't he make the change of habit from intermittent exercise to regular exercise?

Robert's determination to exercise is probably Chimp driven. His Human definitely wants this to happen and his Chimp emotionally wants it to happen **BUT *it doesn't want to commit to it***. There are always excuses and rationalisations for not exercising, and Robert is aware that these are excuses and rationalisations. How does he form a commitment plan to elicit change?

Step 1: *Involve the Chimp*

The first step is to establish what his Human wants and thinks is a good plan and to involve his Chimp with this plan. Is exercising every day a realistic proposition? It might be to his Human but is his Chimp really going to do this? Only Robert can decide. Making unrealistic demands of the Chimp is a recipe for failure. A possible compromise that his Chimp might work with is to exercise four days a week and to start with shorter sessions and to build up once the habit of exercising is established. This plan is far more likely to be accepted by his Chimp.

Step 2: *What does it take?*

When I work with people on making a commitment screen, I break down the list of 'What does it take to commit' into essential, significant and desirable.

Essential   Without the 'essentials' the plan definitely can't work

Significant The 'significants' are those things that will definitely help to raise the chance of success

Desirable   The 'desirables' are those things that will make things more pleasant

As we are all unique, only you can decide what is in each list. For example, Robert might say that exercising with a friend is an essential factor because he knows that without someone else exercising with him, his Chimp is very likely to stop. Some people might say that exercising with a friend is significant, whereas others would say it is only desirable.

Step 3: *What will stop me?*

Things that might stop the plan from working can be divided into hurdles, barriers and pitfalls.

Hurdles   Things we can't avoid and have to jump over

Barriers Things we can get around with some planning

Pitfalls Things that are ultimate plan breakers

Examples of each of these could be:

Hurdles
- Some days will be uncomfortable
- Your Chimp might need managing
- Minor injuries can happen

Barriers
- Being tired might mean adjusting your training
- Time pressure might mean better time management
- Limited opportunity to train might mean being inventive for exercising that day

Pitfalls
- Going on how you feel rather than what you have to do
- Seeing a poor training session as indicating failure
- Losing sight of the benefits of training

Commitment

What I need	What might stop me
- Essential	- Hurdles
- Significant	- Barriers
- Desirable	- Pitfalls

Changing habits

Stating the obvious - to change a habit or instil a new habit needs work!

Here are some suggestions on how to form constructive habits:
Use the Triangle of Change to decide on whether you are in a position to succeed.

Rewards and suffering: reflect on the advantages and rewards of change, and the consequences or suffering if you don't change. Use this reflection to clarify and strengthen the rewards or suffering in your mind. Let your Chimp reflect on *how it will feel* if it does change and *how it will feel* if it doesn't change. This will be speaking to the Chimp in its own language.

Reflection involves exploring what your beliefs are. Have you got some unhelpful Gremlin sitting in your Computer? Is the Chimp giving unwelcome feelings that are stopping you from carrying out the healthy habits that you want to establish? For example, is there a Gremlin that is saying, "We can always start tomorrow", and this prods the Chimp into feeling that it doesn't quite have the energy to begin now?

Psychological mindedness: use your psychological mindedness to think about taking responsibility for change. We are all capable of change, because change is nearly always dependent on decisions within the mind. Don't allow your Chimp to make excuses or to hijack you. Recognise that the mind is a machine and you can manage it.

Commitment: draw up a commitment plan for change and ensure that you have consulted your Chimp. Make it easy and pleasant for your Chimp to join in with change. Many people do this by getting someone to challenge them. Most Chimp's don't like being told that they can't or won't be able to do something. This often drives the Chimp to prove them wrong.

Specifically, learn the skill of rejecting any unhelpful emotions offered by the Chimp and disengage with any unwelcome thoughts. Work with commitment rather than motivation. Distract your Chimp while you get on with your new behaviour.

Beliefs: consider the beliefs you hold that might underpin your habits, including the beliefs that give you your self-image; replace unhelpful ones.

Seeing a habit as a choice: we can view habits as being a choice between the Human and the Chimp. The choice is then down to: Do I want immediate gratification or do I want to have longer-term satisfaction? Many people find it helpful to see habits as being a relaxed choice made by their Human and a rejection of the Chimp's preference.

Why are habits sometimes so difficult to change?
- Scientific points

Once we establish a pathway in the brain that creates an automatic behaviour or a belief, we use this as a default mechanism. These pathways become re-enforced over time. The brain does this by myelinating the pathway neurones. This means coating them with a substance that lags them, which has the effect of speeding up the transmission of the message being sent. For us to stop defaulting to this pathway, we have to create a new pathway and re-enforce this. This means practicing a new alternative habit, in order to extinguish an old one.

Neurones within the brain will form and also break connections with other neurones depending on whether a pathway is being activated or has become redundant.

New habit being re-enforced by practice and myelination

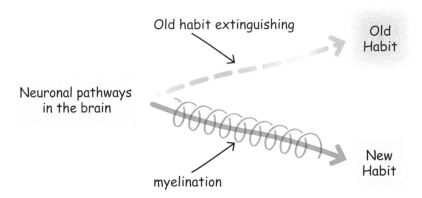

Unit 11 Reminders

- Habits can be formed on self-image
- The Triangle of Change is composed of:
  - Suffering or reward
  - Psychological mindedness
  - Commitment
- Perceived success can be very different for Human and Chimp
- Changing habits usually needs reflection and a plan

Unit 11 Exercises

Focus points and reflective exercises

1. *Checking your self-image*
2. *The Triangle of Change*
3. *Habit formation*

Focus 1: *Checking your self-image*

As discussed in this unit, your self-image will often dictate how you behave. Therefore, it is an important factor for influencing habits.

Exercise: *Define your self-image*

Try and define your self-image. How do you see yourself?

If you have not previously thought about how you see yourself, then try asking the following questions but make sure it's the way your Human sees you and not the way your Chimp sees you! **Warning:** some of the questions are hard hitting:

- Am I a proactive person?
- Am I a problem-solver?
- Do I see myself as a person who primarily works with logic or emotion when making decisions?
- Am I someone who works with immediate gratification or am I wiser than this?
- Do I see myself as someone who gets on with things as soon as they need doing?
- Is my Chimp a winner in life or a loser? Am I a winner or loser in life in my eyes? Which do I want to be? (That's a tough one!)

If you answer the questions with your Human then the answers will be what you want them to be. This is the self-image that represents the real you before the Chimp interferes. If you can use this self-image and see yourself as this person, it can drive you to fulfil this image when confronting habits.

To encourage you: please try to be positive because interference can be managed, it's only a Chimp!

Focus 2: *The Triangle of Change*

Exercise: *Forming a new habit*

Apply the Triangle of Change to a habit that you either want to establish or want to change. Think carefully through the rewards and benefits of change and also reflect on the suffering or detriment to you or others of not changing. Apply your psychological mindedness to take responsibility for habits. Question whether your Chimp is making excuses for not establishing the habits that you want to establish. Remind yourself that habits are decisions made by either you or your Chimp. You always have a choice.

Draw up a commitment plan. I would suggest you don't make this too heavy because it will only put your Chimp off. Try to keep things simple and relaxed. The more intense it becomes, the more the Chimp will agitate and try to take over, usually by disengaging. The Chimp is very likely to become active in an unhelpful way if you see the new habit as being a struggle, something to win or as an abnormal state that you are trying to achieve. [207]

Evoking vigilance
- Scientific points

Domesticated animals, such as horses and dogs, pick up on the mood that Humans are in. If we are relaxed, regardless of what is happening around us, then the animal is likely to relax with us. If we become intense or worried, then this is transmitted to the animal and the animal will become alerted and concerned. Our inner Chimp is exactly the same. If we, as a Human, remain calm and relaxed our Chimps will settle. However, if we approach life or situations with apprehension or see them as a battle, then our Chimps will naturally become alerted and animated. How we perceive situations will evoke appropriate emotional responses for that perception. [227]

Focus 3: *Habit formation*

Linda is trying to establish a new habit of always getting things done as they arise, instead of putting them off. She has tried several times to make this a habit but each time she tries she very quickly reverts to her usual habit of putting things off.

Exercise: *Working out why habits don't change*

Suggest five reasons why Linda cannot establish her new habit. After you have thought of the reasons, check with the suggestion answers.

Answers

Clearly there could be lots of reasons, so here are a few of the commonest ones:

1. **Reward or suffering:** Linda doesn't see this habit as being that important. She doesn't really believe that she will gain that much from doing it, only that it might feel good. She doesn't believe that there are any significant consequences to leaving things until later.
2. **Psychological mindedness:** Linda hasn't accepted that it is within her power to manage her mind and take responsibility for this.
3. **Commitment:** Linda has made no real plans to form her new habit. She hasn't thought about what it will take to do this nor has thought about what might stop her.
4. **Self-image:** Linda believes that she is just one of those people that can't focus or stick at things. This excuse becomes a self-fulfilling prophecy.
5. **Chimp alerted:** Linda perceives the habit to be a big challenge and has alerted her Chimp to this. The Chimp has then seen the task as overwhelming and is fighting back to let Linda know that it isn't possible to achieve.
6. **Engaging the Chimp on its terms:** when Linda is ready to act, her Chimp begins a conversation with her and Linda engages with this conversation. Instead of being ignored, the Chimp draws Linda into a discussion that Linda loses.
7. **Linda is being unrealistic:** although many things can be done immediately, some things can't. Linda is not distinguishing between the two and is therefore condemning herself unfairly, and this is putting her off trying.
8. **Linda is being too harsh on herself:** Linda is being too critical regarding how she is already performing and she actually does do things immediately, but is asking for perfection. She might attend to things immediately but some things need to be done in stages and take time.

Unit 12
Processing and managing life events

Unit 12: covers *how to accept and make sense of life events and move on from them. Some life events are fairly easy to move on from but others can prove very difficult. This unit begins with the easier ones.*

What it means to process life events

Processing a life event means that we have come to terms with what has happened and are moving on. Therefore, processing life events or information means:
- Accepting the reality of the facts that you are faced with
- Understanding and making sense of these facts
- Working with the facts to go forward

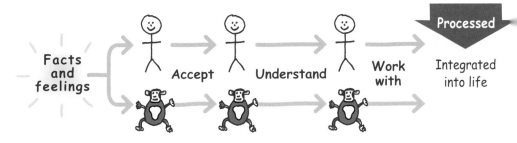

Once we have processed a situation we can move forward because it has been integrated into our lives. If we don't process information, then we get stuck and start receiving emotional messages from our Chimp that all is not well. [228] [229] [167]

For example, if you have attended an interview and you felt that you had done really well but didn't get the job, you might keep going over the scenario. You might feel it was unfair, or dream about what could have been, or start thinking maybe you didn't present well and so on. In other words, you can't seem to get over it. This is because processing the information hasn't taken place.

Nearly all life experiences can be processed. If we did process all experiences, it would lead to a much more peaceful life. The processing exceptions are emotional scars, which we might have to live with.

Processing occurs at two levels:
- Simple processing
- Complex processing

Some experiences respond to simple processing, whilst others need complex processing.

The difference between simple and complex processing

Simple processing occurs when there is little emotional consequence from an event and complex processing occurs when there are lasting emotional or practical consequences from an event. [230] [231]

Examples of experiences that respond to simple processing:
- If I trip and scrape my knee, it might hurt but it is of little emotional or practical consequence
- If I am criticised on social media but it's clear that the criticism is ridiculous and it doesn't particularly bother me
- If I have a disagreement with someone, which initially leaves a bad feeling
- If I miss a train or it is delayed but I will be fine once it is sorted

Examples of experiences that need complex processing:
- If I am an elite athlete and I trip and break my leg and this ends my career
- If I am criticised on social media and I feel it has unfairly damaged my reputation and it really bothers me
- If a serious injustice has happened to me
- If I have had a serious loss in my life

Simple

Just let me have a scream!

Complex

I need some time and some help.

Key Points
- *Simple processing is when the Chimp just reacts in the moment and might complain for a while, but soon recovers.*
- *Complex processing is when the Chimp needs time and help to come to terms with something significant.*

This unit will cover simple processing and the next unit, unit 13, will cover complex processing.

Simple processing: *Acceptance seen as a skill*

The first step to processing anything is *acceptance* of the reality that is in front of you. Imagine having the skill to immediately accept any unpleasant experience that you encounter, and deal with it constructively. This could save many hours of distress or agitation from your Chimp that wants a different reality and is not accepting the situation. By accepting situations immediately, it would also help you to reach solutions and move on more easily. This is why *acceptance* is the first step to processing (or constructively dealing with) any life event or situation. [232] [233] [234]

> **Key Point**
>
> *Acceptance of things that can't be changed, and of circumstances that have to be worked with, is a skill that can be acquired.*

Example: *The spilt milk*

This is an easy example to show the process of acceptance. We will keep returning to this example throughout this unit to demonstrate the various aspects of simple processing.

Imagine you have gone to the fridge to get some milk. As you pick up the carton of milk it slips from your grasp and spills onto the floor. The ideal processing might be as follows: *I did not intend this to happen, but it has. I need to clean up the milk. Once I have cleaned up the milk everything will return to normal.* Therefore, you, as a calm human being, would process the information appropriately and deal with the situation. (We will come to reality shortly!).

The problem we have with this example is that, for most people, it is not the way the mind works. Instead of calmness there will be an emotional reaction from the Chimp that ranges from being slightly upset to an embarrassing complete loss of emotional control. It is likely that the carton will be blamed for being impossible to hold and victim mentality will ensue, "Why me?" Nobody should have to witness our Chimp's reaction. We can change this reaction, if we learn to manage the way the mind works.

So how does the mind process information?

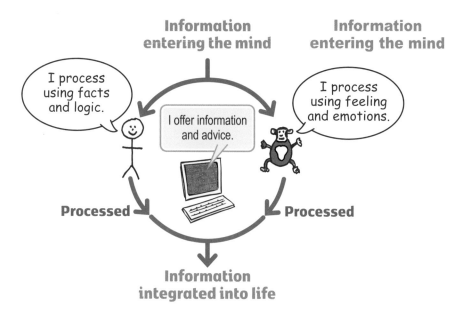

We can think of the mind as processing the information down two separate tracks: Human and Chimp. The mind must successfully process the information down both of these tracks and not just down one of them; otherwise we can't integrate the information into our lives. [235] [236] [237] Both tracks carry out their processing by conferring with the Computer to help them.

Note: The Computer can help immensely with this processing, ***but only if it has helpful beliefs stored in it***. We will look at how to do this shortly, after we have first considered the two tracks.

The Human processes the information using a logic-based approach and the Chimp processes the information using an emotion-based approach. [238] [8] [7] [3] [48]

How the Chimp processes events

> **Key Point**
>
> *The two tracks are different and both are essential for processing information.*

Let's return to our spilt milk example. There will typically be an emotional reaction first, before logic kicks in. [40] [41] [17] [239] A natural emotional reaction from our Chimp could range from distress to anger or frustration. Although this might be a natural reaction, it's not a very helpful one. As we are all unique, you must recognise what ***your unique Chimp*** will do, and then work with this. It might not even react. Let's assume the Chimp is typical in voicing some expletives and then chuntering for a while.

What we can acknowledge is the need to express emotion and not to hold it in. Once our initial feelings have been adequately expressed, our Chimps will settle down and allow us to deal with the situation. This is because simple processing by the Chimp can be done just by expressing emotion. [7] [3] [8]

If expressing emotion doesn't process the situation, then the Chimp will turn to others to seek understanding and acknowledgement of its distress, along with some approval. The Chimp wants other people to justify the emotions it feels. In simple processing these two aspects are all that the Chimp needs in order to accept and understand the situation and to move on.

The Chimp using simple processing

With simple processing, the Chimp usually moves on when it gets tired of listening to its own moans or complaints, and often gets distracted with other things!

Note: If you feel your Chimp is not processing the information, then it helps to turn to someone who will listen and give your Chimp the understanding and approval it needs. Alternatively, you could give your own Chimp some understanding by acknowledging that what it feels is normal and understandable.

How the Human processes events

The Human processes events by using facts, logic and rationality. [238] [8] Events are given some perspective, and facts are established and become the working basis for moving forward. This processing is much easier than emotional processing because the facts are usually relatively easy to establish.

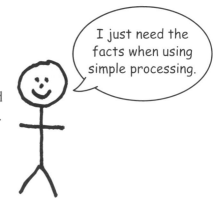

The Computer comes to the rescue!

We have already looked at the way the Chimp and Human struggle to communicate with each other because they speak a different language. The Human speaks with facts and logic, whereas the Chimp speaks with emotion and feelings. The Computer can now come to the rescue! The Computer acts as an interpreter. It can communicate clearly with both Chimp and Human because it speaks both languages.

The problem the Chimp has, is that it can't work with facts. However, if the Human can programme the Computer with facts and the Computer interprets these for the Chimp, then this will help the Chimp to process information.

Brain connections
- Scientific points

One of the complexities, in the way that the brain is connected, can give rise to some difficulties in understanding it. Generally, people think of the brain as having specific areas that do specific tasks, and specific pathways that do specific things. Although there is some truth in this, the brain doesn't always work this way. The brain acts more like a group of people. Some people have specific roles but can turn their hand to other roles, if need be. All people tend to get involved if an opinion or decision is to be made that will affect the entire group. The brain acts in this way. Therefore, sometimes it's easier to think of all areas being connected to all other areas, and when it comes to specific tasks, some areas drop out and leave it to the rest. Therefore, even though there are specific Human and Chimp teams in the brain, they often share areas to work with. The orbitofrontal area of the brain (the lead for the Chimp) and the dorsolateral prefrontal cortex (the lead for the Human) area of the brain are both connected to most parts of the Computer circuits (for example, the cingulate gyrus and hippocampal formation). This means that unless one of the lead areas of the Human, Chimp or Computer take over a task, all of the brain can get involved and we end up feeling confused or as if we are going in circles, as we switch between Human, Chimp and Computer. [240] [241]

The Computer as the advisor

Both Chimp and Human always look into the Computer for advice or help before they make any decisions or act. [243] [243]

In our example of the spilt milk, both the Chimp and Human will search the Computer for any beliefs attached to the spilt milk before they act.

Let's say that prior to the accident the Human has spent some time reflecting about minor accidents or setbacks. The Human has drawn up some conclusions and turned these into beliefs. These beliefs include the following:

- Most accidents or setbacks are easy to sort out
- Accidents are mishaps, not done on purpose, therefore there is no blame
- Getting upset or emotional only makes things worse
- Acting immediately can quickly put the problem right
- Whenever you can, it's better to smile than get upset

The Computer will offer these beliefs to the Chimp immediately the milk is spilt. The Chimp will process the event and not react to it in an emotional way. The person can therefore mop the milk up and peace continues!

The interesting point for us is the moment the Chimp consults with the Computer. [17] This is because the Computer is about to guide the Chimp on the best way to react. If the Computer has no programme to deal with unexpected incidents then the Chimp will have free range to react in any way it wants to. Clearly, this is often unhelpful. However, if the Computer is programmed to guide the Chimp with reality, and some perspective, then the Chimp is very likely to either not react or to react in a more measured and constructive way.

Programming the Computer

The Computer can be programmed to maintain perspective but the Chimp might still want to react emotionally. We have previously looked at ways of expressing emotion effectively. We can now build on this concept and programme the Computer to express emotion on behalf of the Chimp. This programmed emotional reaction will help the Chimp to process any event, particularly those with unwanted outcomes.

The example of the spilt milk represents many incidents that happen to us daily. It is relatively trivial and often accidental but can evoke a Chimp reaction.

Some simple constructive programmed emotional reactions might be:
- Laughing at the situation or seeing the amusing side of it
- Allowing one giant expletive
- Letting someone know, in order for your Chimp to get a sympathetic response
- Smiling and calling out a simple phrase such as, "It happens"

Processing and managing life events

Seeing the amusing side is more constructive than becoming frustrated.

Key Point

Programming the Computer with constructive emotional reactions helps the Chimp to process emotion.

Three questions to consider

Always remember that you are unique; I can only offer ideas and suggestions. It is up to you to find what works for you and this means it sits comfortably with your values and personality. For example, some readers will be happy using anger as an emotion to communicate with, whereas others will avoid this. Generally speaking, anger is an unhelpful emotion because it can be detrimental to health, especially if it is long-term anger. Also, anger being projected at another person, or even at yourself, can be damaging to communication and relationships. [244]

Returning to the spilt milk, try imagining that it was you who spilt the milk. Let's make matters a little worse, and imagine it has seeped into the gaps in the tiles on the kitchen floor and you have no more milk.

Question 1: "What emotional reaction would you like to have?". If you are going to manage your emotions then you need to programme the desired emotional reaction into your Computer ***before anything happens***.

Question 2: "Will your programmed reaction help you to have a better quality of life and less stress?".

Question 3: If your answer to question 2 is "yes", then my final question is: "Why are you waiting?".

To help you, we can consider some emotional reactions that can be programmed into your Computer.

Programming an emotional response

Here are some suggestions of emotional responses or reactions under 'probably helpful' or 'probably unhelpful' headings. The Autopilots are helpful beliefs, whereas the Gremlins are unhelpful beliefs, even though these Gremlins might have a ring of truth to them.

Probably helpful (Reaction with Autopilot)	Probably unhelpful (Reaction with Gremlin)
Emotion or action (Programmed beliefs that could help to enforce this)	**Emotion or action** (Programmed beliefs that could help to enforce this)
Sense of humour (See the funny side)	**Irritation** (This shouldn't be like this)
Tolerance (These things happen)	**Blame** (There is guilt attached to this accident)
Disappointment (It isn't great, but it's not the end of the world)	**Anger** (Anger is the best way to sort things out)
Acceptance (Let's start sorting it out)	**Frustration** (I can't accept reality)
Resignation (My starting point is where I am)	**Despondency** (It's all too much)
Patience (Time will move things on)	**Feelings of helplessness** (I am unable to cope)
Positive approach (Let me be proactive and find an immediate solution)	**Feelings of stupidity** (I shouldn't make mistakes)

You always have a choice of what type of reaction to place into your Computer. After programming the Computer, practising your chosen reaction within your imagination helps to embed it. Remember, if you don't programme your Computer, your Chimp will!

Moving into solution-finding

Example: *Late for the train*

Emma has not thought about programming her Computer in readiness for any setbacks. She has booked a ticket for a train that leaves at two o'clock. The train is taking her to meet a friend for a catch up, with no special importance attached. Emma gets ready for the train and is about to leave the house when a neighbour calls to let her know that she is throwing a party on Saturday and Emma would be welcome to come along. Emma knows she needs to avoid wasting time, but doesn't want to be impolite, so she chats before explaining that she has to go for the train (Is Emma being polite or lacking assertiveness?). Emma is now under a little bit of pressure to hurry and her Chimp is appropriately alerted. She has a ten-minute drive to the station and jumps into the car and sets off. The car stalls, which it has never done before, and then stops. She is now faced with a car that needs to be dealt with and the realisation that she is likely to miss the train.

Without programming her Computer for any setbacks, her Chimp now takes charge. Which of the following do you think is most likely to happen with her Chimp in charge?

a. Emma smiles and says "I really must buy a better car next time"

b. Emma says "I love challenges in life"

c. Emma's windscreen steams up!

> **Key Point**
>
> Remember that when the Chimp takes over, it only reacts and isn't looking for solutions.

Solution finding is your job, ***but you have to get into Human mode in order to do this***. A pre-programmed Computer can place you back into Human mode. If Emma had programmed her Computer with a constructive emotional reaction, ready for any 'disaster', then she would be much more likely to accept the situation and go into solution finding mode.

Rejecting 'natural' emotional reactions

Example: *The laptop glitch*

Mike is working on his laptop, when suddenly it crashes and he loses the last thirty minutes of his work. He is about to lose his composure. The first Chimp reaction is to see this as catastrophic. Expletives are abundant and even unknown expletives are invented. For some reason, Mike attacks the laptop and takes the delusional stance that the laptop clearly had a choice about whether to crash or not. It is now described as a useless piece of junk and anger increases as the laptop refuses to engage and wilfully taunts him. The laptop's clear lack of intelligence and idiocy is beyond Mike's Chimp's comprehension.

It could all have been different. Mike could have chosen a different reaction and emotion, other than frustration and anger. For example, he could have programmed his Computer to always respond to frustrating situations by seeing them as just a setback and a challenge to overcome. It could be that he still wants to let his Chimp out first, but probably for only a brief moment, rather than letting it take over. He could then follow this up with a plan to move forward. The emphasis here is that Mike must pre-programme *and rehearse* which emotion he wants to express, before it all happens.

> **Key Point**
> *We don't have to engage with an emotion that comes naturally; the emotion we choose can be more effective in managing a situation and far more productive.*

Being let down or betrayed

Example: *Rob and the promotion*

Rob and his friend, Andy, work in the same company doing the same job. Rob notices that a senior position is being advertised and tells Andy that he will be applying. Andy doesn't tell Rob but he speaks with the line manager about the new job and expresses an interest. The line manager asks if Andy will step in temporarily and Andy agrees. Rob's sense of betrayal is eating him up.

Rob has options: he can either allow his Chimp to keep focussing on the problem or he can process the situation and move on. Rob's Chimp will need to express itself and be heard by someone who cares. It will help if Rob's Human accepts the reality of the situation and forms a plan. His plan could be *to rethink his original beliefs* about Andy and the type of friendship he had with him. At some point in life all of us feel let down or betrayed by a friend or colleague.

Some truths that might resonate could be:

- The friend I had, is not the friend I thought I had
- Everyone gets let down at some point
- Being bitter is only going to hurt me
- Life doesn't always go the way that I want it to
- Not everyone is like this
- Sometimes people make mistakes

Try not to allow your Chimp to overstate your mistakes

Sometimes, we cannot process an event because we are allowing our Chimps to overstate mistakes that we make. This means not living with blame, guilt or beating yourself up. These are destructive coping strategies from your Chimp. They are not rational. A better approach is to forgive yourself, if you have created a problem, apologise and then learn from any mistakes. Once the Chimp accepts that mistakes are rectifiable by taking responsibility, it will be able to begin the processing.

Inappropriately taking personal attacks to heart

Although it can be difficult, try not to allow your Chimp to read personal attacks into comments people make. Sometimes, a distressed person is reacting to their situation and not to you; so don't take it personally. If there is a deliberate personal attack, it reflects more on the person making it than on you. [245] People who make critical remarks or comments generally do this to everyone. It's important to decide whether you want your Chimp to be happy by trying to gain approval from everyone, or your Human to be happy by giving self-approval or approval from those who know and love you. Is trying to please someone who can't be pleased a wise thing to do? Processing unkind remarks means making a decision about what you want to do with them.

Gremlins preventing the processing of comments

- You must have approval from everyone.
- Any remarks about you must be true.

Gremlins preventing the processing of comments

For example:
- You must have approval from everyone
- Any remarks made about you must be true

Unit 12 Reminders

- Acceptance is a skill that can be acquired with practice
- Information is processed down two tracks; Human and Chimp
- The two tracks operate with different methods and timescales
- We can learn to understand processing and help to make it easier
- Emotional responses can be programmed into the Computer

Unit 12 Exercises

Focus points and reflective exercises

1. *Acceptance as a skill*
2. *Choosing your emotion to help process experiences*

Focus 1: *Acceptance as a skill*

Whenever things don't go the way that we want them to, it's natural that our Chimps react. It isn't usually helpful and merely delays any positive actions we could take. The period of time that it takes the Chimp to react, and usually express negative emotion, can be reduced significantly if we learn to accept unwanted situations and immediately work with them. This is similar to programming your Computer with a "What's the plan?" response.

The subtle difference is that we are now taking *an approach to life* whereby we learn to immediately accept what is in front of us, without our Chimps reacting. We are learning the skill of going with the flow of life.

Exercise: *Pressing the pause button*

When something doesn't go according to your plans, try programming your Computer to make you take a deep breath or pause for a few moments. This allows your Human to take charge and at this point focus on acceptance.

Focus 2: *Choosing your emotion to help process experiences*

Our Chimps tend to offer us the same emotions even when they are dealing with different situations. For example, some Chimps always become anxious and some always moan. What the Chimp needs are more helpful alternative emotions to select from.

Exercise: *Alternative choices of emotion*

List the emotions that you recognise that your Chimp tends to use frequently. Ask yourself if these are helpful and appropriate. Then decide which emotions you would like to experience instead of the typical ones the Chimp offers. By naming the emotions that you wish to express, it will help you to manage your emotions much better and to process situations more easily. Once you have decided on the emotions you would like to experience, you will need to practice these so that they become a habit and replace the usual unhelpful Chimp choices.

Unit 13
Managing significant life events

Unit 13: covers *how to process significant life events that might have left us emotionally scarred. We will consider how to manage common feelings associated with change and loss, and how to manage emotional scars.*

Complex processing of information

We use 'complex processing' when we try and come to terms with serious events that could have lasting consequences. Spilling some milk is easy to get over but coming to terms with a significant loss is not so straightforward. When we experience situations that have a more serious impact on us, the Chimp needs more help to process them. [246] [247] The simple process of expressing emotion and being given understanding by others is insufficient. These emotionally tough situations usually involve some form of loss, error or injustice. They often involve a significant adjustment to the person's life. [248] [249] [250]

Two examples

In order to show how this works, we will look at two examples of situations requiring complex processing. Following this, we will look at the emotional stages that we commonly go through during this processing and how to recognise and manage them.

The first example: *Jean and the neighbour dispute*

Neighbour disputes can be extremely distressing and make people ill. This is a true example, as many of my examples are, but with the names and details changed. Several years ago, Jean's neighbour planted a hedge between Jean's garden and the neighbour's garden. The hedge was growing heavily on Jean's side of the fence. Jean decided to trim the hedge but only on her side of the garden. When she had finished, the neighbour became very angry and said that trimming the hedge could kill it. The neighbour then began a vendetta against Jean, almost as a punishment for what she had innocently, and legally, done. Jean could not prove that her neighbour had been behind many of the attacks on her property, such as the poisoning of her goldfish pond and lawn killer sprayed on parts of her lawn; nor could she explain the effects on her of being ignored or occasionally shouted at by the neighbour.

Jean had tried all avenues of reconciliation but the neighbour was not going to change their position. How does she now process this event, which is on going?

Her human would reason something along the lines of the following: "I will use facts and logic and come to a conclusion". Here are the facts, as she sees them:

- The neighbour is not a pleasant person and is unlikely to change
- What they are doing is wrong and they are unlikely to stop
- They will never apologise and they will never put things right
- I could go to court but the stress and cost just aren't worth it
- Injustice will happen, but it's up to me to accept it, when I can't change it
- If I can't manage the situation then I must take myself out of it
- I have a choice to move house and cut my losses

Jean can obviously decide not to let her neighbour's actions take away her happiness. Wouldn't it be great if we could do just that? For most of us, this is so far from reality. Now let's bring in Jean's Chimp.

Jean's Chimp will work with feelings and plan to act on these. Here are her Chimp's reactions:

- I will not accept this; no matter what it takes, I will keep going until I get justice
- If I can't get justice, I will get revenge
- I might get upset, become angry or despondent but I need to keep going and win
- I would rather suffer than concede because I am not in the wrong
- I have rights and expectations
- I will rally other neighbours to get support
- I see this as a win or lose situation, and I have to win

Some of the Chimp's thoughts are absolutely correct and could be acceptable. Fighting for justice might be the right thing to do, regardless of how long it takes. The point of this example is to clarify the two approaches and to check their appropriateness. Any plans will have consequences. When Jean is in Chimp mode, there is no peace of mind because her Chimp cannot process the situation, as it will remain focussed on the problem. [251] [252] There are no solutions from the Chimp, just strategies to 'win'. Winning or losing will both have their own long-term consequences, but the Chimp can't see this. How can Jean process this situation, if she wants to move on?

Jean will process the Human track by rationalising and using her facts and logic. Her Chimp will need help because this is a complex situation. It won't be solved with logic alone or by just allowing the Chimp to vent its feelings and receive some reassurances from others. [83] [253]

In the real situation that this example was based on, 'Jean' sold her dream house and moved. After some time, she told me that it was the best thing that she had ever done and she was really happy again with pleasant and helpful neighbours. In effect, her Chimp 'lost' but Jean won! Her Chimp needed to grieve, which will be described later in this unit.

The second example: *A relationship break-up*

Being involved in an intimate relationship is something that almost everyone experiences. Sadly, a break-up can be very painful. Imagine someone who is involved in an intimate relationship and their partner has decided to end it. The person doesn't want to break up but their Human knows that the relationship is over. Rationally, the break-up might make sense, and so in theory, the person should be able to move on quite quickly. We know that this just isn't the way it usually works. This is because the Chimp **cannot process emotional information quickly**, frequently gets stuck and becomes unable to move on. [254] As the Chimp cannot work with reality, it will try to reject reality and make things happen the way it wants or expects them to happen. This will lead the Chimp to go through various emotional stages, as it struggles to accept the facts. [246]

Emotional stages during complex processing

When the Chimp tries to process an emotionally complex situation, it can go through a number of stages. These stages of emotional processing can be experienced in any situation where there is difficulty coming to terms with something. [247] [248]

Here are examples of situations where we might experience the stages:
- An injustice or perceived injustice
- Failing to achieve; in exams, gaining a promotion or achieving a sporting success
- A loss of something that is valued
- A missed opportunity or poor decision, with implications
- A tragic experience that could have been avoided
- Consequences of being in the wrong place at the wrong time

Rather than go through each of these experiences individually, we will look at some common emotional stages experienced, by using the 'grief reaction' as an example. [249]

The Grief Reaction

We can experience a grief reaction when *any change or loss has occurred in our lives*; it is far more common than you might think. [250] We can also experience anticipatory grief, when we fear that a loss or change might happen. [251]

Here are some examples:

- Loss of a role or job
- Loss of a routine
- Loss or break-up of a relationship
- The death of someone close
- A change in body image
- A change in self-identity
- Moving house
- A serious illness occurring

Brain connections
- Scientific points

The grief reaction represents the Chimp trying to come to terms with reality. As we grieve, we see both the Human and Chimp processing grief assisted by the Computer. [252] [246] Areas of the brain, including the anterior cingulate gyrus and the nucleus accumbens, become active and register pain. The Human parts of the brain, such as the dorsolateral prefrontal cortex, will put words to the pain and loss and this will help to process the grief. [253] [254] [255] [256]

The way our mind deals with the grief reaction is divided

The Human processes grief by working through facts and logic.

The Chimp processes grief by experiencing and working through emotions.

The Computer can help both Human and Chimp to work through grief, depending on what beliefs it has in it.

Gremlins of **unhelpful** beliefs will hinder or prevent working through grief.

Autopilots of **helpful** beliefs will assist or promote working through grief.

The Human can be helped to process grief

Talking through what has happened will help the Human to come to terms with grief and accept it. The Computer will help with this process by offering helpful beliefs, truths and perspective. This rational part of dealing with grief might be difficult but nowhere near as difficult as our emotions. [142] [257]

What are the typical emotions that a Chimp goes through?

The grief reaction is well recognised and described in various ways by many therapists. [257] [258] Grief usually takes months or years to work through and can lead to adjustment problems. [258] As the Chimp tries to process what has happened, typical emotions experienced are:

Denial: *A refusal to believe it is true*

Bargaining: *Trying to see how things could be changed or could have been different. The sentence from the Chimp often starts with "if only".*

Yearning: *A longing for what was*

Anger: *Often randomly directed*

Disorganisation: *The reality dawns and distress and grief set in*

Organisation: *A new life emerges with acceptance and energy*

These emotions occur because the starting point for the Chimp is not with reality but with a position of what it wants to see. In the eyes of the Chimp no change should have taken place. Knowing that the Chimp cannot accept reality helps to us to make sense of many of the emotional 'stages' that the Chimp will experience. These emotions cannot be rushed through but always take time. Not everyone will experience these stages, which might not follow an order and can repeat.

The first four stages: denial, bargaining, yearning and anger are seen during the period leading to acceptance.

Stages leading to acceptance
- Denial
- Bargaining
- Yearning
- Anger

Once we have come to accept the reality of a situation, a disorganised stage begins followed by an organised stage.

Typical emotions experienced during grief

Although these stages of grief have been identified, grief is a very personal experience and dealing with change or loss has no 'normal' pattern. [173] Each of us must deal with grief in our own way. However, as some feelings are common, it helps to recognise and know how to deal with these.

Dealing with common feelings associated with change and loss

Denial

In the very early stages of grief, a common experience is denial. This is when we just cannot believe or accept what has happened. We keep challenging the facts and carry on as if they were not true. When presented with the facts of a situation, the Human will settle, but the Chimp typically does not. Denial can last for minutes or years. When denial lasts for years, it is more of a protective mechanism to prevent emotional pain than a refusal to accept the situation.

Example of denial: *An inability to take the facts on board*

Years ago, I was running a clinic in hospital. A patient had received some terrible news and it had been clearly explained to them that sadly they would not live more than a few more weeks. It seemed as if they had understood. However, at the end of the consultation, the patient then asked if it was wise for them to book flights to visit relatives in the USA at Christmas. It was February at the time. When denial is this strong it is best to allow the person more time to process the information they have received. It's not unusual for someone in denial to be unable to accept the obvious. This is not a conscious decision by the person not to listen; it is a mechanism to try to prevent emotional pain.

Example of denial: *The inability to accept a breakup*

I worked with a young woman, whose partner had left her. She had not managed to grieve because she was stuck in denial. This denial had been continuing for several years. All of the facts were obvious and clear to her friends, but she did not accept them. The young man she had been with, had not only found a new partner but had married the new partner and also started a family and had two young children. The person I was working with continued to rationalise that her ex-partner had made a mistake and he would eventually find this out. This is quite a severe example, but if you search your own experiences, you might find many subtle examples of it happening to yourself. The important point is that it is normal, and almost expected. It comes from the Chimp because this is its way of handling what is happening.

Bargaining

Bargaining is seen when our Chimp begins sentences with words such as, "If only…". It's an attempt to turn back the clock and stop the change or event from happening.

Bargaining occurs because the Chimp always believes that what it wants to happen will happen. Therefore, when things don't go according to its expectations, it will try to bargain to change things. In grief, the Chimp works with this approach in order to process the emotions it is experiencing. Bargaining can help the Chimp to close off avenues of futile hope; therefore allowing a Chimp to bargain is a worthwhile exercise.

Example of bargaining: *An error of judgement*

Richard was driving his car and had to pass a cyclist. Instead of slowing down, he thought he had time to get around the cyclist but misjudged the speed of an on-coming car. He managed to avoid the cyclist, but the on-coming car swerved and hit the back of Richard's car. When the situation had been dealt with and Richard was on his way again, his Chimp began bargaining, in order to process the emotions he was feeling. This would range from self-recrimination to a blame game. He could blame the cyclist, the other driver, the road for being too narrow and so on. As he processes the situation, he might well 'bargain' with: "If only I had been five minutes later", "If only I had slowed down", "If only the road were wider" and so it goes on. If your mind is going to go down this route, then go with it and help the Chimp by discussing the futility of bargaining. It's still worth allowing your Chimp to bargain, if it won't move on, so that you can help it to lay the incident to rest.

Example of bargaining: *The end of a relationship*

We experience a lot of bargaining when relationships are ending. This time the bargaining is in real time as the event is unfolding. So typically, we might hear one of the couple saying, "If I do this" or "If you do that" believing that it will all be better again. It could be a reasonable suggestion because it might help to save or build up a relationship, but often we are not recognising that this is mere futile bargaining and part of a grieving process. The reason that this is often futile bargaining is that it is one-sided with just one of the couple interested in trying to go forward with the relationship. Once we realise that the bargaining is not achievable, it might be time to help the Chimp to move on with questions such as, "Where do I go from here?".

Yearning

Yearning for what used to be is another emotional process seen within grief. [259] We might have to express these yearnings on several occasions before these emotions begin to settle and don't keep dominating our lives. By expressing yearning, the Chimp can begin to accept what has happened. To yearn for the past is very common and it often accompanies bargaining. Whenever we reminisce about the past, yearning can be replaced by enjoying the memory from the past but living in the here and now.

Yearning can help us move on

Example of yearning: *The change or loss of a role or job*

Our Humans know that life is dynamic and nothing is static. Changes are often imposed upon us and can cause a lot of grief. At work, we hear ourselves saying things such as, "Why can't we just go back to what we were doing", "Why do we need these changes", "It was really good when I worked that way" or "It'll never be the same". If the changes are non-negotiable, all of these statements might be true, but it won't bring back what used to be. Allowing the Chimp to yearn will help it to eventually process the situation and move on.

Example of yearning: *Ageing and acceptance*

We all age but not always gracefully or with acceptance! Our physical bodies and minds mature as we go through life. Each morning we wake up as a different person from the day before. Accepting changes that occur to us physically and our inability to do what we used to, is a skill in itself. [233] Yearning for youth or previous abilities can become very destructive. Learning to move away from yearning and into accepting yourself, as you are today, sometimes has to be worked on.

Acceptance is a skill

Anger

Anger is a common stage of grief. [173] The anger experienced might not be rational, *so there is not much point in trying to rationalise it away*. This anger can be directed almost anywhere. It might be directed towards the person himself or herself or to someone else connected with the loss or change. It is rarely justified, and it is frequently out of proportion. Anger is a substitute emotion. The more appropriate emotion would be sadness or sorrow, but these emotions can be too painful to experience or manage, so anger is substituted instead. Anger can also present as irritability.

Example of anger: *A devastatingly sad situation*

I worked recently with a man whose son had taken his own life. Clearly this was a devastating experience for the father. During the next year he told me of an uncontrollable anger with the world. He knew this was not rational because it was often aimed at total strangers. He could experience anger when just seeing a stranger smiling. Whenever we feel that an injustice has occurred or when we feel that something could have been avoided, anger is a common emotional reaction. Anger is then easily displaced into other areas of our lives. For this man, it helped to understand that anger is a common, though irrational, experience during grief. He went on to explore his beliefs about his son and what had happened. This enabled him to start seeing the situation from a different perspective. With time, his anger subsided and gave way to acceptance.

Anger can be irrational during grief

Example of anger: *Suppressed grief*

I worked with a young man who wanted help to manage anger issues. Anger is a symptom not a diagnosis. The symptom of anger can appear for many reasons and it is important to find and treat the cause of the anger. Sometimes, grief is the cause. Relatives described the young man as frequently becoming intolerant and aggressive. During discussions, it transpired that a very strict aunt had raised him. Her method of dealing with him was to demean and mock him. His reaction to this was suppressed anger. His grief arose from the loss of his mother and his perceived 'loss' of a loving and caring upbringing. Grieving for a *perceived loss* is when we grieve for something we never had.

When we worked through his childhood experiences, he was able to process them. We also looked at his anger as being natural but unhelpful. Along with other aspects of therapy, we were able to diffuse the anger and replace it with constructive emotions and a different interpretation of himself and the world.

Many of us carry emotional scars that will affect us from time to time. However, if we are suppressing emotions from any event in our life and have not addressed them, then they will typically keep breaking through. [177] Often, we don't even realise where our emotions are coming from or why we might appear to be overreacting or becoming angry.

Disorganisation

Disorganisation can be spread across the whole of the grief reaction, but it usually appears after the reality of a loss has been accepted. Denial, bargaining, yearning and anger give way to a sense of despondency. Disorganisation is the lack of ability to keep things together. It often involves:

- Forgetfulness
- Numbness
- Loss of energy
- Inability to prioritise
- Lack of motivation to do anything

> **Key Point**
> *Low mood, insomnia and loss of appetite are common during the disorganisation stage.*

Example of disorganisation:
The aftermath of an affair

Sarah has just experienced a relationship break up. She had been with her husband for nearly ten years, but he had an affair and decided to leave her. She felt that they could have worked it out and was devastated by his leaving. She attempted to change the situation and went through phases of denial, bargaining, yearning and anger. Finally, she has accepted that the marriage is over.

This state of realisation and acceptance typically brings a collapse in our ability to function. It can bring a loss of confidence in many areas of our life and show symptoms more characteristic of depressive illness. Therefore, Sarah might have trouble sleeping, becoming very weepy, and lose all interest in others and any social activities. Although these symptoms might appear to be a depressive illness, this is not the case. They represent a stage of disorganisation, where any effort might seem exhausting and there feels little point to doing anything. It is a passing phase and the Chimp just needs time to process the realisation. It is not usually helpful to push people in this position to socialise or make efforts to move on. Compassionate support is mainly what is needed. When they are ready, they will pass into a stage of reorganisation.

Example of disorganisation: *Loss of routine and purpose*

Elite athletes often find it difficult to adjust to the loss of their lifestyle when they retire. The abrupt change in daily routines of training and focus are lost. Their disorganisation can be quite severe and can be accompanied by a change of self-image. It can be the same for someone who retires from work. It can appear to be as if the person has depression. Elite athletes can even experience a similar sense of loss and disorganisation following success at a major championship. Anyone who has completed a project or challenge can feel disorganised when it is completed. Sometimes, a simple remedy to prevent this low mood can be to plan beyond the event; even a planned holiday can make the difference.

Depressive features caution

Many people will experience what appears to be a mild form of depression when grieving. They could have:

- Low mood
- Poor sleep
- Loss of appetite
- Feel physically exhausted
- Periods of weepiness

These are all 'normal' experiences following a loss. If, however, you are at all concerned, or the symptoms really are stopping you from continuing with your life, then you should see a doctor to get a check-up: ***depressive illness can occur*** in the setting of grief or loss.

Reorganisation

Reorganisation is the time when the symptoms have been processed and you begin to pick up again. It is a time when the loss has been accepted and a new chapter in your life is being written.

Two very important points regarding reorganisation

1. There are no time limits for when this should happen. It depends on many factors. One is whether the processing of the grief has gone smoothly. Another is whether a new lifestyle is able to emerge. It can also depend on exactly what has been lost or changed and the specific impact of that loss on the individual. Many people find the first three months are the hardest. Most find that after the first twelve months have passed, they are moving on with their lives.

2. After a grief reaction, we might be left with an 'emotional' scar. This means that any stage of grief can revisit us from time to time; this is a normal experience.

It's normal for emotional scars to resurface

This is so painful.

Emotional scar

We need a plan in place to manage emotional scars.

How can we help to process change or loss?

There are a few things that we can do to help the grieving process, whatever the change or loss is. The first suggestion is to go with the events and feelings as they present. Try not to fight any emotions you might experience. This doesn't mean giving up, but instead work with them. As we are all unique, it is important that you treat yourself as a unique individual. Don't compare what is happening to you, with what is happening to others. Work with the way that your mind is presenting to you.

I can recall an unhappy case where two parents were trying to come to terms with the death of their child. Each was grieving very differently from the other. By working in their own way, they learnt to understand how different the process was for each other.

> **Key Point**
>
> *Any grieving experience you have will be normal for you.*

Key Point

It helps to talk your emotions through. If possible, try to find someone who will listen and not judge.

Keep talking if it helps

Usually by talking through a situation or event, we continue to process the emotions involved. [88] [81] [260] [143][261] For example, I worked with a professional man whose role at work had been adversely altered. He was extremely distressed by being disempowered and desperately wanted his previous role re-instated. It's not an unusual problem. His Human could rationally accept that the decision had been made and would not change. It accepted facts such as: roles change with time, companies don't always deal with change well, others make the decisions and it is out of my hands. However, his Chimp still had to process the change and it would not respect reason and logic. By going through his feelings and emotions several times, and by allowing him to keep expressing these, without judgement, he reached a point whereby he began to laugh and remarked that he was fed up of listening to his Chimp. If you let your Chimp keep on complaining or wittering, it often gets exhausted and decides that it has moaned enough. Be warned, some Chimps can complain and witter for a long time. There is a point where you might have to confront your Chimp and ask it how long it wants to keep this up!

Repeatedly talking about the same event can help to process it

Re-living emotions can help

In treating post-traumatic stress disorder, re-living the emotions experienced at the time of the event can help with the processing of the situation. By 're-living emotions', it means going back in time to the event and then re-experiencing the same feelings that were experienced at that time. This re-living of emotions helps the Chimp to process the event. It does this by allowing the Human to put into the Computer some rational beliefs to accompany the emotion. [246] The Computer then interprets these beliefs and speaks to the Chimp in its own language. For example, if someone had been trapped in a building where a fire started and had escaped, they might suffer from intrusive memories and emotions. Some rational beliefs would be:

- This is an experience from the past and is over now
- It is unlikely to happen to you again
- You did escape and are unharmed

The next time the emotions appear, the Computer will remind the Chimp of these beliefs. Re-living emotions is best done with a qualified therapist, as it can clearly be traumatic while the processing is taking place.

Pathological Grief

In a few cases of grief, the grieving process halts and the person gets stuck. We term this 'pathological grief', because we see it as not being 'usual' or acceptable. People can get stuck for many reasons and if this happens, or aspects such as guilt appear, then it is wise to consult a professional for help. If you are in any doubt seek out help. [250] [262] [263]

Writing a new chapter in your life

Some people find it helpful to see their life in chapters. When a change has occurred, they round up the chapter in their mind and see tomorrow as a new chapter. This way of looking at life can help to draw clear lines for seeing the past, the present and the future.

An ever-changing world

One of the strongest Autopilots that we can have in our Computer is a truth that our worlds are forever changing. Our Chimps find this truth difficult to accept and therefore react to any changes that they don't approve of, which can lead to a grief reaction. Accepting and managing change is a Human role. The Computer can manage the Chimp by introducing the truth that the world inevitably changes every day.

Creating Autopilots for an ever-changing world

Unit 13 Reminders

- The grief reaction has recognised 'stages':
  - Denial
  - Bargaining
  - Yearning
  - Anger
  - Disorganisation
  - Organisation
- Working with and through these stages helps grief to be managed
- Hidden loss is worth searching for and addressing
- Professional help should be sought when the emotional response is severe

Unit 13 Exercises
Focus points and reflective exercises

1. *Helping the Chimp through stages of a grief reaction*
2. *Recognising that the past is the past*
3. *Your lifeline exercise*
4. *Explaining how Chimps and Humans process information*

Focus 1: *Helping the Chimp through stages of a grief reaction*

By far the largest factor that will help the Chimp to manage a loss, regardless of what stage it appears to be in, is to talk about the loss. By talking we allow expression of emotion but also start to work through phases such as yearning and bargaining. [88] [81] [260] [143][261]

Exercise: *Managing a loss*

Whatever the loss is that you have experienced, try to set time aside to talk this through with a friend. It isn't always necessary to understand the feelings associated with a loss because they are not always rational. The important point is to express your feelings. Discussing more rational thoughts will follow expression. It is important to deal with grief or loss in your own way, because we all do this uniquely. Try to remind your Chimp to be your best friend and not your worst critic when managing loss.

Focus 2: *Recognising that the past is the past*

Although change and loss can be very painful, eventually, the pain will subside. Try to see past events as belonging to yesterday and not to today. All of us carry emotional scars into our future but these scars don't have to prevent us from enjoying our future.

Exercise: *Putting the past into the past and living in the present*

Try to detect any events that are still troubling you, but belong to the past. Sometimes…painful memories can be managed by simply recognising that they are not currently happening and that time has moved on. Every day, when we get up in the morning, we are not the same person that we were the day before. It is helpful to recognise this and to leave your past life and past self behind. Of course, we want some memories, but recognising them as memories, can help us to live in the present.

Focus 3: *Your lifeline exercise*

Exercise: *Accepting the reality of life*

A useful exercise to gain some perspective for helping to process events in your life is the 'lifeline' exercise.

Simply draw on a piece of paper a timeline graph, as shown in the diagram, to represent your life from your birth to the age of 100 years.

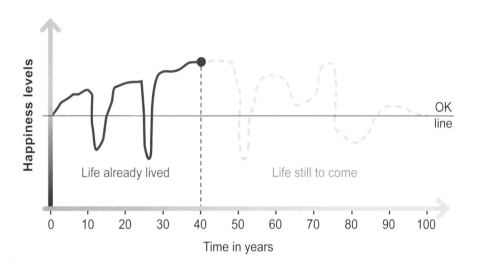

In the example, the OK line represents neither happy nor sad.

Our example, shows a person who has had an up and down life so far. They have predicted their future years with the dotted line.

On your graph, put your life so far with the highs and the lows. Then try and predict the future by drawing a line to show what you might predict will happen.

For most people, this predicted line shows both highs and lows. Nobody draws a line that is consistently high and nobody draws a line that is consistently low because life doesn't do this to any of us. Life gives us both highs and lows in different portions.

The point of drawing this line is to help you to accept the reality of life and thereby process events more easily. There will be future highs and there will be future lows. When the lows happen, they will pass and when the highs happen they are there to enjoy, as they will also pass. Could we influence the amount of highs and lows we have? Quite evidently, we can all develop a more positive approach to life and to any setbacks. We can also learn to live in the moment and enjoy it more. The lifeline graph can help by giving us a reality check on expectations. We always have a choice in how we manage what life brings our way.

Focus 4: *Explaining how Chimps and Humans process information*

We can fail to process information if we don't recognise the need for both Human and Chimp systems to act. This can leave us wondering why we can't seem to get over some events.

Exercise: *Advice to a friend*

Imagine that one of your close friends, Archie, tells you that they are struggling with a problem and would like your advice on it.

The problem: Archie said that at work there had been an investigation into the theft of packs of notebooks and pens. The thief was never caught. About twenty people work in the company. By chance, Archie overheard that a colleague had spread malicious rumours about him being the thief. When confronted, the colleague denied spreading the rumour, but Archie is sure it was them.
Archie says, "I can't prove my innocence and I can't prove their guilt" and asks why he can't move on. Before you read the suggested answer that follows, try and explain how you would help Archie to process the event and to move on.

Suggested answer

Archie needs to appreciate that he will need to process the event in two very different ways: The Human way and the Chimp way.

The Human can process the event by stating some facts, such as:

- Sometimes in life, we **CANNOT** win
- Sometimes an injustice **DOES** occur, and we can't do anything about it
- Sometimes we **CANNOT** prove a truth because it becomes one person's word against another
- Reasonable people will understand your situation and be fair
- We cannot make people's minds up for them
- You know the truth

The Chimp is likely to need:

- Time to work through this painful experience
- To express some strong emotions, such as anger
- To express its feelings repeatedly until it feels it is heard and understood
- To go over the details of the injustice until it feels exhausted going through them
- To hear some truths from the Computer that have been repeated regularly, such as:
  - Most people suffer an injustice at some point in their lives
  - What you are experiencing is normal
  - You can move on from this, often after going through the same stages as a grief reaction
  - You have people on your side and you are not alone

Unit 14
Working with reality and truths

STAGE 5: *will look at in detail at how we set up an effective Stone of Life and our Troop. These two aspects are critical to stability. The Stone of Life will always settle the Human and usually settle the Chimp. If the Stone of Life fails to settle the Chimp then it will use the Troop.*

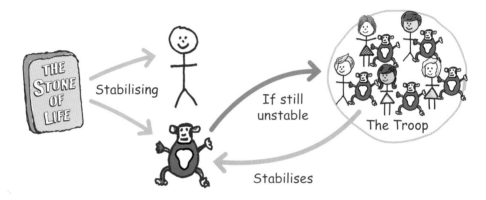

Unit 14: *will cover the first feature on the Stone of Life: Working with reality and the 'Truths of Life'. This is the first large step towards emotional stability.*

Changing our default position

If left alone the mind will default into Chimp mode and remain there. If the Chimp is fully relaxed, then it will allow the Computer to take over and run our lives. [235] [236] The problem is that the Chimp is rarely fully relaxed, and it constantly takes the steering back off the Computer. The Computer can only remain in charge if it is programmed.

What we are going to do next is to learn how to change our default position from Chimp mode to Human mode. In other words, we will set off each day in Human mode and default back to this mode throughout the day, rather than setting off each day in Chimp mode and defaulting back to Chimp mode. This way, in times of 'danger' the Chimp will still alert us and take over, but the rest of the day we will remain in Human mode.

The USUAL default position

A more HELPFUL default position

Is it going to be a struggle every day?

When dealing with the Chimp on a day-to-day basis, it can become exhausting because the Chimp never stops commenting or expressing emotion. It can feel like a permanent battle to keep the Chimp at bay and move into Human mode.

Rather than doing battle with the Chimp, it would be great if we could **prevent** most hijacks from ever occurring in the first place. This would make mind management much easier.

The good news is that managing your Chimp doesn't have to be a struggle every day. The reality is a paradox in itself: the more we relax the more influence we have in managing our Chimps; the more we take the Chimp on, and tense up, the less influence we have. This is because when we relax, we are sending a message to the Chimp that all is well and therefore the Chimp stops looking for danger. [262] [48]

Those who work with dogs know that if a dog senses the owner is uneasy, then the dog becomes uneasy and will go into protection mode. However, even in the face of uncertainty, if an owner remains calm then the dog will settle. Fussing a dog and telling it not to worry when walking near heavy, noisy traffic will only alert the dog to believing that something must be wrong. If instead, we ignore the traffic, then the dog will take its lead from us, and settle down and accept the situation.

Our Chimps are similar. If we make a big deal of something, then the Chimp will join in. If we retain perspective and refocus, then the Chimp is very likely to follow. [263] So how do we relax ourselves and thereby our Chimp? By using a preventative approach!

Preventative stance

As doctors, we have always diagnosed and treated patients. During the 1970's and 80's preventative medicine gained prominence. We realised that it was better to prevent a problem from occurring in the first place than to wait until there is a problem to treat.

So far we have considered how to manage the mind when the Chimp has hijacked us. Now we can consider *how to prevent Chimp hijacks* from happening. Battling with your emotions and thoughts can be exhausting and is called 'arm-wrestle the Chimp'.

Do you really want to arm-wrestle your Chimp?!

Example: *Penny and low self-esteem*

Penny's Chimp says that she feels she is not as good as everyone else and this makes her feel like an onlooker to the world. She experiences a lack of confidence, poor self-image and feelings of sadness and emptiness. Penny's Chimp engages with these feelings, which results in further feelings of despair. Penny tries to block the emotions and control them by trying to make herself happy but it just isn't working. She is arm-wrestling her Chimp.

Penny could have prevented the emotions from occurring in the first place, by choosing to let her Human decide on her self-esteem. The Human will look to her values to determine how she perceives herself. It's true that Penny might not have the talent, skills or looks that others might have. It's whether she wishes to allow her Chimp to measure her self-esteem on these things.

The brain's operating process and how to improve things

The key to preventing Chimp hijacks and remaining in Human mode is hidden in the Computer. The Computer can restrain both Human and Chimp and even take over. [264] [217] [265] Hence, programming the Computer is our focus.

Key Point

Programming the Computer well is the key to mind management.

The 'Oracle'

We have to consult the computer

The Computer is key to mind management

Working with reality and truths

Example: *The talk*

Alice is about to give the staff briefing. Her Chimp is worrying that she won't be well received and that she might make a mess of things. The Chimp cannot take over because Alice has programmed her Computer effectively.

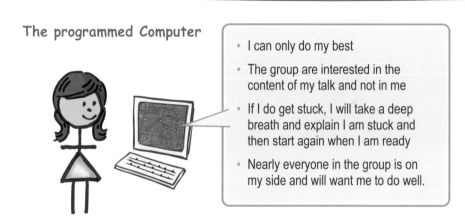

The programmed Computer

- I can only do my best
- The group are interested in the content of my talk and not in me
- If I do get stuck, I will take a deep breath and explain I am stuck and then start again when I am ready
- Nearly everyone in the group is on my side and will want me to do well.

Her Chimp must consult the Computer and is reassured that all is well, so it doesn't react.

This example shows how Alice programmed her Computer with truths that settled her Chimp down. It was for one specific situation, giving a briefing to some staff.

This all sounds great, but we have a problem. If we have to have a programme specifically for every single situation we encounter, then we would have to think of thousands and thousands of truths and this just isn't practical. Some common situations we regularly encounter might benefit greatly from having a specific 'truth package', so I am not saying we shouldn't do this, but the question is, *"Can we have a package that will help to settle the Chimp in every situation?"*

The answer is "YES"!

> **Key Point**
>
> There is a way for the mind to deal with virtually ANY situation.

The Stone of Life: *The ultimate mind stabiliser*

Contained in the Computer system are three components that can create a stable mind, if they are programmed correctly. These three components are:

1. Reality and the Truths of Life
2. Values
3. Perspective (your Life Force – what it all means to you)

We can put them together to form the 'Stone of Life', our ultimate reference point for all situations. If we get these three components correctly programmed, then no matter what situation we meet, we can manage our Chimp. It is critical that the Computer is programmed correctly and that it is **YOUR** programming not that from someone else! The details on your Stone of Life will be unique to you.

The ultimate stabiliser

Key Point

The Stone of Life is the stabilising force that will keep you in Human mode and also return you to it if you are momentarily hijacked.

We will now consider the three components of the Stone of Life in more detail, starting in this unit with Reality and the Truths of Life.

Working with reality

There is a very important difference between the Human and Chimp when it comes to reality. We have briefly covered this concept earlier but need to explore it in more detail and then utilise it.

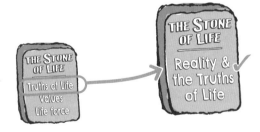

The Human's *starting point* is to work with:

- What presents itself
- The facts
- The many truths of the way life runs
- Hopes rather than expectations

The Chimp's *starting point* is to work with:

- What it wants to be in front of it
- What it expects to happen
- What it believes should happen
- An expectation of fairness

The difference between these two approaches to life is massive. If you are in Chimp mode, your entire approach will be centred on what should happen. If it doesn't happen as you expect or want it to, then the Chimp will react adversely. If you are in Human mode, you will work with reality. [266] [267] [268] [269] You will accept the reality of whatever is happening and work with it, even if it is not what you want.

The starting point for going forward

Reality: I start with what **is** happening.

Expectation: I start with what **should** be happening.

Expectations versus reality

The Chimp effectively writes the script in advance of how any day or situation will unfold. When it doesn't happen the way the Chimp expects, it can then react with emotions such as frustration, anger or despondency. [270]

To add to the Chimp's distress, the Chimp doesn't believe that it can deal with any consequences. Therefore, anything that isn't going according to plan becomes a source of frustration and stress and even fear.

Chimps and children find it very hard to deal with any consequences of life events. Adult Humans can work with anything that life brings to their door.

Example: *Unexpected item in the bagging area*

Irene is going home and looking forward to watching the final episode of her favourite programme on the television. She is running slightly late but has time to get some items from the supermarket.

She is in a good place and in a relaxed Chimp mode, which is not necessarily a bad thing. She has no particular emotions and there are no obvious indications that she is in Chimp mode, but the Chimp is in charge. Her Chimp has already written the script of how the supermarket shopping will go. It sees the self-service area as clear, with no queues, and a quick and easy way to purchase her items. However, when she gets to the self-service area, she sees two people in the queue, which the Chimp sees as a horrendous number of people in the queue. Immediately her Chimp begins to get ruffled because this isn't the way it pictured the check out.

The Chimp starts to single people out who clearly are not going quickly enough. The Chimp has already begun to become irritated because it can't work with the reality of the situation.

Finally, a little agitated, Irene reaches the scanner. She starts quickly putting her items through the scanner. The Chimp is already whispering in her ear, "You are going to miss your programme because of these stupid people". Suddenly the scanner beeps and she hears it say in a well-spoken voice: "Unexpected item in the bagging area". As there is no unexpected item in the bagging area, her Chimp bares its teeth. It cannot cope with reality, but only with what it expects and wants to see. She calls for the assistant, who is helping an elderly woman to operate her scanner. The assistant smiles, as does the elderly woman. The Chimp has now totally lost it because these people are not taking her distress seriously. Who knows what the Chimp might do next? Whatever it is, it might end up explaining it to a judge.

Let's rewind and begin the scenario again to see the difference between the Chimp and the Human in action.

Irene is going home and looking forward to watching the final episode of her favourite programme on the television. She is running slightly late but has time to get some items from the supermarket.

She is in a good place and in Human mode. She has no particular emotions and there is no obvious indication that she is in Human mode, but the Human is in charge. Her Human has no expectation of how the shopping will go but it hopes that it will be uneventful, and it plans to use the scanner at the self-service area to save time. When Irene gets to the self-service area, she sees a small yet insignificant queue of two people. Her Human accepts the reality of the situation and rationalises that the queue will eventually go down. She reaches the scanner in a calm and collected manner.

She quickly starts putting her items through the scanner. The Chimp is already about to whisper in her ear, "You are going to miss your programme because of these stupid people". However, the Chimp doesn't do this because before it whispers it looks into the Computer and sees a programmed truth; life may not run the way I want it to, and when it doesn't, I will be able to deal with any consequences. The Chimp settles because this fact is true, whatever happens Irene will cope. Suddenly the scanner beeps and a voice says, "Unexpected item in the bagging area". As there is nothing unusual in the bagging area, Irene looks for assistance. She calls for the assistant, who is helping an elderly woman to operate her scanner. The assistant smiles, as does the elderly woman and Irene rationalises that "It is disappointing not to see the last episode of my programme". Then she says to herself, "I might still make it and if I don't, then there will be a way to watch it later". She smiles back at the elderly lady.

***Can we attain this response as a reality?* YES!**

Programming the Computer to work with reality

So how do we stay in Human mode?

We need to programme the Computer with some Truths of Life in order to remain in Human mode and to stay working with reality.

Here are some common sense reality statements:

- Just because I want something to happen doesn't mean that it will happen
- Not everything will go the way that I want it to
- Getting frustrated with reality isn't helpful
- Nearly everything will work out in the end
- This situation probably won't be important in a week's time
- Learning to accept reality and then working with it, can be very constructive
- Adult Humans can deal with anything in life
- Setting off each day in Human mode will help me to work with reality

The Computer getting into reality mode

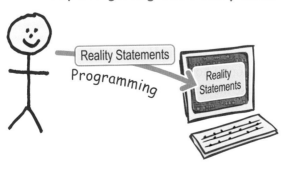

It might be worth re-reading the above list. Allow the statements to sink in. If you rush through them, their meanings might not be fully appreciated and applied.

Try to visualise situations where you could have switched mode, so that you can see the benefits of working with reality. In Human mode, we can still make plans, when reality is doing us no favours. ***Being proactive is still necessary*** to ensure that being in Human mode is an advantage.

Key Point

It is important to appreciate the advantage of setting off in Human mode, and working with reality. Without this appreciation, it is unlikely that you will make the effort to acquire the skill.

Key Point

Working with reality is not the same as accepting life and doing nothing about it.

The difference between 'reasonable' and 'realistic' is important

Reasonable does not mean realistic!

Expectations can be reasonable or unreasonable. Unreasonable expectations of situations or people are obviously unhelpful. However, even reasonable expectations can create problems. Reasonable expectations must also be realistic (that is, they are based on reality), otherwise we will end up back in Chimp mode. For example, *it is reasonable* to expect everyone to be thoughtful, ***but it isn't realistic***. The Chimp demands that things are reasonable, the Human hopes things will be reasonable but operates with realism. Therefore, when someone isn't thoughtful, the Chimp becomes irate, whereas the Human remains calm and deals with the reality.

If we are in Chimp mode and have reasonable expectations *but are not working with realistic expectations*, then it is likely that we will have unhelpful emotional reactions. If we are in Human mode with reasonable expectations but also work with realistic expectations, then it is unlikely that we will have any emotional reaction. The Human can accept that sometimes we can't win and sometimes tasks are impossible to achieve, and we can only do what we can do.

Here are some examples to demonstrate the difference between reasonable and realistic.

Example: *Interacting with a stranger*

John is at a gathering where he is meeting a lot of new people. The atmosphere is vibrant, and everyone seems to be on their best behaviour. At this point, his Chimp has set its expectation that anyone it meets will be civil, pleasant and engaging with him. ***This is a reasonable expectation***.

After meeting a few people, he meets Marion, who is not having a good day! Marion confronts him quite abruptly and is rude and dismissive towards him. If John is in Chimp mode, he will get an emotional reaction and might act on this. His Chimp could personalise the situation and see Marion's behaviour as an 'attack' directed at it.

If he is in Human mode and working with reality, he will not react. This is because although it is reasonable to expect a civil interaction, the *reality* is that we do meet rude or unpleasant people. His Human can accept this. He will not personalise the 'attack', but rather accept it, and then deal with it appropriately. He could, for example, be assertive and explain that he does not wish to be disrespected. He could choose to enlighten Marion about her approach, or he could just simply excuse himself and walk away.

Chimps and Humans both have reasonable expectations, but the Human adds reality.

The key principle is that the Human accepts reality as it is and then has a plan to deal with it in a dignified manner. He has also not taken the situation personally, even if it were intended as a personal insult.

Example: *The 'thoughtless' teenager*

Teenage brains are not fully developed and show many traits that are typical of the undeveloped brain. [271] [272] The brain is actually *at the right stage of development for the age of the person*, which means that it will function differently from the adult brain. One adolescent feature is that teenagers are easily distracted and lose focus. [273]

Zak is a typical teenager and has been asked by his mother to remember to feed the family dog, while she attends his parents' evening at school. Zak clearly wants to do this and has every intention of doing it, especially since his mother is getting feedback on him from school.

When his mother returns, she finds Zak watching the television and the dog looking hungry. Zak's mother is very reasonable in expecting that her son will remember to feed the dog. However, is it realistic to think that a teenage brain will definitely remember? The fact that he forgot might be disappointing, but Zak's intention was not to forget. Of course, there is a need to address how he might manage better in the future, but we are working with a teenage brain!

Being realistic about people and life helps us to stay in Human mode.

So how could Zak's mother have responded in Human mode? She could have expressed her disappointment and she could have asked Zak how he might remember in future. For example, he could set an alarm on his phone to remind him as soon as he is asked to do something. What Zak would be doing is learning how to compensate for his brain's limitations, which is what most of us do. However, it's up to Zak's mother how she wishes to see things.

Key Point

Distinguishing between what is reasonable and what is realistic helps us to prevent a lot of unnecessary negative emotion.

Programming the Computer to stay in reality mode

It is helpful to start each day by programming our Computer ready to accept whatever comes our way, and how to deal with it. This exercise can take just a few minutes. We can rehearse going through any difficult or important conversations and plan how to present ourselves. We can also decide on what outcomes we want from various situations and the best way to achieve them. We can pre-empt the script that the Chimp is likely to try to write. It is a skill to accept what is in front of us rather than to become frustrated by it. Again, this doesn't mean rolling over and accepting the unacceptable. It also doesn't mean being passive towards situations or people. At the end of this section there will be exercises and help in how to programme your Computer.

The 'Truths of Life'

The Truths of Life is simply a term used to describe the beliefs, based on reality, which Humans work with. These Truths are formed from evidence acquired through learning or from experience. Truths can be statements of *fact* or they can be a statement of *experience* or a mix of both. Truths might only be true for the individual because they can be based on experience.

For example, you might hold the belief that taking risks is not sensible. Alternatively, you may hold the belief that taking risks is a necessary part of being successful. Either statement can be true.

You have to work out what rings true for you and have that firmly embedded in your Stone of Life, as your reference point.

Truths are what we have to *ACCEPT* and base our logical plans on, to help us move forward. They clearly help us to stay working with reality.
Most people can work out around ten fundamental Truths that they have to remind their Chimp of, on an almost daily basis, so that it is settles down. Some examples of Truths might be:

- Not everyone will understand me
- I can only do my best in life
- Everything will pass
- Not everybody's opinion matters
- Reflection is a great promoter of change
- Opinions are not facts
- Something is only as important as I allow it to be

Key Point

It is crucial that your Truths resonate with you; otherwise they are not really Truths to you.

The reason the Truths of Life are so important is because they can settle the Chimp in virtually all circumstances.

Example: *The cancelled flight*

Juan is going to an overseas business meeting on the 1st June. He has packed his bags, but on the 1st June he is told that the flight won't be leaving until the 2nd June because of 'difficulties'.

Unless he is prepared, his Chimp will naturally take over. It starts with protests and accusations and even paranoid interpretations of what the airlines are doing to it. It complains bitterly that they have ruined his business trip and feels they really need to know how hard he has worked for this opportunity.

His Chimp can't work with reality.

If Juan had programmed his Stone of Life with some Truths, it's much less likely that he would have reacted emotionally. For example, he could have had the following Truths in his Computer:

- Life doesn't always go the way I expect it to
- I need to find all of the facts before I respond to a situation
- It will probably all work out in the end
- I can always look to find positives in any situation

By preparing the Stone of Life with Truths, Juan could save himself from a lot of stress and potentially embarrassing Chimp behaviours. He is also helping his own health. Whenever we allow our Chimp to hijack us with a negative emotional reaction, we can create an increase in unhealthy hormones and neurotransmitters in our brain and body. [48] [155] If these Truths had been in the Computer then the Chimp would have seen them before it reacted. An unmanaged Chimp can put you into a poor state of mind for the rest of the day.

Working with reality and truths

A frequently asked question:

What is the difference between an Autopilot, a grade A hit and a Truth on the Stone of Life?

Autopilots can take two forms:
- A constructive and helpful behaviour
- A constructive and true belief

Autopilot beliefs that are very helpful can take two forms:

1. Grade A hits
Very helpful Autopilots of belief that resonate *within a specific situation or setting* are called grade A hits.

2. Truths of Life
Very helpful Autopilots that *could apply to nearly all situations* are Truths of Life found on the Stone of Life.

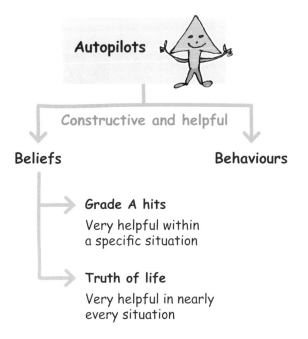

Example: *Greg and the leaking water pipe*

Greg has a leaking water pipe in his house. It is dripping through the ceiling. Here are three true beliefs that he holds:

1. **An Autopilot:** The belief that the damage will be repairable. The Autopilot helps him, but it doesn't settle him too much because it is still leaking.
2. **Grade A hit:** Greg believes that the plumber, who is on his way, can mend the leak. This truth settles Greg much more because it holds a solution.

3. **Truth on his Stone of Life:** The belief that 'everything in life will come and go and be resolved'. This is a Truth that Greg believes and this he can apply to any situation.

Example: *Jocelyn and her golf tee off*

Every time that Jocelyn attempts to tee off, her Chimp tries to take over. Her Autopilot for her golf is: '*Look where you want the ball to go and fix on this.*'

A Truth on Jocelyn's Stone of Life is: '*If you want to succeed in life then focus on the process of succeeding and not on how you feel*'. This Truth helps with whatever she is doing in life.

Stabilising yourself with Autopilots and Truths of Life

The Truth can change!

Not everyone believes the same things to be true. For example, take the following statement: *All people are friendly, if you take the time to get to know them.*

Some people believe this to be true, but others might disagree. Therefore, what someone believes to be true is unique to that person.

Truths can even change! For example, if someone used to believe that *all people are friendly, if you take the time to get to know them,* but has now changed their mind, then that Truth becomes untrue.

Our Autopilots are therefore unique to us because they are what we believe to be true and what helps us.

Unit 14 Reminders

- Preventing the Chimp from reacting is easier and better than dealing with it when it has already reacted
- The Stone of Life is the key to success for staying in Human mode
- The Stone of life must be appropriate for you and specific to your unique life
- The Stone of Life is composed of three features:
  - Reality and the Truths of Life
  - Values
  - Perspective
- Being in Human mode can have some major advantages
- Reality is 'working with what is in front of us' and not 'working with what we want to be in front of us'
- Reasonable and realistic are very different words; realistic helps us to stay with reality
- The Truths of Life need to be worked out and reflected on if they are to have any effect in helping us
- Truths are specific to the individual

Unit 14 Exercises

Focus points and reflective exercises

1. *Don't arm wrestle your Chimp!*
2. *What is your understanding of 'expectation'?*
3. *Reasonable and realistic*
4. *Your Truths of Life*

Focus 1: *Don't arm wrestle your Chimp!*

Arm wrestling the Chimp effectively means that someone is fighting with their emotions and trying to control them. They struggle to step back and recognise that these emotions *are a message from their Chimp* and they are best worked with. **Recall:** emotions are there as a prompt to do something, rather than to engage with them or try and control them.

Exercise: *Advising others on using emotions*

Try this exercise before reading the comments below the exercise.

What advice would you give the following people to help to move forward, all of who are struggling with their own specific situation?:

Paul – frustrated that he can't find the right girlfriend
Jennie – upset that her mother doesn't appear to care about her
Archie – annoyed at being called lazy by one his friends
Bertha – anxious about not getting good quality of sleep

Comments

In all of the cases we are saying the same thing: Don't engage with these emotions but rather use them to prompt you to form a plan to solve the problem. What we do, as a friend, is to move into Human mode and come up with a plan to help them to go forward. We effectively ignore the emotion and use it to act.

How easy it is to give others advice! When we look at other people's problems, it often seems fairly straightforward on how to manage things. This being the case, why can't we give ourselves the advice and take it? Unless we exercise the Chimp and then move into Human mode to find a solution, we are likely to get stuck. Next time you find yourself in a similar position, try and give yourself the advice that comes so easily!

Consider whether you are engaging or using your own emotions. Try to look at situations where you might be arm-wrestling your Chimp, trying to battle with and control emotions, rather than using them to form a plan and move forward.

Focus 2: *What is your understanding of 'expectation'?*

'Expectation' can be interpreted in a number of different ways. For example, "I expect to see the train arrive in five minutes". When we say this, we can experience different emotional responses, depending on how we are interpreting the word 'expect'. This is made clear, if we look at three different uses of the word "expect" and then see what type of emotion is likely to occur if the train didn't arrive in the next five minutes.

1. **I expect** (it is predictable that) the train will arrive in five minutes
   Emotion: Fairly neutral just surprised
2. **I expect** (I hope that) the train will arrive in five minutes
   Emotion: Disappointment
3. **I expect** (it should/ought to be that) the train will arrive in five minutes
   Emotion: Frustration or anger

The Human uses the first two types of expectation, whereas the Chimp uses the third type. The Chimp goes through life with the expectation that things will happen as it thinks they should happen.

Exercise: *Working with a helpful expectation of life*

Consider what hidden 'should' expectations your Chimp has put into your Computer. For example, if you meet a partner or friend, has your Chimp put in a 'should' expectation of the type of greeting that you expect from them? Has it put into your Computer that they should be in a good state of mind and they should be attentive to me?

To find some hidden 'should' expectations, always consider what your beliefs are underpinning any emotions of frustration or annoyance.

Focus 3: *Reasonable and realistic*

We have talked in this unit of how the Human approaches life by being realistic, as opposed to the Chimp working with what is reasonable. There is nothing wrong with wanting people and objects to be reasonable when you work with them but it is often not realistic.

Exercise: *Working with a realistic outlook*

Think through and recognise the subtle difference between reasonable expectations and realistic expectations in your own life and the implications of this. Try to find and address any reasonable but unrealistic approaches from your Chimp that are creating stress in your life.

Focus 4: *Your Truths of Life*

Exercise: *Work out your Truths of Life*

Try to think of specific areas that unsettle your Chimp and work out some truths around these areas. Then test them out to see if they are effective for you.

For example:

- *There are always good people around to lean on*
- *I can always create my own sunshine.* This might be useful if you feel you are not in control of many areas of your life
- *People are unlikely to change but I can change my approach towards them*
- *Effort and results don't always go hand in hand.* This might be useful if you always try very hard but things don't always work out.
- *Sometimes the bad guys win.* This could be relevant if injustice has happened to you.
- *There will always be good times ahead*
- *When the goal posts move, just keep kicking the ball*

Use your imagination and find Truths that really resonate with you.

Unit 15
Establishing peace of mind and happiness

Unit 15: *will cover how to find and establish values, which will unlock the door to finding peace of mind. We will also look at achieving happiness. Values are the basis for the second large step towards emotional stability.*

Defining values: *A comment on the word 'value'*

The word 'value' can be used in two distinct ways:

1. 'Value' is sometimes used to describe things that are important or valuable to you. An item is 'of value' to you.
2. 'Value', meaning an ideal or moral guide to your conduct.

We will be using 'value' to mean what is the right thing to do, morally or ethically, as this is the meaning of 'values' on the Stone of Life.

First, we will look at some examples to clarify the difference between the two uses of the word and then we will look at *why it is very important not to muddle the two up*.

Examples of 'items of value':

- Having fun
- Being content
- Being creative
- Having personal space
- The relationship with my partner
- My car
- Family
- A family keepsake
- Health

What this list is showing, is that these are important to the person because they believe they will bring quality of life and make both their Chimp and themselves feel *secure and happy*. Therefore, it is wise to know what these things are, to invest in them and to make them a priority in life. The list can include; objects, people, practices or habits and some states of mind.

Examples of 'value' meaning an ideal or moral guide to my conduct:

- Respect for others
- Justice and fair treatment
- Loyalty and faithfulness
- Compassion
- Kindness
- Honesty
- Integrity
- Respect for diversity
- Equality without prejudice
- Promotion of peace and goodwill

These values form a moral code to guide your behaviour in all situations. These values are based on society and personal beliefs. What these values represent is a way of knowing that we are doing the right thing. All of them are judgement-based and give us a moral compass. They are all measured by behaviours we display and actions that we take.

Why is it important not to muddle up the two lists?

What is 'valuable to us', and what our 'values' are, create two different lists. The first is our '*happiness list*' and the second is our '*peace of mind list*'. What is valuable to us gives us happiness. What is morally correct gives us peace of mind.
The 'happiness list' is a list of what we feel is valuable to bring quality of life and give us happiness.

The 'peace of mind' list defines our moral code. It will be our Human's inner strength, particularly in times of trouble. By living by our moral values, we will find peace of mind, knowing that we have done the right thing. These moral values are the values inscribed on our Stone of Life.

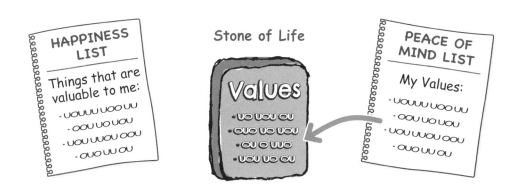

Surely, 'happiness' and 'peace of mind' are the same thing?

These two states of mind usually go together but they don't always. Most people would not be happy unless they had peace of mind, but here are some examples to show that we can have one without the other.

Happiness without peace of mind

Example 1: *Jane and the student*

Jane was a teacher who was very fond of one of her students, Alice. Alice struggled with the work but set her heart on passing the exam. On the day of the exam, Alice became nervous and underperformed. She failed the course by just a few marks. Everybody else on the course passed. Jane could not bring herself to fail Alice. She lifted Alice's result by a few marks and passed her. Alice was elated. Jane was happy that everyone had passed and especially that Alice was so happy in herself, but Jane didn't have peace of mind. This was because she chose to go against her own values and moral code of integrity when she changed the marks.

Example 2: *Lorna and the money*

Lorna borrowed some money from her Aunt. Lorna promised to pay her Aunt back and felt very uncomfortable about owing the money. One week, Lorna miscalculated how much money she had and overspent. Consequently, she fell behind in her payments. This made Lorna feel bad because she was going against her own values by not honouring her debts. After talking with her Aunt, Lorna felt better because her Aunt understood and asked Lorna to pay back what she could, when she could. As the relationship between them had been repaired, Lorna was happy again. However, because she still owed the money she did not have peace of mind and wouldn't have until she had put things right in her own eyes.

Peace of mind without happiness

Example 1: *Raymond's heartbreak*

Raymond's girlfriend, Deborah, had decided that their relationship had come to an end. She told him that it was over and that they must both move on with their lives. Deborah now feels she has peace of mind because she is finally being honest with Raymond, in line with her values. However, she isn't happy because she knows how much emotional pain the break-up is causing Raymond.

Example 2: *Graham and social media*

Graham had been criticised heavily on social media for something that misrepresented him. He had received some harsh comments. This had upset him and he was not happy. However, when he looked within himself, he saw that he had done the right thing and had lived out his own values. He therefore felt peace of mind, despite his unhappiness.

Are there overlaps between the two lists?

Morals are specific to the person. As our moral code is specific to us as individuals, some items might be on either list; it depends on the person. To decide which list an item belongs to, you can answer the question: "Has this item got a moral judgement attached to it?".

If, in your opinion, it has a moral judgement attached, then this will be a value on your '*peace of mind*' list and will guide your behaviours. If, in your opinion, it hasn't got any form of moral judgement attached to it then it will be on your '*happiness list*', as something you desire to have.

For example, 'being honest' is a value and would be on your peace of mind list because it is the right thing to do. Being confident would be on your 'happiness list' as something valuable to have. This is because lacking confidence isn't wrong, it's just unhelpful. What about 'being generous'? This is now up to you. It might be on both lists or it might not feature on either of your lists!

Consider the case of two vegetarians. One has their diet on the happiness list and one has their diet on the peace of mind list. This is because the first vegetarian might be choosing this diet because they don't enjoy eating meat. This puts their vegetarianism on their happiness list. The second vegetarian doesn't eat meat because they feel it is morally wrong to kill and eat animals. Therefore, they would put vegetarianism on their peace of mind list because it is the right thing to do.

What about telling lies?

Surely, telling lies is wrong and honesty would appear on the peace of mind list? Well it would for most of us, but there are some people who believe that lying is fine and part of life. They would not think being honest is a value worth having.

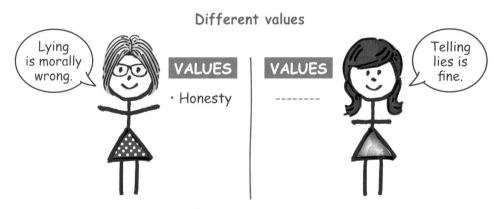

Different values

What about approaches to life?

Some people approach life with a risk-taking attitude; others would choose a more defensive and less risk-taking stance. Is the way we approach life a moral position or a desirable attribute? For example, where do we place courage?

This would be up to you. If you believe that having courage is important, but accept that if you don't display courage then this is not wrong, then courage becomes a desirable characteristic and goes on the 'happiness list'. If on the other hand you believe that it is wrong not to show courage, then this would go on the 'peace of mind' list. The way that you approach life could be a value or be valuable.

Example: *Company values*

Many companies describe their core values. These are often a mixture of true values and desirable characteristics or behaviours that they would like to see in their staff. For example, most people see honesty, integrity, equality, helpfulness, patience and respect as moral values. On the other hand, collaboration, flexibility, persistence and accuracy are generally seen as desirable characteristics or behaviours that are less likely to generate a moral judgement, if someone fails to display them. Therefore, what a company calls its 'values' often include ideal characteristics and behaviours they want to see in their staff. These give standards that staff can measure themselves against, but they might not be 'values' in the moral sense. [274] [275]

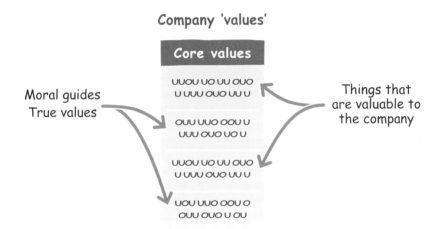

Happiness: *Forming happiness lists*

When working with someone, I usually recommend that they write out two happiness lists: one for immediate happiness and one for longer-term happiness (something to look forward to). The immediate happiness list is composed of things that you can do very quickly in order to improve your chances of being happy.

For example:

- Have a cup of tea
- Phone a friend
- Stroke the cat
- Go for a walk
- Look outside and see nature
- Think of things I am grateful for that I might be taking for granted
- Think about people who care about me

A longer-term happiness list of things to look forward to might look something like this:

- A holiday
- A night out with friends
- Buying some new clothes
- Watching a film
- Catching up with family
- Progressing with a project
- Reading a book

These lists are to help you to get into the right frame of mind for both Human and Chimp. It's important to make sure that there are items on each list that will work for both of them.

Don't let someone else decide on your happiness or peace of mind

Who is in charge of your happiness? If you are in charge, then you will have the final say on what you want to base your happiness on: your opinions, your decision-making and what you think is important. If you allow your Chimp to be in charge, then it will forget its happiness list and hand your happiness over to others: the opinion of others, their approval and what others think is important.

Key Point

Your Chimp will hand over your happiness to others: you can take it back.

If your Chimp has hijacked your happiness and handed it over, take back control of your own happiness and allow yourself to decide what is right and OK for you.

You have the option to look within yourself to establish your happiness based on your own values and beliefs. This will allow you to go out into the world and enjoy yourself.

The same applies to your peace of mind. Allow yourself to decide on your peace of mind and don't hand your peace of mind over to others.

Key Point

Your happiness and peace of mind belong to you and not to others.

Example: *Byron and his lists*

Byron wants to be happy and have peace of mind, so he draws up his two lists. Please remember that as everyone is unique, we will all have our own two lists and only you can decide what is on your lists.

Here is Byron's *'immediate happiness list'*:

- Put some music on
- Reflect and be grateful for being with my partner
- Have a coffee and ten minutes time out
- Think about a memory that I always find amusing
- Call a friend

Here is Byron's *'long-term happiness list'*:

- Visit the local Jazz club
- Develop a positive approach to life
- Joining an evening class
- Have a go at making some home made wine
- Plan time to sort out my music collection
- Download a collection of new movies

These lists clarify for him where he needs to make the effort in order to be happy. Some things he can directly put into place, such as making himself a cup of coffee. Some things he will have to work on, such as developing a positive approach to life. What he is saying is that, if I put in place all of the things on my happiness list, then I will have the best chance of being happy. He can't guarantee some of them, such as developing a positive approach to life, but he can work towards attaining them.

Now we will look at Byron's moral and ethical ideals about what is right and what is wrong. These are his personal moral statements.

Here is Byron's *'peace of mind list'*:

- Respect everyone
- Be honest
- Be generous and not greedy
- Put my family first
- Be compassionate
- Always work hard
- Always be grateful

These values are guides to live his life by. They are not there for him to punish himself! Nor are they there to make him feel under pressure and to feel like he has failed, should he not manage to live by them every day. They are there to let him know what is right and what is wrong and to guide his future behaviours.

The Truths of Life are beliefs that we have worked out by experience or have been taught. Values are ideals that we believe will give us moral guidance and shape our behaviour.

We have to live **WITH** truths	BUT	We choose to live **BY** values
We have **no** choice		We have a choice

Values contribute significantly to defining someone.

The Chimp and Human both possess values, but they could be very different. We are only focussing on the Human values, because these contribute to stability of mind.

Key Point

Peace of mind is achieved by establishing and living out your values.

Forming a 'peace of mind list'

The importance of values is often underestimated. We all hold values; it's just a case of finding out what they are. Finding your values might not be as easy as you think and might take some time. It's important to make sure that these are your values and not those of others, no matter how much you might respect them. Often people form their values from the society that they live in or from their parents. This is fine provided these values are truly in tune with that person. [276] [277] [278] [279] [280] [281]

The easiest way to find your values is to think of a moral belief and then an action that would demonstrate that belief. For example, I believe that everybody should be shown respect. My value is that I will show respect to everybody. One way I can demonstrate this, is by listening to their opinions and acknowledging them. I don't have to agree, but respect is demonstrated by listening.

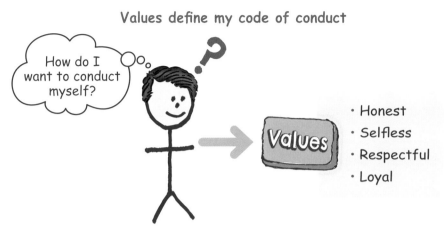

If you do not live out the values that you hold, it can cause you to have severe inner conflicts. This will lead to a restless and uneasy mind. For example, if someone holds a value that being honest is important and then lies, they will usually start feeling guilty. The inner conflict will be resolved when they tell the truth. Otherwise, they might attempt to resolve the inner conflict by rationalising their actions and somehow justify to themselves that it was fine to tell a lie. The strength of the inner conflict depends on the conscience of the individual and how important that value is to them. One of the difficulties with values is that we often hold them unconsciously and are not fully aware of them. This means that we might not be aware that we are going against them at a conscious level, but it doesn't stop the inner conflict from happening. It means that this can result in us feeling unsettled without knowing the cause. [279] [282]

Example: *The neighbour dispute*

Maria believes that everybody deserves respect. She has come into conflict with a neighbour. Maria justifies her lack of respect toward this neighbour by saying that the neighbour is in the wrong. Maria's behaviour isn't in line with her values. Maria recognises this, but if she continues to disrespect the neighbour she will continue to feel unsettled. Sadly, she will believe that the neighbour is the cause of her being unsettled, but in reality it is her own inner conflict arising from not living by her own values.

Values as a way of living life

If we remind ourselves of our values, we can live our life by them. Our values can help us with decision-making. We can also use values to manage the Chimp. For example, managing the eating drive by using values is a very different way from using restrictive dieting.

Example: *Denzel, the diet and self-respect*

Denzel's Chimp has a strong eating drive, which he struggles to manage on a regular basis. His weight yo-yos up and down. When his weight goes up, it bothers him a lot. He could learn to use his values to change his behaviour. One of the values he strongly holds is to have self-respect. Instead of seeing his eating as a weight problem, he has come to see over-eating or eating the 'wrong' food as showing disrespect towards himself. If he can remind himself about showing self-respect before he chooses his food, then his choices can be based on his values. If we truly hold a value, and remind ourselves of it, it is difficult to go against it.

Example: *Self-esteem and values*

Values are part of the character of a person and help to define them. If we live by our values, then we can look into the mirror and be proud of who we are. Knowing who you are and living by your values brings strength, peace of mind and raised self-esteem. Values can remind of us of principles that are important in life.

Example: *Values used to manage stress from criticism*

Many forms of stress can be managed and removed by looking to our values. For example, everybody receives criticism from others at some point. This is especially true on social media. We have a choice of how we want to manage criticism. We can try to justify or defend ourselves. This usually has no effect because the person or people who are criticising are usually fixed in their opinions, regardless of the facts. Our defence might be rejected, and this leaves us in the same position.

As an alternative way of dealing with criticism, we could first look to see if there is some validity in the criticism and act appropriately. Then, we could look to our values and assess whether we are living by them. If we are, then peace of mind is usually restored.

People who are destructively critical on social media demonstrate a lack of morals and show impulsive Chimp reactions. It helps to recognise this and not to take any comment personally. The comments they make merely reflect on themselves.

Values in conflict

It is common for our values to come into conflict with each other, producing an inner conflict. If you realise that this is happening, then it helps to know that our values have a hierarchy. This means that we can resolve a problem, where two values oppose each other, by deciding which one is the more important. [279]

Example: *Values in conflict*

Ian holds two values that have come into conflict: loyalty to his parents and always helping others in difficulty. Ian's parents have had several clashes with the neighbours. The neighbours have a fifteen-year-old son who has caused damage to Ian's parents' garden, stealing garden ornaments, and polluting the pond and causing several of the fish to die. The boy has been caught on a security camera, but the parents will still not accept responsibility for their son's deliberate acts of vandalism. Ian's parents are very angry with the neighbours, especially the boy.

When Ian is driving to see his parents, he sees the neighbour's son at the side of the road. The son has had an accident and fallen off his bike. The bike is very damaged. The son sees Ian, recognises him and flags the car down. The son tells Ian that he is able to walk home but feels shaken and asks Ian to give him a lift home.

This scenario will divide opinion. There are no rights or wrongs, but values come into play. Ian himself must decide which of his values he thinks is the more important. Does he stay loyal to his parents (knowing they might be angry if he offers help) or does he help an adolescent who is distressed? Only he can decide and obviously there could be consequences whatever he does.

Example: *Misplaced loyalty*

Sadly, during my working life I have come across several cases of people who have very low self-esteem and misplaced loyalty. A severe example of this is often seen in abusive relationships. The abused person might be staying with their partner out of misplaced loyalty. The abusive partner is unlikely to be truly sorry, won't stop being abusive and fails to seek help. The abused person often stays because they have put their own self–respect below their value of loyalty. Often, it is only when an abused partner leaves the relationship that they can see what was happening. They then realise that they do deserve better and can happily survive without the abuser. Self-respect is a value worth putting above misplaced loyalty.

Balancing work and family life

The problem of getting a work-life balance is a common one. Practicalities involving finance and securing a long-term future are bound to play a part. One way of resolving the situation is to try giving advice to your own child, as if the years have rolled by and they are now in your position. Sometimes, we put practical things in front of quality time with our family, and forget to look at both our happiness list and our peace of mind list for guidance. Most work-life balance situations just need some time spent on them to sort out a happy compromise but this isn't always the case, as with the next example.

Example: *Geraldine and the cake shop*

Geraldine has always wanted to run a cake shop. Her dream is finally realised and her business opened two years ago. The problem she is facing is that despite two successful years of hard work, it seems she is now on a necessary treadmill of work, just to keep the business afloat. Geraldine is married with two young children and she is becoming distressed. She works long hours and feels her marriage and her role as a mother is being compromised. She has tried to manage the situation in numerous ways but there always seems to be shifting ground with problems arising out of the blue. The more attempts she makes to solve the emerging problems and balance her work with home life, the more despondent and unhappy she becomes. It appears to be a no win situation.

It is **not** a no win situation. The reason that Geraldine is distressed, and also seeing this as a 'no win' situation, is because she is not accepting reality. She only sees a 'win' situation as her being in two places at once and giving both her home and work life her full attention. If she can be realistic about the needs in both her home life and her work life, then she can redefine what 'winning' means. A solution would be to sit down and accept that none of us can do justice to two things, which between them demand more time than we have. If a balance can't be found because it is impossible, then one of the two must be compromised and this has to be accepted. It's just a decision on which is most valuable to you along with which marries with your values. (You could say that Geraldine can't have her cake and eat it. – sorry!).

Clearly, there is a point for some people, where it will dawn on them that they have given enough time and effort to try to obtain a work-life balance. Their own situation doesn't seem to allow a compromise. Geraldine's situation is one of these. Only she can decide on the way forward. She will 'win' if she accepts reality and makes her choice.

Values as the basis for an enduring relationship

Lasting relationships are built on values. Two people will have a great chance of an enduring relationship provided they share the same values. If they do not have the same values, the chance of the relationship lasting diminishes. It also helps if the two people have their values in the same order of importance.

Common interests, physical attraction and other aspects might initially attract us to a person. However, they might not be sufficient for a lasting relationship, if the person we are attracted to does not share our values. It's wise to check out a person's values if you want to know if a relationship with them is likely to last!

Unit 15 Reminders

- Things that are valuable to us form our happiness list
- Values form our peace of mind list
- Values are ideals based on morals that we wish to live by
- Values are often unconscious, but can be worked out
- Living by our values prevents inner conflicts
- Values can come into conflict with each other
- Values help us to make decisions
- Lasting relationships are based on shared values
- Your happiness and peace of mind belong to you and not to others

Unit 15 Exercises

Focus points and reflective exercises

1. *'Happiness list'*
2. *'Peace of mind list'*
3. *Checking values and self-esteem*

Focus 1: *'Happiness list'*

Exercise: *Forming your 'happiness list'*

This is a straightforward exercise that can make a great difference to the quality of your life.

Forming your immediate and long-term happiness lists is really about knowing what will make you happy and ensuring that ***you make it happen***. The happiness lists will give you things that you can do immediately and things that you can do in the long-term. Planning is only the start of this exercise; making the plans happen is the important part.

Focus 2: *'Peace of mind list'*

Exercise: *Forming your 'peace of mind list'*

Finding your values can be made easier if you follow the guidelines in this unit. Take some time to work out your values. It might not be as easy as you think, so you might have to have more than one go at doing this. Remember that a value is a moral code to let you know you are doing the right thing. Values lead to behaviours. Remember that the list will only work for you, if you make sure you implement it.

Focus 3: *Checking values and self-esteem*

Exercise: *Living by your values*

The main function of values is to bring about a moral code to live by. If you base your self-esteem on living out your values, it can bring not only peace of mind but also raised self-esteem.

Ask yourself, if you are really living by your values. Try to reflect back on your day and see if you can detect anywhere where your values were compromised. There might not be a right or wrong way of approaching some problems, but there might be a better way, in line with your values. It's helpful to encourage yourself when you recognise that you have put your values into good use.

Unit 16
Keeping events in perspective

Unit 16: *will complete our understanding of the Stone of Life by looking at the third aspect: The Life Force. This will bring perspective.*

Keeping perspective: *The 'Life Force'*

Keeping perspective is about being able to see everything in relation to what life is all about and what at the end of the day is really of importance to you. If, at the end of our lives we look back, most things will take on a different perspective. The 'Life Force' represents what we think life is all about. This Life Force gives us a reference point to keep all aspects of life's experiences in perspective and can stabilise us emotionally.

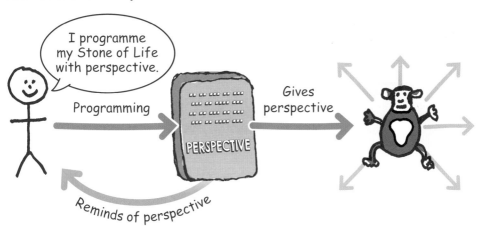

To find your Life Force, simply answer the question:
"What advice would you give yourself, if you were to start your life again?"

The deathbed advice scenario

Another way to help you to find your Life Force is to imagine being 100 years old and in the final minute of your life. Your great granddaughter arrives and asks, "What should I do with my life?" how would you answer?

Giving perspective

When you have done this, ask yourself if you are taking your own advice.

When people answer the question about giving advice, they typically say something like:

- "Don't worry about anything – it isn't worth it"
- "Most things work out, and those that don't, will change with time"
- "Don't worry about what other people think"
- "Make the most of your family and friends"
- "Take every opportunity"
- "Live the life you want to live"
- "Get over things quickly"

It's all great advice and if we took it then we could have a fantastic life. So now we reach a different question. Why can't we take our own advice?

By now it's probably becoming obvious why we can't do it. Our mind won't let us. [283]

The Human in our mind can take the advice and act on it. It can see the bigger picture. In contrast, the Chimp cannot see beyond what is happening in the here and now. The Chimp is preoccupied and worries about day-to-day living. It worries about being seen as good in the eyes of others. [284] It worries about how it will cope if things go wrong or if it fails at something. If you reflect on the way your Chimp is so ridiculously short sighted, you could come up with many other things that the Chimp does to stop you from living your life, as you want to.

If we have to accept that our Chimps cannot keep perspective, how do we manage this situation? The rules of the mind are clear; the only way we are going to manage the Chimp is to rely on the Computer.

How to keep perspective

If we are to manage our Chimp and retain perspective, we need to work out precisely what to put into our Computer. As always, only you can decide what you will put into your Computer. What follows are suggestions that you might like to try. These are Autopilots that will automatically provide behaviours or beliefs to immediately take over *before* the Chimp gets a chance to act. The starting point is to programme your Computer to learn to recognise that perspective is being lost. If we can recognise a loss of perspective, then we have the insight to allow the Computer to bring in our automatic behaviours and beliefs.

Example: *Edward and the failed contract*

Edward is a tradesman who is self-employed and has just submitted a bid for a contract. This contract would see him into a financially secure position for a few years. Without securing the contract, Edward will struggle to pay his mortgage, put his marriage under strain and it will challenge his ability to even feed his three children. He has just heard that he did not get the contract. Unless we have been in this position it's difficult to imagine the feelings that Edward could be experiencing. It's a serious situation, so we can't just dismiss it. However, we can retain perspective, which will help.

Once Edward has recognised that perspective might be lost, the first thing he could do is to accept that any emotions or thoughts from his Chimp are normal and to be expected. He could exercise his Chimp by letting his emotions out and then turn to his Computer. Alternatively, he could take a deep breath and turn straight to his Computer for help.

Either way, the Computer will offer some truthful reassurances to help keep perspective, provided it has been programmed.

Edward has programmed into the Computer two main themes:
1. **See the bigger picture** – This is formed from a number of truths.
   Edward has included:
   a. Everything we experience in life will change with time
   b. Problems do get solved and solutions can be found
   c. We never know what is around the corner
   d. Help is always available, if we reach out

2. **Reminder of his deathbed advice**
   a. Place your focus on what really matters
   b. Make the most of today
   c. Worrying doesn't help anything but action does help

Keeping perspective will help, but might not take away the Chimp's unease or suffering. It will vary from person to person. If Edward's Chimp doesn't settle then he might want to exercise it regularly, that is to let it express itself out loud. This allows his Human a chance to progress with solutions and action. [81] [88]

Seeing the bigger picture

Keeping perspective by looking at the bigger picture puts a situation into context with an overall longer-term view. When we are in Chimp mode, we cannot retain perspective and lose sight of the bigger picture. [285] [178]

Example: *Marek's Flat sale*

Marek is very keen to sell his flat. He wants to move because the noise from a nearby bus station is too much for him. It has stressed him for months. He put his flat on the market at a price he would like, but the only offer he has received is slightly lower. The fact that his flat has been on the market for several months is also stressful and Marek is finding that it is now affecting his health. Despite this, he is holding out for his 'hoped for' price. Effectively, he has lost the bigger picture. What Marek can't see is that the potential financial gain from waiting doesn't outweigh his own peace of mind. What price does he really put on his health and quality of life? Could he manage on the offered price and accept that the 'lost' money is worth it, and could be seen as paying for his own happiness in being able to move on.

The principle in this scenario is very common. How often do we forget the bigger picture of what we are trying to achieve and focus on minute detail instead? It is so common for people in this situation to be able to move on quickly once they step back and see beyond the fine detail. This principle can also apply to relationships, as shown in the next example.

Example: *Kim and her daughter*

Kim has been divorced from her husband for four years. Her husband had an affair and when this came to light Kim decided to leave him. She has settled into a single life and lives with their five-year-old daughter. Her ex-husband left the other woman and has since had several more relationships. Kim dislikes him and feels he is a poor role model for their daughter. Her ex-husband rarely turns up to see his daughter and so Kim has now refused to allow him near the child. This is all very understandable, but let's look at the bigger picture. Is Kim protecting her daughter or taking her anger out on her ex-husband by limiting visits to the child? What does she really want for the child? Kim's answer is almost certainly "I want the best for my child". Most therapists would tell her that the best for the child is to have two parents in the picture and for her daughter to have some relationship with her father. Some might disagree, reasoning that he is a poor role model for his child. The only way that Kim can resolve this is to step back and to see the bigger picture. By bringing perspective into what she wants for her daughter and what her daughter needs in the long-term, Kim will be able to make appropriate decisions.

The bigger picture and long-term vision

The Human always has perspective and time awareness. [286] When we are in Human mode, we see the bigger picture and tend to take things in our stride because perspective gives us a long-term vision. As the Chimp does not possess perspective, it operates with a short-term vision and therefore overreacts in situations of stress.

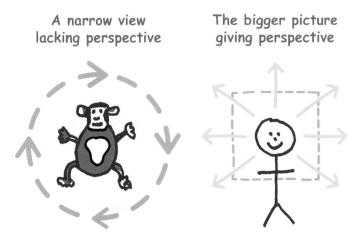

Example: *The effective business leader*

Businesses can suffer if the Chimp hijacks them with its lack of perspective. This is seen, for example, with plans that give short-term gains but long-term losses, a lack of succession planning or inappropriate performance indicators.

When a leader operates with a short-term vision, they effectively create unwanted behaviours in their staff. With a short-term vision, staff will become more preoccupied and stress over small and often unimportant details. A leader with a long-term vision will settle their staff, who can then see trivia as trivia and deal with it in a stress-free way.

Example: *The sales team*

Imagine a sales team is asked to sign up new recruits for a programme and they will be rewarded on the number of new recruits they sign up. The company might measure this to show success. However, the evidence might show that new recruits don't remain within the programme. Therefore, this short-term measure of success might eventually give the company a poor name for running a programme with a high dropout rate. A better measure of success would be how many people are retained over time.

Perspective when defining a successful life

If we changed the deathbed advice scenario slightly, we could find another way to retain perspective. Imagine again that you are one hundred years old and on your deathbed with just one minute to live. This time your great granddaughter's question to you is: "How will I know that I have been successful in life?"

What would you say?

Success?

How will I know that I have been successful in life?

This is not an easy question to answer and requires some thought! Two of you will try to answer: your Chimp and your Human.

It's likely that your Chimp will look outwardly to achievements to answer the question about success:

- Did I make a lot of money?
- Did I get lots of qualifications?
- Did I impress a lot of people?
- Was I the best in my field?

It's also likely that your Human will look inwardly to self in order to answer the question about success:

- Did I achieve a happy life?
- Did I go through life with peace of mind?
- Did I spend my time doing what I wanted to do?
- Did I live a constructive and positive life?

I don't know what your Chimp or Human might say. It could be very different to these. What is crucial is that whatever you have said, you have now got a guide as to what is important to you and where you should spend your time and energy. Why waste your energy on unimportant things?

How you define success at the end of your life will bring perspective to what is really important to you. What do you want to spend your time and energy on during your lifetime?

Perspective on Success

Example: *The wayward asteroid*

Some time ago, a man I was working with shared several legitimate worries that his Chimp was having about his life.

I asked him to imagine that an asteroid had been detected and it was heading straight towards Earth. It would definitely collide with Earth in four weeks time. Which of those worries is most important now? He answered with a smile "none of them", and then added, "Why would I waste my time and energy on worrying". After thinking, he said, "but there isn't an asteroid hurtling towards us".

Actually there is. That asteroid represents the reality that our own personal life on Earth will one day end. It might not be in four weeks time, but it is certain it will come to an end (at least as of the time of writing this book!). We can face this constructively and make the most of our lives, knowing that they are transient. Alternatively, we can deny our temporary situation and continue to worry throughout our time here. You always have a choice about whether you want to keep perspective or not.

Confidence and perspective!

Confidence is a choice.

It is based on how strongly you can believe that you will, either:

The Chimp puts its confidence in how strongly it believes that it will *achieve* its best. Therefore, if it believes it can definitely do something, it will be 100% confident. However, if it doesn't know if it can achieve something, then its confidence can be very low. This approach by the Chimp raises a number of problems. The Chimp tries to reconcile the following:

The Chimp believes to be really confident, it must:
- Have a guarantee that it will succeed
- Not show weakness to others or to itself
- Achieve its optimum performance every time

None of these are realistic. You can never guarantee that you will always succeed. We can never influence how others see our performance, however we do. No one ever achieves their best performance every time. Therefore, in Chimp mode we are rarely confident.

The Human puts its confidence in *trying* their best *and* knowing that they can *deal with any outcome*. Just a note on trying your best: trying your best means exactly that. It means it was impossible to have put more effort in, but despite this, you were not perfect. There could still have been areas you might improve on or things you could have done differently looking back.

The Human believes to be really confident, they must:
- Know that they gave it everything, their best effort
- Know that they are proud of themselves
- Accept that they can deal with failure or a less than perfect performance

Of course, we all want *to achieve our best* but this can never be guaranteed, but *trying your best* can always be guaranteed. We can never do better than our best effort. Therefore, if we approach challenges in Human mode and reassure the Chimp that we can guarantee giving everything we have, even if we get it wrong or make some errors, the Chimp will settle. It settles, because it now has a guarantee of success, defined by giving everything you have: your best effort.

Rationally we can never do better than our best effort and to keep on trying to improve in the future.

Example: *Chloe, confidence and perspective*

Chloe failed some of her exams at school and felt that she had underachieved. This haunted her for years. She always felt that the exam results undermined her self-esteem. This led her to be lacking in confidence and having a reluctance to try new challenges. She decided to look at what she was basing her confidence on. Did she give her best effort? Yes she did, but she also made many mistakes and underachieved. In Human mode she could accept that best effort and best achievement don't always go together. Continuing in Human mode, Chloe could take an adult stance and let her Chimp know that she can deal with this disappointment. Finally, she can bring her exam results into perspective. They really are not that important to her now and certainly don't define her. By bringing perspective to their relative importance, Chloe was able to see herself in a different light and raise her self-esteem along with her self-confidence.

Perspective and sentimental items

Example: *Rod and the lost wedding ring*

Rod got married twenty years ago. His wedding ring symbolised the amazing relationship that he has had since then. While travelling, his car broke down and he had to sort the engine out. He got the car going again but had to stop to clean his hands. While washing his hands in a public service area, he took off his wedding ring but forgot to put it back on. He left the service area before he realised what he had done, and despite returning and searching, the ring was never seen again. He was distraught. Some possessions represent something of major significance to us. We can then lose perspective and not recognise the item for what it is, rather than what it represents. Others looking on have perspective and can see that the ring was just a band of metal but Rod has lost sight of this. Rod still has a great marriage and although the exact ring can't be replaced, it is in effect, a sentimental piece of metal. It isn't easy for most of us to separate an item from what it represents to us. However, we can lose perspective in our sadness, if we don't recognise it is only a representation. We *have not lost* our memories.

Key Points

You always have a choice on what you base your confidence on:

- **Human:** *doing my best and managing outcomes.*
- **Chimp:** *achieving my best and falling apart if I don't manage.*

Operating with the Stone of Life

We have now covered the three major components on the Stone of Life. Before we complete this unit, we can give an example of how these three components operate. After this, we have a challenging situation to consider.

Helping to keep perspective and using the Stone of Life

Example: *The ice cream scenario*

The tale of Freddie and the ice cream cone is an example of how to gain perspective by having your Stone of Life up and running.

Picture it: Four-year-old Freddie has just been given an ice cream. The ice cream van drives away and Freddie drops the ice cream cone. It's all over! So Freddie's Chimp screams the place down. His Chimp needs to express its desperate emotions and if we don't help Freddie to gain some perspective, he might still be telling people how awful it was, twenty years later.

What we could do is to allow Freddie to cry, and after this, reason with him.

In other words, exercise his Chimp and then box it with logic. Therefore, we apply facts, such as:

- The van will come again
- It's only an ice cream
- It's not the end of the world
- We can't change it

Effectively we are moving him from Chimp to Human. We are acting as his Human because a four-year-old can't do this. [287] [288] He can't gain perspective or believe that he can deal with the situation.

Now consider situations in your own life and experience. How many times do we 'drop the ice cream' and allow our Chimps to keep on over–reacting? To get back into Human mode we can legitimately state some facts to ourselves, such as:

- I do have a Human part to my brain – so let me use it
- I am not a four year old – I am an adult and I can deal with anything life throws at me
- I can gain perspective by seeing the bigger picture
- I can deal with consequences, no matter how serious they appear to be

The secret is to recognise 'the ice cream situation' and to manage it immediately.

Here are some examples of where the 'ice cream situation' could apply:
- Failing a test
- Being criticised by someone
- Experiencing an unfair situation that you can do nothing about
- Not getting the outcome that you want
- Not being on time (for whatever reason)
- Making a decision that turned out not to be the best

Now we can consider an alternative scenario, where we *prevent* the Chimp from hijacking us in the first place, by using the Stone of Life.

If we 'drop an ice cream' with the Stone of Life in place, the Chimp part of our brain will look into the Computer for advice and find the Stone of Life. The Stone of Life will unconsciously and rapidly remind the Chimp of reality, truths, values and perspective that we have already prepared on the Stone. This settles the Chimp and prevents it from reacting.

The Stone of Life can bring reality, truths, values and perspective to any situation. In Freddie's case, if he were older and had now put into place a Stone of Life, the Chimp would look to this before it could hijack him. It will see perspective and also many truths. It will effectively be boxed before it has chance to react. Freddie might not be aware that his Chimp has been reassured and boxed by the Computer, but what he will experience is calmness within the situation.

The Stone of Life must be pre-programmed with truths, which strongly resonate with you otherwise it won't work.

> **Key Point**
>
> *If your Stone of Life is in place and you live by it, then you will find that there are no over-reactions to life's ups and downs.*

Stabilising mechanism

A challenging situation to consider

The deathbed advice scenario can be used to offer some perspective but there is a challenge that it throws up.

What is the best advice you can give?

Example: *The imaginary great granddaughter and the deathbed scenario*

Someone went through their deathbed scenario with me. She said she would give her imaginary great granddaughter some sound advice. It would include making sure she was happy, taking every opportunity, not worrying about things or what others might think and so on. This was all really great advice. If her great granddaughter follows this advice, then she will have a very happy and fulfilling life. However, there is a major flaw!

What is the point of giving this common sense advice if none of us can take it? Some very fundamental advice is missing and this fundamental advice is what this book is all about.

The most crucial Key Point of them all

The reason we can't live life the way we want to, is because we have a machine in our head that works very differently to us and is stronger than we are.

This machine consists of the Chimp and the Computer. It must be managed if we are to have the best chance of being happy, successful, and confident, and leading the life that we want to lead.

A problem with the deathbed scenario...

The best advice to her great granddaughter, might be to tell her that:

It is your own mind that could stop you from being you and leading the life that you want to lead. It is so important to learn to understand and manage your mind. You are going to share your life with a Chimp!

If her great granddaughter can grasp the fact that she will have to share her life with this machine, and there are advantageous ways of doing this, then she can gain a better quality of life. She can also be successful, no matter how she defines success.

This advice will help her to:

- Work with reality
- Establish and live with the truths of life
- Establish her values and live by them
- Always keep perspective

In other words, travel through life in Human mode and base her journey on her Stone of Life.

Keeping perspective by re-setting the brain with laughter
- Scientific points

What seems like a serious situation can often be defused by seeing the funny side of things. If you can laugh at yourself or the situation, it brings perspective to the picture [289]. Whenever we laugh or display a sense of humour, we re-set the brain into Human mode because we disarm the Chimp. [290] [291] The Chimp is programmed to detect danger and causes for concern, so by laughing at a situation the Chimp will lose its power. Then, perspective can be brought into play from the Human.

Learning to laugh at yourself or seeing the funny side of a situation can be programmed into your Computer as an Autopilot. Not taking yourself too seriously can also help to defuse stressful situations. When we laugh, levels of the stress hormone, cortisol, decrease and are replaced by higher levels of other more positive transmitters and hormones such as dopamine and endorphins. These relax us and make us feel good. [292] [293]

Unit 16 Reminders

- Perspective can be gained by having a clearly defined Life Force
- Your Life Force can be defined by looking back on your life and stating what advice you would give someone who is setting off in life
- Perspective can be gained by programming your Computer to automatically see the bigger picture whenever an unwanted event occurs
- Defining success can bring perspective
- Perspective can be gained by reminding yourself that you are sharing your mind with a machine
- The Stone of Life can remind us of key Truths that maintain perspective at all times
- We can reset the brain with laughter

Unit 16 Exercises

Focus points and reflective exercises

1. *The one minute 'Life Force'*
2. *Creating your own Stone of Life*
3. *Setting off each day using the Stone of Life*

Focus 1: *The one-minute Life Force*

Exercise: *Explaining The Chimp Model to a child*

Try to imagine that you are sitting with Florence your great granddaughter. She has told you that she wants to be happy in life and be successful and she has shared her dreams with you. You have offered some advice. Now try to explain to her, in one minute, why our minds might be the limiting factor for our success and happiness! If you can explain how the machine that we are born with might have different ideas to yourself and also how we can then manage it, you will have done well and shown that you have understood the course so far!

Specifically, try to explain how to manage the machine so that you can live the life that you want to live and be the person that you really are.

Don't be concerned if you can't explain it! Some of us are better at explanations than others but please have a go.

If you find this exercise impossible, then here is a cheat sheet for you!

The mind has two thinking parts, one is you and the other is out of your control and thinks for you.

This out of control part of your thinking mind is the same as the one a chimpanzee has, so we will call it your Chimp.

Your mind has a storage memory system that will remind you and your Chimp what the best way is to live your life and how to deal with situations. Both you and your Chimp put information into this: some information is helpful and some isn't helpful.

To manage your mind and get the best out of yourself, you need to:
- Accept, manage and work with your Chimp
- Tidy up your Computer to have only helpful beliefs and advice in it
- Make sure you know who you are, and separate yourself from your machine

If this makes sense, have another go at explaining the basics in one minute. Add any other important pieces of information that you feel you would want to add.
This will hopefully act as a reminder for applying the principles to yourself.

Focus 2: *Creating your own Stone of Life*

Now that we have completed the three main components of the Stone of Life, you will be able to create your first draft of your own unique Stone. I am saying first draft because it is good to keep refining it. The importance of this exercise can't be underplayed, because it is the Stone of Life that will support emotional stability and peace of mind.

Exercise: *Creating your Stone of Life*

Work through each component and draw up your Truths of Life, values and Life Force.

Remember that your Truths of Life need to be statements that truly resonate and mean something to you. Any Truths that you add, that don't really mean something to you, will weaken the impact of your Stone of Life. When you consider the Life Force, think carefully about what advice you are giving because this is what you will need to follow, if you want to keep perspective.

Focus 3: *Setting off each day using the Stone of Life*

Exercise: *One minute reminder*

Assuming that you have formed a good Stone of Life, try to allocate just five minutes each morning to go through and reflect on the components. This will help to bring these to the front of your mind as your basis for living. By doing this, it is likely that you will set off each morning in Human mode. By reminding yourself of your Stone of Life contents for just a minute or two at intervals throughout the day, you might find that your Chimp will begin to settle into a new pattern of behaviours.

Unit 17
External support

Unit 17: *will cover establishing, maintaining and employing external support. External support is the safety net for emotional stability when self-management falters.*

Managing a distressed Chimp using external support

A major part of mind management is managing the Chimp. So far, we have looked at how we can do this in various ways. The Stone of Life is the ultimate stabiliser for the Chimp but what happens if this fails to settle the Chimp and all other methods also seem to be failing?

The Chimp naturally looks outwardly when it is functioning. It also does this under stress to find support. We have altered this by programming the Computer and managing the Chimp internally. Therefore, if our internal mechanisms for managing the Chimp are not working, for whatever reason, we can work with external factors. For example, if we experience a very stressful event and are unable to process it, or we are going through a grief reaction, then the Chimp might not be able to engage until it has worked through the situation. At this point, the Stone of Life might not be accessible to the Chimp and it will turn to 'the Troop' for stabilisation.

We are now operating in a way that engages the Chimp's natural drive to find a supportive troop. Ideally, we would like the Chimp to be managed internally in our mind because we can have significant control over this, but we can use the Troop as our safety net. It is therefore very important to make sure that the Troop is in place and fully functioning. This unit will look at how we can do this.

External support
- Scientific points

Where does the drive for the Troop come from? For survival, chimpanzees rely heavily on being part of a troop. [294] The drive to remain within the troop is extremely strong because a single chimp is vulnerable to attack from prey animals, such as a leopard, or from other Chimps that are not part of its troop. We can't always transfer what we see in animals and relate it to human beings. However, our in-built troop drive seems just as powerful. We search out companionship and form groups readily. Our problem is that we sometimes make poor choices and don't recognise unfriendly chimps! [295] [296]

There will always be some individuals who are exceptions to this rule but generally we are gregarious creatures. We seek others for protection, reassurance, procreation and many social and emotional reasons. These are all needs that both the Chimp and Human share. How they go about obtaining these needs might be very different. [299] [300] [301] [302]

External support (continued)
- Scientific points

The natural emotional support during human development moves from parental, to peer to individual. Children usually depend emotionally on parental figures until around puberty. At puberty, the typical teenager will turn to the peer group. During the later teenage years, individual stability develops. Under stress, we usually revert to a previous stage of emotional support. Adults turn to friends; teenagers turn to parents.

When teenagers turn to each other for support, it can raise problems. This is because support is driven by popularity and approval from other adolescents, who themselves are struggling. A teenager leaning on other teenagers to try to form a basis for stability is a very precarious system! [299] [300] [301] [302]

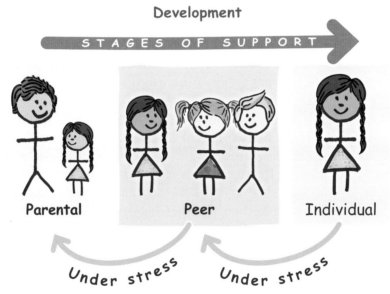

Establishing your support

Support from other people, that gives rise to security, is fundamental to helping our Chimps become emotionally stable.

This support must therefore be chosen carefully. The Troop is the person or group of people, who we can completely rely on. They are friends who will stand by us, whatever is happening. They will be the people, or the person, who when our back is against the wall, won't let us down or walk away. Effectively, they will put us first. We might have many friends but only a few will be part of our Troop. [303]

In our search for this support, our Chimp and Human take very different approaches. The Chimp will search for emotionally based and often superficial characteristics, whereas the Human will look for deeper characteristics.
Here are some examples of what the Human and Chimp typically look for in people, when deciding whom to admit to their Troop.

I admit someone who...

- Pleases me
- Shares common experiences or similar interests
- Shares a common background
- Looks attractive physically
- Can provide financial or other security
- Makes promises that please me
- Have impressive talents or status

- Shares common values
- Has integrity
- Demonstrates compassion
- Shows understanding
- Demonstrates tolerance
- Is reliable
- Is selfless

Clearly some of the characteristics might overlap between what the Human and Chimp are looking for. For example, the Human will obviously want someone who pleases them, but without the other essential features, this won't put that person into their Troop.

Key Point

The Human chooses characteristics in people that will fit the purpose of the Troop.

Research shows that personal relationships that endure are founded on values, and not on more transient things, such as common interests. The Human chooses characteristics in people that are likely to lead to a long-lasting and constructive relationship. The Chimp chooses characteristics that are much less likely to lead to long-lasting or constructive relationships.[304] [305]

Example: *Kurt and the craft group*

Kurt enjoys craftwork and works with wood. He has joined a craft group and made several friends within the group. Kurt's Chimp has picked out two friends, who he feels he has a close relationship with: Andrew and Terri. His Chimp sees these two people as part of his Troop. Andrew is a dynamic individual full of energy and he seems to give Kurt a good feeling when he is with him. Terri flatters and compliments Kurt very easily and this makes Kurt feel good about himself. There is nothing wrong with any of these characteristics but do these two people share more reliable characteristics? Millie is a quiet member of the craft group and although pleasant, Kurt's Chimp doesn't see her as part of his Troop.

Kurt has suddenly found himself in difficulties with his finances and has become depressed. He knows he could have managed things better but really needs some understanding and guidance. He tells both Andrew and Terri about his problems. Within days, Kurt finds that Andrew has told several members of the craft group details about Kurt's financial problems and has been very critical of the way Kurt is running his life. Terri listened but then walked away and rebuffed Kurt when he tried to discuss the situation with her. She told him that he needed to get his act together. Both people don't share his values, nor are they demonstrating characteristics that Kurt's Human would like to see. However, Millie overheard about Kurt's problems and gave him support and understanding with compassion. Letting your Chimp choose your Troop can be fraught with difficulties.

Kurt

Example: *Summer and her father*

Summer has a reasonable expectation that her father would be part of her Troop. This reasonable expectation is coming from her Chimp. The reality is that she does not share his values and finds that he is unreliable and she thinks he is selfish. By expecting him to be part of her Troop, she is setting herself up for disappointment. There is a difference between a Troop member and someone who is friendly but outside the Troop. Friendly Chimps are great to have around, as long as we don't invite them into the Troop and then try and rely on them.

The 'psychopath'
- Scientific points

You definitely don't want a 'psychopath' in your Troop!

The statistics show that these people occur at the rate of about 1 in 150 individuals. At first, it was believed that they displayed anti-social behaviour because of the way that they were raised or the experiences they had been through, often thought to be inflicted upon them by society. Therefore, the term 'sociopath' was given. As scanners and neuroscience advanced, it was revealed that some anomalies in the structure and functioning of the brain were present in these individuals. A pathway of neurones called the uncinate fasciculus was found to be much smaller and had less connecting neurones than the rest of us.

The 'psychopath' (continued)
- Scientific points

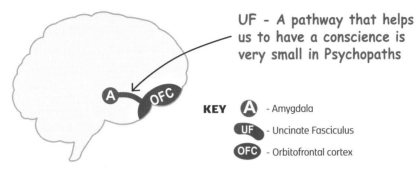

Therefore, it was believed that there was a genetic component and people were born this way. The term 'psychopath' became common language. The truth behind the cause of the unwelcome behaviour is probably a mixture of the two. Therefore, strictly speaking, we know them now as 'dissocial personality disorders'. However, all three terms have become interchangeable.

All agree that the hallmark of the dissocial personality disorder is a lack of conscience. Along with this are many other recognised features, such as: a lack of compassion or empathy, a disregard for the law, a user of people and a self-interested lifestyle.

How do people avoid being taken in by a dissocial personality disorder? This isn't easy but the best way of avoiding emotional damage is to check on a person's past behaviour. The biggest predictor of a person's future behaviour is to look at their past behaviour. Once the background facts are established, it's wise to act on them.

Most people are not psychopaths, but many people display repeated destructive or unhelpful behaviours. Consider these carefully when selecting your Troop members. [306] [307] [308] [309]

Maintaining your Troop

Investing in those who mean a lot to us is often neglected. Frequently when working with people, I hear stories of how relationships with close friends or partners are lost because of a lack of attention and effort in maintaining the friendship.

Key Point

Familiarity can blind us to the privileges of a great friendship.

This passage can remind us to stop and reflect on the friendships we have, and invest some time into strengthening them.

Questions that you could ask yourself about those in your Troop could include:
- Have I let this person know what they mean to me?
- Have I thanked this person for what they have done and also for just being themselves?
- Have I asked myself what I could do to support them?
- Have I checked with them that they are in a good place?

Employing your support

We all have life situations that will create some stress or unsettle us. At this point we hope that we can stabilise emotionally by turning to our Stone of Life. Naturally, there will be times when the Chimp does not appear to settle using the Stone of Life. It is in these times that we have the Troop as our safety net. The members become the settling influence on the Chimp. They can do this by either getting us back into Human mode by talking through a situation rationally or they can be understanding and offer some comfort by their Chimps joining us in sharing the problem.

If you have become emotionally unstable but have not reached out to your friends, ask yourself how you would feel if one of your friends were in your situation but did not reach out to you. Almost everyone says that they would be upset that their friend didn't reach out because that is what friends are for. Many add that it would make them feel good to be able to share their friend's problem and to support them. Reaching out allows our friends to be able to demonstrate their friendship.

Independence is a great thing to achieve but it is a fine line between independence and stubbornness or reluctance to receive help. Being a friend is not just about supporting others but also allowing others to support you. When you do share concerns with others, it will settle your Chimp by feeling understood and supported. By sharing concerns, you will also have chance for your Human to reflect, in a supportive setting.

Key Point

Sharing concerns with a member of your Troop will not only settle the Chimp but will also give your Human chance to reflect.

Unit 17 Reminders

- The Troop is the Chimp's stabilising factor, if the Stone of Life doesn't settle it down
- The Chimp and Human have different ways of choosing the Troop
- The Human chooses characteristics in people that will fit the purpose of the Troop
- Long-lasting personal relationships are based on common values
- Maintaining your Troop takes effort
- Employing your Troop means recognising when you need to reach out

Unit 17

Exercises

Focus points and reflective exercises

1. *Checking on your Troop selection*
2. *Maintaining your Troop*

Focus 1: *Checking on your Troop selection*

Your Troop size might be just one person or it could be several. By establishing the members of your Troop, you might save yourself future disappointment by inappropriately turning to the wrong person for support. Please note, ***what is not being said***, is that anyone outside of your Troop is not worth relating to or having a friendship with. ***What is being said***, is that the Troop is where you can turn to, should your Chimp become unsettled and not emotionally stabilised by your Stone of Life.

Exercise: *A check on the Troop*

Take time to check that you have the right people in your Troop. Is someone missing or is there someone in there who probably shouldn't be? Consider the Human and Chimp selection criteria for selecting their Troop members. Check the people closest to you to see how they compare with these lists. Has your Chimp chosen someone for the wrong reasons or missed someone?

Focus 2: *Maintaining your Troop*

Maintaining your Troop relationships is very important. It is self-evident that, the more we invest in any relationship the more that relationship is likely to build.

Exercise: *Strengthen your friendship*

Having established your Troop, try to be inventive and think of ways that you could strengthen your friendship with members of your Troop. Some questions have already been offered in this unit, here are some more suggestions:

- Make time to ***listen*** to your Troop member. Listening strengthens any relationship.
- Involve yourself in a joint experience, where you can get to know them better.
- Reaffirm their values by discussing various topics.
- Ask about experiences that have made an impact on them, in order to appreciate what has helped make them into the person that they are today

Once you have plans to strengthen your friendships, put them into action!

Unit 18
Managing stress

STAGE 6: *addresses stress management, how to create an optimal environment and reviews recuperation.*

Unit 18: *is about managing stress. By understanding your mind's natural reaction to stress, you will be able to engage this and stop stress before it builds.*

The stress reaction in three stages

The body and mind are constantly trying to keep us safe and in a steady state. [310] If the body becomes physically stressed, such as becoming dehydrated, then the body recognises this, and we become thirsty and search out fluid. Once we drink, the body returns to a steady state. [311]

Similarly, the mind has a system to return us to a steady mental state. [312] [313] If the mind becomes aware that we are not in a good place, then the mind sends out an alarm and demands that we act to put things right. If we act and put things right, the mind will return to its steady state. [314] [315]

The stress reaction can be considered as having three separate stages to it:

1. The alerting stage
2. The resilience stage
3. The stress stage

The alerting stage

The first thing the mind does is to alert us that something needs addressing. It does this by releasing a lot of hormones and transmitters (chemical messengers) throughout our body and brain. [316] These substances usually make us feel uncomfortable and alert us to take action. The alerting stage is therefore an alarm system within us.

The resilience stage

This alerting stage rapidly transforms into a resilience stage. The mind moves into the resilience stage by releasing resilience hormones. This is our opportunity to act. At this point, our mind is activated to take on whatever is challenging us.

If during the resilience stage we manage to deal with whatever has alerted us, then we will return to a steady state of mind. However, if we do not manage to deal with the situation satisfactorily, then the resilience stage will transform into the stress stage. [317] [318] [319] [320]

The stress stage

The stress stage occurs because we have been unable to use the resilience stage to manage the situation. Stress hormones have overtaken and blocked the resilience hormones. These stress hormones now create stressful feelings within us. The stress hormones can have various effects on us, such as: creating anxiety and worry, causing physical symptoms and making us feel exhausted. [321]

If we fail to address the stress stage, then this temporary stress will change into a more permanent long-term stress. Long-term stress, also called chronic stress, is very damaging to our health and well-being. [322] [323]

The stress reaction
- Scientific points

In the alerting stage of the stress reaction, the Hypothalamus releases as many as 40 different hormones. These hormones begin a cascade of reactions. The main two significant hormones that result from this are cortisol and noradrenaline. Cortisol and noradrenaline initially alert us and we experience unpleasant stress symptoms, such as a dry mouth, a racing heart and general unease. As cortisol builds, we experience these symptoms more and more.

The body now moves into the resilience stage. Cortisol is formed in the adrenal cortex of the adrenal gland. This is a small structure just above our kidneys. This same structure now comes to our rescue and releases a different hormone: dehydroepiandrosterone (DHEA). This hormone reduces the cortisol levels and gives us chance to sort out the cause of the stress. At this point, if we recognise that we have control of the stress then we can act. [324]

However, if we don't sort out the stress then the level of cortisol overpowers the DHEA and we return to the stressful state again.

In our model, this equates to the Chimp alerting the Human and the Human having chance to sort the problem out. If the Human doesn't sort the problem out, then the Chimp will overwhelm the Human and remain in a stressed state. [325]

Managing the stress reaction

There are significant points during the stress reaction, where we can intervene or promote behaviours that will help us to return to a steady state of mind. These interventions can help us to avoid reaching the stress stage.

Intervention points

Alerting stage		• Recognise stress potential • Spot the warning symptoms • Don't engage with the symptoms • Actively move into the resilience stage
Resilience stage		• Recognise the need for action • Clarify where the stress is coming from • Form a plan • Revise or refine the plan if it isn't working

Before I go into details, I would like to give a gentle reminder! We all know that most things in life need some dedicated time, effort and commitment to make them work. Managing stress is definitely in this category. Therefore, I recommend that you consider reading the following passages as *a working blueprint to put into practice* and not just as an academic exercise. The ideas and information that follow are to offer suggestions for your plans, but I am sure that you will be able to add some of your own ideas.

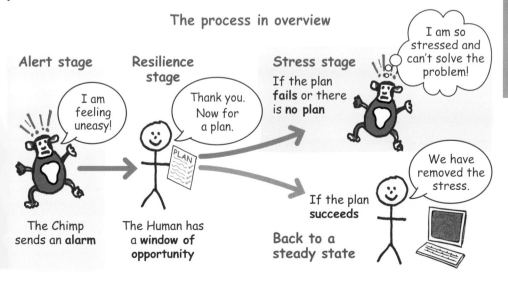

The process in overview

The stress reaction: *Stage 1* - *The alerting stage*

Recognising an alert

Not every alerting reaction that we experience needs addressing. Some situations pass quickly. Other situations can result in a stress state if we don't recognise the possibility that things can get worse, unless we act. Therefore, it is important to spot the first stage that could lead on to stress if we don't address any alert as it happens.

For example, if someone makes a rude remark about you or causes offence, you are likely to feel some kind of reaction. If you don't address this reaction in a constructive way, it is possible that future interactions with the same person could be jeopardised and lead to feelings of stress. If at the time you address the rude remark or offensive comment in a helpful manner, then future interactions with the person could be positive and stress-free.

Example: *Harriet and the error*

Harriet works as a shop assistant. A customer arrived and bought some goods. Harriet accepted the payment but made an error in recording the sale. This error resulted in significant work for Harriet's colleagues.

In this example, Harriet will have had an alerting stage where she realised that she had made the error. If she recognises this stage, then she can use the uncomfortable feelings to prompt a helpful response. If she doesn't formulate a plan to remove the feelings, but just allows them to subside, she might be storing up future stress.

For example, her Chimp might interpret the mistake to mean that she is not competent. The Chimp could then put into the Computer an unhelpful Gremlin stating: "I am incompetent at my job". Therefore, Harriet needs to address the mistake. Otherwise, it is likely that this Gremlin will appear in the future, whenever she is asked to do something where there is room for error. This in turn could lead to low self-esteem or a lack of confidence.

All of this could happen if Harriet doesn't stop and move into the resilience stage, with a plan of action, when the first alerting stage is experienced. Her plan could be to immediately address what it means to make an error and how this doesn't imply anything, other than an error had been made.

A platform for stress!

Managing stress

The alerting stage is therefore a wake-up call to move into the resilience stage and form a constructive and effective plan of action. By forming a plan, we can stop the stress process.

Symptoms of the alerting stage

The symptoms experienced will depend on how threatened your Chimp feels. These symptoms are the same symptoms that you could experience if you end up in the stressed stage. However, in the alerting stage, they will be helpful to you because they are temporary and can guide you into action. Typical symptoms could include: [326]

- Upset
- Unease
- Being on edge
- Anxiety
- Worry
- Feelings of panic
- Fear
- Increased heart rate
- Irritability
- Frustration
- Anger

> **Key Point**
>
> *As soon as you recognise the alerting stage is present, don't engage with emotions but move into the resilience phase and form a plan.*

As you can see from the list, your mind can alert you in numerous ways. It doesn't matter which of the emotions your mind sends you, the message is always the same. That message is: move into the resilience stage and *form a plan*.

What happens if you don't acknowledge the symptoms or if you ignore their message and allow your emotions to take over?

Not acknowledging the alerts given in the first stage means that you will enter the resilience stage without employing the resilience hormones that nature is sending you. It is then more likely that the symptoms will take over, allowing them to upset you, and you will miss the chance to act on the message that they are conveying. [327] [318] [328]

Not acknowledging the alert stage

"I'm feeling uneasy!"

"Maybe if I ignore my Chimp, it will settle down."

A danger of missing an alert → **Stress!**

The stress reaction: *Stage 2 - the resilience stage*

After the mind and body has alerted us to the fact that something is not going as desired, we enter a stage where our mind tries to help us. The mind does this by converting the negative symptoms into energy for action. We release a hormone that heads up the resilience phase. This is a window of opportunity to form a plan and ensure that we don't enter the third stage of the stress reaction. If we form an effective plan, then we will revert to a relaxed state of mind. (329) (330) (327)

How do we form a plan?

The first step is to clarify exactly where the alert is coming from. What exactly is behind the alarm that something is wrong?

The cause of the alerting phase is sometimes obvious but sometimes it is disguised and can easily be missed. Our plans can then be ineffective.

Example: *Sharon and the hidden Gremlins*

Sharon works as a teacher in a primary school. External authorities are assessing the school and everyone can see that the headteacher is feeling under pressure. The headteacher has visited Sharon's class unannounced and pointed out several concerns about Sharon's running of the class. Sharon was not given chance to respond and the headteacher left.

The real problem here is unlikely to be about Sharon's teaching; it is to do with the headteacher. The headteacher is likely to continue to criticise Sharon, whatever Sharon does. The problem is to address the headteacher's stress. However, if Sharon doesn't recognise the real problem, she might form an ineffective plan that could lead to further stress. For example, her plan could be to question her own ability to teach. In assessing herself, she might turn to her Computer to check on her beliefs. If the Computer has a number of Gremlins in it, then these can cause her to behave in an unhelpful way. Here are some possible Gremlins that Sharon could have:

- A belief that she is a fraud and is being found out (imposter syndrome)
- A belief that she cannot implement assertiveness
- A fear, that if she speaks up, there will be repercussions that she cannot deal with
- A belief that nothing she does will change things
- A belief that the headteacher is personally attacking her

This example is to demonstrate that sometimes our plans are based on *a wrong assumption of what the cause of the alerting phase is*. If we then make incorrect assumptions, they can be strengthened by hidden Gremlins. If we don't find the real cause of the stress and don't *also address the Gremlins*, we will continue with these beliefs and enter the stress stage.

Our plan of action must be effective in addressing the real cause of stress if we are to return to a peaceful state of mind.

It's important to find the real cause of stress

Once we have clarified the underlying reason that we are being alerted, we need to have a plan that will solve the problem that has arisen. What we don't want is a plan that merely avoids the problem or creates a problem of its own.

Example: *Ayaan and his Chimp's plan*

Ayaan has three children in their teens. He gets along well with two of them but has difficulties with the third. This daughter, Emma, appears to challenge everything he says and is very moody. He recognises that this is alerting him, and he has entered the resilience stage.

Sadly, his Chimp has taken the lead in forming a plan of action. Please remember the Chimp doesn't look for solutions, it just tries to remove the problem or uses avoidance to escape the situation. Ayaan's Chimp has decided that the best plan is for him to avoid his daughter and to ask her mother to deal with any difficult situations. Although this might remove him from the situation, it is not solving the underlying problem. It is also laying down foundations for further problems.

He is not forming a healthy relationship with his daughter. Additionally, she is not learning to interact appropriately with others or to take responsibility for her own moods. A true solution would be for Ayaan to form a plan of how he can communicate in a constructive way with his daughter and help her to learn to manage her moods. His Chimp's plan is very likely to fail and result in a stress stage setting in.

> **Key Point**
>
> *The plan formed during the resilience stage must be effective and result in resolution of the cause of the alert; otherwise we will enter the stress stage.*

Working with the Human during the resilience stage

In order to return to a steady state of mind, the Human needs to take over and form the plans during the resilience opportunity. [269] [331]

The Human can use the Stone of Life to begin the stabilising of the mind. The three key elements that the Stone of Life will bring are:

1. Accepting and working with the facts and the reality of the situation
2. Gaining strength by working with your values
3. Bringing perspective to the situation

The plan will first establish the facts and accept what can and can't be done. The Human will look inwardly to see if your conscience is clear and that you have done the right thing according to your values. Then the Human will look at the bigger picture to bring perspective.

Planning the resilience stage

Example: *George and the criticism*

George has been unfairly criticised by a few people on social media. The comments made were untrue and unkind. He feels upset and distressed by the situation. His Chimp has gone into a bad place and is unsure what to do. It wants to fight back or run away but feels it can do neither.

At this point, George can clearly see that he has entered the alerting stage. The resilience stage is emerging as George's Chimp tries to find a plan of action, which would be to just avoid the problem but not find a solution. It is at this point that George needs to move into Human mode and begin to construct a plan of action.

Unfair criticism has moved the Chimp into the alert stage

First, he gathers the facts. Here is what George believes to be some of the facts:

- Some people can be unkind, and I can't change that
- I have a choice about whether I want to listen to them
- I can't stop people believing what they want to believe
- If my Chimp won't settle, then I can put out a statement of truth
- It isn't worth trying to get unkind and unreasonable people to change their minds because they won't
- This untruth might stay forever in some people's minds
- The people who love me know the truth
- It is my choice where to put my focus and thoughts

George now adds perspective:

- Most people will forget very quickly what was said
- Most people don't care
- In a few weeks' time everything will have moved on
- Looking at my whole life, this is trivial
- Everybody suffers unfair criticism at some point in their life

Finally, George looks inwardly to his values:

- I know the truth
- I have lived by my values
- I have done the right thing
- I will hold my head up

What if George hasn't done the right thing?

What if George looks inwardly and finds that he has done something wrong and some of the comments made were true? Then George can still move into Human mode to make the plan to return into a steady state of mind. This time he might have the following, as examples for the plan:

Facts:
- I have done what I have done
- I can't change the past, but I can change the future
- I can apologise and make amends
- I can try to put things right
- I can accept responsibility

Perspective:
- Everyone makes mistakes
- Everyone does something that they wish they hadn't done
- Mistakes can be amended or compensated for
- This isn't my first mistake and it won't be my last
- If it isn't too serious then get a sense of humour and laugh at yourself

Values:
- I can accept responsibility and be honest
- I can say sorry
- I can try to put things right
- If I can forgive others, then I need to forgive myself

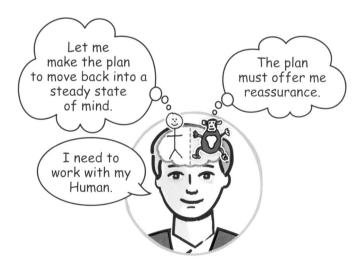

What happens if my plan doesn't return me to peace of mind?

Sometimes, we have to refine our plan or try with a second plan. In the example above, George could find that his plans seem to fail because he hasn't allowed some time for his Chimp to get over the situation. The Chimp might need a few days or weeks before it can move on. If George can accept this, then he can rewrite his plan to include some time for his Chimp to express its grievances before he brings himself into reality.

As long as there is a plan in place that offers a way forward, there will be energy. Energy diminishes when a plan fails and there seems to be no way forward. [327]

A constructive plan might need refining to maintain resilience

> **Key Point**
>
> As long as we have a *constructive* and potentially effective plan in place then resilience will continue.

Stress reaction: *Stage 3 – The stress stage*

If we don't manage to form and execute an effective plan that will take us back to a steady state of mind, then we will move into the stress stage. The transition into the stress stage is seen in the brain by the resilience hormone being suppressed by the stress hormone. [328] [332] [333]

Sadly, what often happens during the resilience stage is that people waste a lot of resilience hormone energy. They do this by allowing their Chimp to complain and worry instead of acting. They then enter the stress stage with what is termed an acute stress reaction, which is a state of stress where damaging hormones are being released. They experience very similar symptoms to those experienced in the alerting stage. These symptoms are usually accompanied by lots of concerns and worries with feelings of being trapped.

Energy is wasted on the Chimp complaining and worrying, instead of using the energy for your Human to act.

Acute stress will be damaging to the body and mind. It typically causes such things as:

- Anxiety
- Panic attacks
- A sense of unease and restlessness
- Disrupted sleep
- Bingeing or comfort eating
- Loss or gain in weight
- Fatigue
- A disrupted social life
- Compromised relationships
- Agitation or irritation
- Tearfulness
- Loss of confidence
- Ruminating about the same things over and over

The list offers some examples and is extensive. Although it might seem strange, many people fail to recognise that they are stressed and begin to accept a whole host of symptoms as part of a normal life for them.

The acute stage of stress has two progressions:
- Moving back to the resilience stage
- Moving into a chronic stress state

How to return to the resilience stage

Moving back to the resilience stage can be accomplished by using the following steps:
1. Recognise that you are under stress and work out exactly what the cause is
2. Accept that there is a problem to address and that it needs addressing
3. Accept that there is a way forward, it just needs some time to work this out
4. Acknowledge that the way forward might include having to accept:
   - Something that is not ideal
   - A compromise
   - Something that can't be changed

It is pertinent to see that the way out of stress involves a lot of acceptance and working with reality and truths. In other words, moving from Chimp to Human.

Once these steps have been followed, you will automatically begin to form a plan for moving forward and enter the resilience stage again.

Learning to use relaxation techniques will help to ease a lot of anxiety and is recommended. However, using relaxation techniques alone, without addressing the cause of the stress or anxiety won't get to the root of the problem.

Key Point

Using relaxation techniques can be very helpful, but they won't address the cause of the stress or anxiety.

What happens if I move into a chronic stress stage?

Chronic stress mostly occurs from two main causes:

1. That we can't see that we are stressed.
2. That we have come to believe that how we feel and the symptoms we experience, are normal. It's not unusual for someone who is in a state of chronic stress to start describing themselves in terms of their symptoms. For example, they might say, "I easily become irritable", "I know I am a worrier" or "I seem to be a control freak".

What they could be describing are symptoms of chronic stress and nothing to do with them as a person.

The chronic stress state is very likely to result in physical and psychological health problems [334] [323] such as:

- A lowering of your immune defences (resulting in colds and other infections)
- Headaches or migraines
- Loss of energy
- Poor sleep patterns
- Various states of anxiety or low mood

Chronic stress symptoms

- Loss of energy
- Irritability
- Poor sleep
- Worry
- Anxiety
- Headaches
- Controlling features

How do I remove chronic stress?

For any individual, chronic stress usually has a number of causes. Each cause must be found and addressed with a specific plan. There is a very important point to consider when removing chronic stress. You need to know how you aim to be, in order to know that you have been successful. In other words, first work out what you think the ideal situation will be and how you will be when the stress is gone. As you have been in a chronic stress state for some time, it can be difficult to appreciate what normality could feel like.

Chronic stress can create its own persistent habits, symptoms or consequences that might need addressing in their own right. These created Gremlins add to the stress.

For example, chronic stress often results in poor sleep patterns. [335] [336] Even after removing whatever it is that created the stress in the first place, you might find that you still have a sleep problem. Therefore, it can help to form a list of created problems that have resulted from chronic stress and address these problems individually.

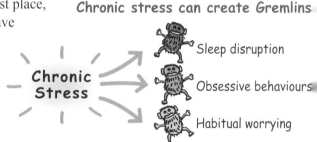

Unit 18 Reminders

- Neuroscientifically stress can be seen to form in three stages:
    1. The alerting stage
    2. The resilience stage
    3. The stress stage
- Each stage can be recognised, and we can learn to move between stages
- The key point in the resilience stage is to form an effective plan
- Chronic stress needs to be recognised before being managed

Unit 18 Exercises

Focus points and reflective exercises

1. **Managing the stress reaction**
2. **Checking on the real cause of stress**
3. **Checking for and removing chronic stress**

Focus 1: *Managing the stress reaction*

Forming a plan, the moment that you recognise that stress is appearing, is the most important part of managing the stress reaction.

Exercise: *Recognising and managing stress*

Try to develop a habit of forming an immediate plan that involves both the Human and Chimp. In other words:

1. Planning how your Human can go forward with a logical and factual approach
2. Planning how you can manage your Chimp to go forward by expressing feelings and emotions; *then looking forward to better feelings*, when resolution of the stress has happened.

Focus 2: *Checking on the real cause of stress*

The trigger incident that causes stress is commonly underpinned by Gremlins. Common underlying causes include:

- Lack of assertiveness
- Low self-esteem
- Unrealistic expectations of self

Exercise: *Questioning the cause of stress*

The next time your Chimp becomes stressed about a situation, ask yourself why the situation has occurred. A question that might help to find underlying causes could be:

"If someone else were in this situation and didn't get stressed, what feature do they possess that I don't have?"

If you can elicit this feature, then you will know where to put your focus and efforts.

Focus 3: *Checking for and removing chronic stress*

Chronic stress is usually based on unhelpful habits that we have allowed to develop. These Gremlins become the norm and then it is hard to detect them.

Example: *Shirley and the family*

Shirley has a family of three children and loves them. However, she has become exhausted looking after them and finds herself irritable and tired a lot of the time. She hasn't recognised that the situation is stressing her but has rationalised it away, by saying that raising three children is going to be tiring. On closer observation, she can see how she has made it a habit to do everything for her children. This Gremlin is backed up by another Gremlin that says to her "a good mother wouldn't complain about helping her children". These Gremlins are blinding Shirley to seeing that a good mother helps her children to take responsibility for their actions and to learn independence. This means they could begin to help out around the house and see that mothers and fathers are supportive but not door mats! Hopefully, this new role model would help them not to fall into a similar trap if they become parents.

Exercise: *Searching for long-standing Gremlins that create stress*

Averill has tried many diets over many years and can't seem to make any work. She knows what to do but always seems to go back to old ways and finds restricting her food really hard to do. This makes her feel bad about herself because she is overweight and has no willpower. This in turn makes her feel stressed.
What beliefs or behaviours could you help her to recognise and change?

Check yourself for any symptoms of long-term stress that are listed in this stage. Try and see where behaviours or beliefs have become unhelpful habits. Make plans to replace any Gremlins that you find with helpful Autopilots.

Unit 19
Preventing stress from occurring

Unit 19: *will look at how we can prevent stress from occurring and how to remove persistent stress. This unit will give you some practical ways to get you into a good place and feeling stress-free.*

Prevention of stress

Preventing stress from occurring would be the ideal situation. We can consider two specific strategies that could help:

1. Putting our basic needs in place
2. Finding personal trigger points

1. Putting our basic needs in place

If we put in place the basic needs of both Human and Chimp, then we are less likely to become stressed. We will call this our 'basic needs list'. By having a list, you will know where to direct your attention and energy.

Reducing the chance of stress occurring

The basic needs list will have:
- Your Human's basic needs
- Your Chimp's basic needs

We are all unique, therefore, only you can decide what these needs are, but I will offer some ideas. The needs of the Chimp will generally be aimed at areas such as: achievements, self-image, recognition by others and security. The needs of the Human will generally be aimed at areas such as: peace of mind, self-esteem, purpose and satisfaction. Some items could be on both the Human's and Chimp's lists, but probably for very different reasons.

Suggestions for your Chimp's 'basic needs list'

Achievement:
- Undertaking a project – e.g. creating a garden
- Gaining a degree or other qualification
- Reaching a high level on a computer game
- Reaching a high level in a sport

Esteem and recognition:
- Feeling that the achievements give status
- Gaining accolade from others
- Having a role with a job title
- Gaining acknowledgement for special talents, gifts or expertise
- Being able to demonstrate knowledge or skills

Security:
- Creating a place or space that the Chimp can call its own
- Having a well-defined role
- Maintaining a support network of people
- Having some routines in the week
- Having specified areas that the Chimp has control over, with decision-making power

Suggestions for your Human's 'basic needs list'

Self:
- Possessing a good self-image – recognising aspects of yourself that make you feel good about yourself
- Demonstrating self-respect – treating yourself supportively and behaving according to your values
- Having self-esteem – recognising that you are equal to others
- Self-worth – knowing that you bring value to others by virtue of just being who you are
- Having a great relationship with yourself

Purpose:
- Having a constructive and productive role in society
- Knowing your role in contributing towards other's happiness
- Developing yourself everyday

Values:
- Living out your values
- Satisfaction by knowing that you have given the best you can
- Being selfless and altruistic

It is important to recognise that in the Human's basic needs list, most of the items relating to self are *ALREADY FULFILLED*. These are the attributes you already have, if your Chimp or Computer is not interfering. Therefore, this part of the list is just to remind yourself of who you already are.

If we have all of these items on the basic needs list in place, then we are likely to have a sound basis for life. This will help in times of pressure.

2. Finding personal trigger points

Trigger points are experiences, situations or even people that cause us stress. We all have specific trigger points and generally we know what they are. Therefore, it would be wise to identify our personal trigger points and have a plan ready to prevent our mind from progressing into a stress state. [337] [338]

Example: *Harrison being overwhelmed*

Harrison works for a company answering emails. Most days he can't keep up. When he gets home he feels restless, and his mind races with concerns that he will forget to follow up some of the emails. When Harrison does catch up, he is fully aware that he will soon have a backlog of emails once again. How can Harrison manage the stress that he is feeling?

First, he can recognise his trigger points for potential stress. These are not being able to:

- Get through his workload and answer all the emails
- Remember to follow up all emails
- Switch off from work when he gets home

The most important point here is for Harrison to move from Chimp mode. He will then have entered the resilience stage of the stress reaction.

In Human mode he can:
- Establish the facts
- Accept and work with the reality of the situation
- Be logical in his approach to solving his problems

Here are some obvious facts:
- This type of work will always have times when you cannot keep up during busy periods
- Nobody can keep up if the workload is impossible to manage
- You can only work to your best
- Allowing the Chimp to think about work, when you don't want to, isn't acceptable
- There are solutions but these might not be perfect

If Harrison can accept these facts, as the basis to start from, then he can move on to finding practical solutions. Just by establishing and accepting the facts of the situation, he might immediately find a sense of relief.

Moving from Chimp mode to Human mode

There are numerous solutions that could be applied, such as speaking to his line manager to bring to their attention the impossible demands of the job. Alternatively, if he gets no support, he could change jobs. However, let us assume that he wants or has to continue with the job, and nobody is complaining about his work efficiency.

In this case, he will need to accept some facts about his work, such as:
- His work is necessarily like being on a treadmill, which will keep moving along and he needs to decide when it is time to step off and take a break
- As long as he stays in this job, the treadmill will always be there
- He has to change his beliefs about how he feels he is doing
- He will need to get work out of his mind preferably before he arrives home

Let's consider the last point: getting things off his mind. Some practical points will help. For example:

- Writing lists and memos to avoid forgetting important things is a very good way to take pressure off yourself and your own memory. If Harrison writes out his concerns at the end of his working day, then he can effectively leave his work behind him.
- Priority lists can also help. Sometimes it is helpful to address your tasks in an order that will give you peace of mind at the end of the day, rather than the natural order you need to get things done.
- If he gets home and he still can't switch off, he could have *a dedicated time to unwind* by reassuring his Chimp that he has written a list of everything he will need to remember. This will be at work waiting for him when he returns.
- Unwinding means different things to different people: music, quiet time, talking with a friend or going online. Only you can work out the best way for you to unwind.

What if his Chimp still breaks through during the down time? It is important not to engage with any unwelcome emotion or thought from the Chimp. You can reject it, with an explanation to the Chimp as to why the emotion or thought is unhelpful. The emotion or thought then needs replacing with whatever you want to replace it with.

Key Point

Just because your Chimp offers you an emotion doesn't mean that you have to engage with it. You can refuse to engage and choose your own emotion. It is a skill to do this.

Example: *Charity and her trigger point of personal criticism*

Charity knows that whenever someone criticises her, her Chimp will take it very personally and become upset. This usually results in stress. If Charity knows this, then she can prepare for her Chimp's usual reaction to any form of perceived criticism. She can follow the steps that Harrison followed in the previous example.

If she moves into Human mode, she can prepare her Computer to respond as soon as her Chimp hears an unfavourable comment about her. The Computer needs to be programmed with facts, such as:
- Everybody receives criticism at some point
- Not all criticism is correct – we can reject some criticisms
- If a criticism is valid then it means we have a learning point
- Some people are critical by their nature and we can't change that
- The only comments that count are those from people who care about us

Charity can then go further by asking how would she like to see herself respond if a criticism is made? For example, she might choose any of the following:
- Ignore the comment but let the person know that they are welcome to their opinion
- Remain calm and let the comment go without responding
- Be assertive and explain that the criticism is unhelpful
- Talk to a friend to gain some support and express her feelings
- Thank the person and let them know that you respect their opinion

What Charity and Harrison are doing is preparing for the moment that their Chimp will try to react. They are both programming the Computer with Autopilots that resonate with themselves. Only they can do this, and it will help to prevent the stress reaction from occurring.

Important point

Often, when friends state the obvious or offer some common sense, it leaves us wondering why we didn't apply these things in the first place. We can all operate with our Chimp system unless it is challenged. The good news is that we can apply facts and common sense to our own stress state if we move into Human mode and programme our Computer.

Thinking ahead – *Priming the Computer*

Our Chimps cannot think ahead with rational plans, therefore, we need to take the initiative and think through our day. We are then much more likely to respond rationally to things we meet. What we are doing when we take time to think ahead, is priming the Computer so that it is ready to intervene when it recognises situations that it has been prepared for.

Example: *Barry and the deliveries*

Barry works for a courier company. He delivers stock to offices where he knows there are some difficult customers. They often make his delivery unpleasant and drawn out. He also knows that he has deadlines to meet. By preparing himself for his day ahead, he can imagine each drop he has to make and the kind of problems he might face. He can then plan to manage his own Chimp and behave as he wishes to, which is in a cooperative and understanding way. He is aware that his Chimp is intolerant of any hold ups and can lose perspective. To pre-empt this, he can visualise each drop by seeing himself in his van, just prior to entering the offices, letting his Chimp know that he doesn't want any emotional outbursts. By working with his Chimp, he is much more likely to have a relaxed and friendly day.

Why might stress persist?

What if you feel that you have done all of these things and still you seem to be stressed, what then? There are a number of reasons why stress appears to persist. I will go through four of the commonest reasons:

1. Addressing the wrong concern
2. Emotional scars
3. Tiredness and exhaustion
4. Learned behaviour

1. Addressing the wrong concern

Sometimes people experience stress because there are hidden Gremlins that are prodding the Chimp and creating chronic stress. If the person tries to address only the symptoms of stress but doesn't address the Gremlins, then the stress continues.

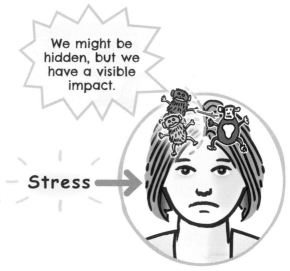

Hidden Gremlins can stress the Chimp

Example: *Joyce and reporting back*

Joyce can't understand why she becomes stressed each week when she reports to her line manager. The manager is very helpful and encourages Joyce, so there seems to be no reason why Joyce should continue to become stressed.

Joyce understands that it is just her Chimp that is doing this. Each week she tries to reassure the Chimp with facts and evidence that there really is nothing to worry about, but to no avail. She tries to reason that her Chimp is irrational and therefore it will do bizarre things that often don't make sense. The stress continues and each week she dreads the meeting. It is now causing her to become unwell. There could be many reasons why this is happening, and Joyce must explore her mind to find the cause. She might need some expert help!

In the end, she finally discovers a Gremlin that is prodding her Chimp. The Gremlin is a belief that should a criticism occur, she will not be able to cope with it, she will lose her job and there will be dire financial consequences. A simple belief, such as this, can keep niggling away at us but be hidden within the mind. Once it is addressed and replaced with an Autopilot, the stress will cease.

2. Emotional scars

We have looked at emotional scars in earlier units. Emotional scars are experiences that we have had, but not fully processed and they come back to haunt us regularly. If we have not constructed *an effective plan* to manage the scars when they reappear, we enter a state of stress. The fear of the scar reappearing is often worse than the actual experience.

Example: *Donna and her father*

Donna had a great relationship with her father until she was 15. At that point, she found out that he had been cheating on her mother and he left the family home to be with the other woman. It has been ten years since the incident and Donna feels that although she has forgiven her father, she still struggles to get over what happened. Donna saw the hurt that her mother went through and felt disillusioned with her father. Whenever she meets her father, she still re-lives the day that he left, feels anger and replays it in her mind. No matter how many times she tries to process the memory of the day he left, she can't do it.

The way forward for Donna is to accept that this is not something that she will get over. Instead, it is something that she will have to come to terms with and work with. If Donna can learn how to manage the emotional scar and accept it for what it is, then her stress is likely to diminish and might disappear. Not accepting that some scars do not go away and can't be fully processed will lead to persistent stress.

Consequence and choices when dealing with emotional scars

3. Tiredness and exhaustion

Experiencing a poor night's sleep is common and we are sometimes tired the next day. Tired Chimps go into action, usually demonstrating irritability and a lack of tolerance. Our Chimps easily become stressed. Clearly, if we are constantly tired or exhausted, we are likely to experience stress, yet the importance of this is generally not appreciated.

If we do not look after our physical bodies, with adequate rest and recuperation, then our mental state is at a much higher risk of becoming stressed. When this happens, persistent stress is a symptom of tiredness, rather than a reaction to any particular situation. The solution is to stop and be proactive in getting the rest that you need. We will look at managing sleep in a later unit.

4. Learned behaviour

Our Chimps are constantly on the lookout for danger, which can turn into an unhelpful habit of looking for something to worry about. This causes constant stress. This habit of finding something to worry about, I call the Mushroom syndrome.

The Mushroom Syndrome

Mushrooms always compete for space and when you pick one another replaces it. If you replace the word 'mushroom' with the word 'worry' then you have the 'Mushroom Syndrome'. Some people grow worries. They are allowing their Chimp to constantly find a new worry to focus on. This habit can be broken once you accept that your Chimp will always find something to worry about, ***if you allow it to***. We can stop this by introducing Autopilots into the Computer. These Autopilots could include:

- Everybody's Chimp will find something to worry about, if they allow it to
- It might be natural for a Chimp to worry, but it's unhelpful
- I can distract my Chimp and remove the focus of worry
- I have a choice on what I focus on; worry doesn't help, but plans do

Settle your Chimp when it cultivates worries

Happiness and contentment to ease or remove stress

We discussed creating 'happiness lists' and a 'peace of mind list' in unit 15. These can clearly be a great help in preventing, easing or removing stress. Don't forget we need to be proactive if the lists are to work!

If you can't manage stress then it would be wise to contact a professional for help.

Unit 19 Reminders

- Putting your basic needs list into action can reduce stress
- Having a plan to manage personal trigger points reduces stress
- It is a skill to choose your emotions
- Stress that persists needs the cause addressing

Unit 19 Exercises

Focus points and reflective exercises

1. *Preventing stress – the 'basics needs lists'*
2. *Personal trigger points addressed*

Focus 1: *Preventing stress - the 'basic needs list'*

"Prevention is better than cure", is a well-known phrase. How can you prevent stress from occurring in the first place?

Exercise: *Forming your basic needs lists*

Look back through the section notes on forming the basic needs list and form your own unique lists. Once you've formed them, ensure that you make it a habit to spend part of your day working from the lists.

If you implement your basic needs regularly, you will find that it becomes a habit for the Computer to keep reminding you to look after yourself.

Focus 2: *Personal trigger points addressed*

Exercise: *Preparing for the day ahead*

Try spending five or ten minutes each morning reflecting on the day ahead of you.

- Before you go through the day, decide what state of mind you would like to end your day in. You will then have a goal to aim for throughout the day.
- Predictable things that you experience in your day-to-day life could have programmed responses put into your Computer by **bringing in perspective.**
- Go through your day and programme your Computer to act in the way you would like to act. If situations arise where the Chimp would otherwise have taken over, your Computer will intervene, and the result is likely to be a much less stressful day. Programming the Computer can be as simple as walking away, counting to ten or biting your lip. The important point is to have some helpful actions ready to follow. First, try this out in your imagination, and repeat it several times, so that your Computer is truly programmed to act.

When you feel you have perfected this, try it out for real. Try not to be disheartened if you don't manage to follow the Autopilot every time. What you are doing is creating a neurological pathway in your brain, and this takes several attempts to establish this. Just keep going!

Unit 20
Managing your environment and lifestyle

Unit 20: *will focus on creating your own supportive and nurturing world within the world. Environment can make a big difference to the way our Chimp feels.*

The world that you live in

We cannot control the world that we live in and this can be a source of frustration and distress to our Chimps. [339] [340] We inevitably have impositions and experiences forced upon us that we do not necessarily agree with. We also share the world with a lot of people who have different morals and attitudes. These people can be damaging through their interactions and comments towards us. How do we survive, what many people perceive as, a hostile and at times threatening world? One way that many people manage to feel more at ease is to distinguish between the external or outside world and the world that they have created. Most of what happens in the outside world we can do little about and it can have no consequences to us.

The world that we create for ourselves, is a much more suitable place to be. If we create our own world, within the world in which we live, and stay within this created world, then we are much more likely to have relaxed Chimps and feel we belong here. Creating your own world, means deciding on who and what is important to you, and then making sure you only work with this. It also means disregarding any events or comments and opinions from people that are not important to you.[341]

Example: *Katherine and her world*

Katherine used to perceive the world as a place that she had to live in and couldn't escape. It seemed harsh with people making constant verbal attacks, leaving her distressed.

Her new perception is now: the external world might be harsh, but if I stay within my world, then I am fine. I disregard events, comments and opinions from those who I do not allow into my world. I distance myself from the external world and view it as an onlooker.

A secure world for you and your Chimp within the real world.

In my world, I only allow my friends to enter and any event in my life is only as important as I allow it to be. It isn't easy to dismiss unkind comments; but I can, if I don't value them.

> **Key Point**
>
> *Something is only as important as you allow it to be.*

Human and Chimp views of the world

Even within our own world, our Human and Chimp see the world very differently. This accounts for the different approaches that they have. The Human might see a society, but the Chimp sees a jungle. This means the Chimp will try to live by the laws of the jungle. The laws of the jungle are certainly not civilised. Our Chimps will search for the right part of the jungle to live in. This would provide essential things to the Chimp such as, security, food and its Troop. Once found, this part of the jungle will have its boundaries determined.

Dominance and violence are two strong working characteristics of the Chimp within a jungle setting. If another unwelcome Chimp enters its boundary then there will be unease and usually trouble. We can see some Chimps using bullying, harassment and intimidation.

Example: *Maurice and his Chimp's boundaries*

Maurice works as part of a small team on a building site. He is the plumber in the team. He has realised that another member of the team has been helping with some plumbing without him being informed.

- **Chimp view:** the way the Chimp will see this is that his territory has been invaded and is under threat of being taken from him. He sees another Chimp challenging his role. The Chimp sees a jungle and will use aggressive and confrontational methods to put things right.
- **Human view:** Maurice's Human will see his team member as being inappropriate and it needs to be talked about and sorted out. This is because Maurice, when he works with his Human, sees a society and knows that society works with solutions and discussions.

> **Key Point**
>
> *In areas involving 'territory', it is always worth asking: are you are in Human mode and seeing a society or in Chimp mode and seeing a jungle?*

Boundary disputes occur at work and at home. Neighbour disputes are often about literal boundaries to their property. Many home boundaries and hedge disputes could be resolved if people worked with their Human and by society rules. If we view the world as a jungle, then we are more likely to move into Chimp mode and allow our Chimp to dictate our behaviour. By viewing the world as a society, albeit with a lot of people in Chimp mode, we can still remain in Human mode ourselves and work with society values.

The right part of the jungle! What can your Chimp tolerate?

As the Chimp sees the outside world as a jungle that it has to enter, it is important that we take our Chimp to live in the right part of the jungle. There are therefore two concepts at play:

1. We need to live in the right part of the jungle (outside world) that suits our Chimp
2. When we are in this part of the jungle, we need to remain within our own world

Therefore, we need to know where the right part of the jungle is. Chimps that are in the right part of the jungle are confident and happy.

The right part of the jungle for your Chimp includes:
- Your environment: home and work
- Your role: at work and in life
- Your relationships

Your environment and role in life are critical to your Chimp's happiness and health. An unhappy Chimp won't tolerate the wrong part of the jungle.

Example: *Jack and his road to freedom*

Jack worked for years in the centre of a big city. He had a good job and there were no specific problems. His Chimp was unsettled and didn't cope with this environment. Jack's logic, from his Human, was to justify staying in the city because it was a good job, with good money and the potential for further promotion. However, the Chimp doesn't work with reason, it works with feelings. When Jack finally gave way to his Chimp he found a more settled and fulfilling life in a rural setting. Happiness is primarily Chimp based: our Human predominantly brings logic to the table but it is the Chimp who brings feelings.

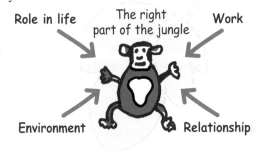

Your Chimp will be happy in the right part of the jungle

Sometimes we have to listen to our Chimps and not box them. There are some things that they cannot and will not tolerate. Sometimes our Chimps won't tolerate a particular job and fighting this will be a losing battle.

The Chimp will let us know which part of the jungle it wants to live in, and this includes its relationships and job; we just need to listen.

Managing behaviour and mood by using your environment

Our environment plays a big part in our health and wellbeing. Creating your own world is not just practical but therapeutic. For example, many people use music or aromas to help them to relax. [342] [343] [344] The setting in which we find ourselves can greatly influence our behaviour.

Nidotherapy
- Scientific points

Although Nidotherapy was introduced to help people with severe mental health problems, the principles can be applied to all of us. Simplified, Nidotherapy aims to help people to cope and to get the best out of themselves by assisting them to create their own supportive and safe setting. It is the setting that is 'treated' rather than the person.

If we place ourselves in an environment that suits us, then we are much more likely to feel empowered, be more relaxed and be more productive. Workplaces and home life have both been studied and demonstrate that the psychological effects of creating your own environment are beneficial to your health. [345] [346] [347] [348]

Several studies have shown that our environment can affect our decisions and mood. For example, restaurants have found that even the colour of the walls and the lighting will modify our behaviour. [349] [350]

Colour and sleep
- Scientific points

All of us respond to colour differently and can attach mood states to colour from experience. However, there are some colours that affect the majority of us in a specific way. The colour blue appears to wake us up whereas the colour red appears to make us drowsy. This might suggest that we are following the colour of the day, with morning and daytime wakening attached to blue skies and red colours attached to sunsets. Restaurants with blue colour schemes apparently sell more food than restaurants with red colour schemes. It's also been found that orange lighting for a few hours before sleep will help to put us into a more sleep-prone state than white lighting. [351] [352] [353]

Your colour schemes can affect not just mood but also levels of arousal.

The question you could ask yourself is: is my environment bringing out the best in me?

When planning your world, remember to be kind to your Chimp and make it a Chimp friendly world.

> **Key Point**
>
> *Create a Human-led environment but make it Chimp friendly.*

Looking after your intellect and your emotions are both needed to keep your Chimp happy. Your environment is not just about the physical setting but also about things such as the company you keep, the work that you do, how you spend your leisure time and the mental stimulation that you have.

Stimulating environments
- Scientific points

Enriched stimulating environments appear to be critical for developing and maintaining a healthy mind. Young children, around the age of four, who have an enriched environment with lots of stimuli, will have increased intellectual development compared to those young children who have a poor environment. [354] [355] [356] [287]

Even animals are psychologically in a better place if they have enriched environments. As we grow older, our mental health and intellect is maintained if we continue to stimulate our minds. Stimulation appears to slow down or ward off mental decline. [357] [358]

> **Key Point**
>
> *A significant part of getting the best out of yourself and becoming resilient involves being proactive in creating your own world.*

Unit 20 Reminders

- Creating your own world can have immense benefits to you
- Something is only as important as you allow it to be
- Seeing the world as a jungle will evoke Chimp activity in your mind
- Seeing the world as a society will evoke Human activity in your mind
- Create a Human-lead environment, but make it Chimp friendly
- Your environment is a significant factor for building resilience

Unit 20

Focus points and reflective exercises

1. **Learning to live within your own world**
2. **Seeing the world as a society, but with roaming Chimps**
3. **Establishing and implementing lifestyle choices**

Focus 1: *Learning to live within your own world*

Over many years of working with people to help them to cope with their emotions and feelings, one of the common factors that causes big problems is taking to heart the opinion of others.

By seeing yourself living in your own world, this problem can be removed. The key to this is to learn to decide on what is relevant and important to you and to dismiss irrelevant or unimportant opinions. It is helpful to replace unkind or critical remarks with kind remarks from friends or strangers, who are being supportive. We can never please everyone; therefore, learn the skill of being selective.

Exercise: *Learning to dismiss destructive or unwelcome opinions*

A starting point would be to first gather opinions that you do welcome and that are constructive and supportive to you. If you have these firmly within your mind, then when an unwelcome opinion comes along it will be much easier to dismiss it. Try and establish some Autopilots that will help settle your Chimp and help you to dismiss unimportant opinions. Here are some suggestions:

- There will always be unpleasant opinions from unpleasant people
- Unpleasant comments are exactly what they are and nothing more
- There will always be people supporting me
- As long as I live by my own values, then I will hold my head up
- I can't change the world, but I can change my world
- Verbal attacks on me are only attacks if I allow them to be important

Focus 2: *Seeing the world as a society, but with roaming Chimps*

Exercise: *Seeing the world as a society*

In order to help you to stay in Human mode try and visualise the world as a society rather than a jungle and go through a typical day that you might experience. As you visualise this, remember that you will meet others in Chimp mode. Try to see these people as roaming Chimps and manage your interactions with them appropriately. As always, begin by asking yourself how you want to behave in any stressful situations and then visualise yourself carrying out these actions.

Focus 3: *Establishing and implementing lifestyle choices*

When creating our own world, we have to think seriously about our choice of lifestyle.

The company we keep, the food we eat and the activities we choose are just part of the environment that we will live in. Therefore, if someone is serious about looking after themselves, each area that creates our environment should be given careful consideration. With good choices, our environment can put us into a good place.

Exercise: *Reviewing lifestyle choices*

Consider all the areas that can make up your world and view them as lifestyle choices. Think about what your needs are and what your Chimps needs are when it comes to how these contribute to the world that you have created.

If you are going to make changes, then small changes are usually the best ones to make because they often stick and reap the best rewards.

Here is a list for you to consider, you might want to add some areas of your own.

Areas that create your world:

- The company you keep
- How you use your leisure time
- Your nutrition
- Your exercise
- What mental stimulation you have in place
- What you do to relax
- Your role in life
- Your job
- Your retreat space

Unit 21
Recuperation

Unit 21: *addresses recuperation, which includes: time out, brief and long rest periods and sleep. We all know how important this is!*

Introduction to this unit

So far, we have specifically focussed on the functioning of the mind. Our physical health can affect the functioning of the mind. [359] [360] The commonest problem for the mind is not getting enough rest or sleep and tiredness can lead to underperformance or even to burnout. When we are tired, our Chimp will find hijacking us very easy to do and we can appear to be a very different person. This is because we can rapidly alternate between Human and Chimp. Therefore, I hope it won't take too much convincing to appreciate how important this unit could be.

Is your Chimp different when tired or rested?

To address the two areas of rest and sleep in detail involves looking at the science. If you would prefer, you can omit the science boxes and focus on the practical suggestions. Some of these practical suggestions might feel like common sense, but being reminded of them can help to ensure we actually put common sense into practice.

Recuperation

Recuperation is arguably the most important area to address. [361] [362]
It could be divided up into:

- Brief rest periods
- Sleep
- Longer rest periods
- Rehabilitation

Why are these areas so important?

There are numerous physical effects on our health when we become tired. [363] [364] [365] [366] Apart from the physical effects, when our Chimp takes over it can affect our perception and interpretation of what is happening. This in turn influences our decision-making and powers of judgement. [365] There is a lot at stake!

Most of us can relate to this when we have not had enough sleep. We might become easily irritated, lose focus or lose the patience to persevere with tasks.

Pause and think for a moment about how much energy and enthusiasm we naturally generate when we are physically rested and rejuvenated. It is common sense to make the effort to put ourselves into a rested state. [367]

Brief rest periods

Brief rest periods are when we take time out for a breather from what we are focussing on. Simple things, such as down time or rest can make a big difference to how the Chimp reacts. [368] [369] As with all aspects of mind management, it is a skill to recognise when your mind is telling you that it needs a short break. It's also a skill to know how much time out your unique mind needs. Brief rest periods during the day, such as coffee breaks or lunch breaks, allow your mind to reset. During this time, the mind benefits from stepping away from its focus and switching off from concentrating. [370] [371]

When relaxed, the mind will move into Human mode. It can then bring in perspective, which has a calming influence. Our ability to retain focus varies depending on the task that we are doing. However, whatever the task, we can always increase our focus regardless of how engrossed we are. Research shows that after a break, there is improved maintenance in executive functioning and decision-making, but without the break they would diminish. [372]

A short break can make all the difference!

Beware of too much coffee!

Example: *Rhonda and her coffee use*

Rhonda has reached her break and is drinking her fourth cup of coffee for the day. She appreciates that this caffeine boost gives her more alertness and the ability to think more clearly. [373] [374] However, she continues her coffee intake until she leaves for home at 5 p.m. What Rhonda claims is that the coffee keeps her focussed throughout the day and that as long as she stops at 5 p.m., she will not be kept awake that night with excess caffeine; but how true is this?

We are genetically programmed differently but for most people caffeine will have a half-life of eight hours. Therefore, eight hours after we first drink the coffee, we will still have half the level of caffeine in our system. There is evidence to show that caffeine can have a negative effect on sleep, if taken too late in the day. [375] Rhonda will have the effects of caffeine fairly strongly for eight hours. She is likely to still be feeling the effects of caffeine after she leaves work at 5 p.m. until 1 a.m. the next day! It's no wonder many caffeine drinks and foods can affect our sleep without us realising and that this can be why we can't fall asleep. Only you can find out what works for you.

Research also shows [376] [370] that breaks from work are more effective when they include a change of focus, such as:

- Moving away from your work or task location
- Taking walks
- Meditating
- Changing your focus to future events
- Taking in nature

If, like many people, you choose not to take these breaks, then consider the risks that could go with it. [377] [378] The risks can include:

- Less productivity
- Affecting others around you negatively
- Having a detrimental effect on your own well-being

Longer rest periods

Evenings, weekends and holidays are when we have the opportunity for longer periods of relaxation, assuming we work regular hours. The critical questions are:

- Is there evidence that taking longer time periods away from work actually helps our wellbeing?
- If so, what can we do during this time that will actually help?

Taking holidays or long weekends does help. [379] [380] [381] However, the extent of the benefit and how long it lasts depends on what we do on the holiday.

If you do take a break, then it should be a complete break from work. If you work during the holiday, then any potential gains could be lost. [382]

Taking a long weekend appears to give health benefits in terms of reduced stress and general well-being, provided you leave the environment of your home and go to a hotel. Staying at home doesn't appear to work as well. The positive effects of the time away from work last for between one and two months. [383] [384]

Longer vacations of a week or more help with general well-being providing the activities on the holiday are based on pleasurable things that result in good memories. [379] When on holiday, having constructive conversations with a partner or a close friend helps, as does talking about your experiences on your return.

Preparing yourself for leaving work by completing or delegating tasks before a holiday also helps, along with ensuring that you will return to a manageable workload.

Although these are common sense ways to promote health benefits when going on leave from work, we often neglect them. [385] In our model, what is happening during rest or holiday periods is that the Human and Chimp are processing and tidying up events in our mind. The Human is able to take over management of the brain and bring perspective and reality to bear on our past experiences and current position in life.

Sleep

Sleep is a necessary function of our machine (body and mind) that we can understand and work with. During sleep, the Chimp and Human communicate, and the Human is able to help the Chimp to process events. [386] [387] It is the reason why we sometimes sleep on a problem then see it very differently the next day.

What is the purpose of sleep?

Although there is much debate about the functions of sleep, it is generally agreed that the following are important aspects of sleep:

- Physical rest
- Maintenance of brain systems
- Memory consolidation and learning

The Human is able to help the Chimp to process events during sleep

Brain activity during sleep
- Scientific points

During sleep, the brain is very active and performs many tasks that appear to have two main functions. The first is general maintenance aspects, such as removal of waste by-products from the brain. [388] The second is improvement work that is centred on learning, and sorting and storing memories. [389] [390] [391] Our sleep patterns and quality of sleep also affect other systems, such as synchronising some hormonal systems and fine-tuning our immune system. It's not surprising then that prolonged mild sleep deprivation has been linked with impairment of our cognitive functioning and emotional instability. [389] [392] [393]

Sleep cycles
- Scientific points

During sleep we pass through different stages of sleep that keep repeating. These repeat cycles of sleep typically take about 90 minutes to complete. [394] The cycles take us into light sleep, then into deep sleep, then back again to light sleep and finally, a state called REM sleep. REM stands for Rapid Eye Movement. After REM sleep, we return to deep sleep and continue the cycle.

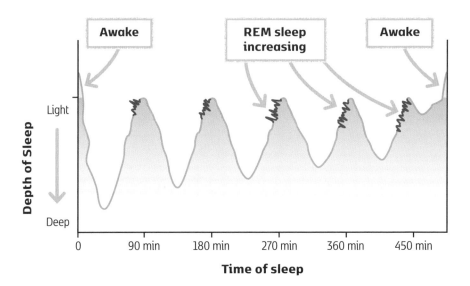

The cycles are repeated throughout our sleep, but deep sleep becomes less, and REM sleep becomes more, as we progress through each cycle.

What happens at each stage?

Drowsiness and light sleep: during this stage of sleep, we experience visual images crossing our mind. Initially our eyes are still moving but as we progress into sleep they cease to move, and our body enters a rest phase. We can easily be awoken during the early stages of sleep.

Deep sleep: brain wave patterns change, and it is difficult to rouse somebody when they enter the deep sleep stage. If we do rouse somebody, then they usually feel groggy and find it difficult to focus. During this stage there are no eye-movements and no muscle activity.

REM sleep: as we rise out of deep sleep and back to light sleep, we enter a very different kind of sleep state called REM sleep. During REM sleep everything changes! Our eyes move about rapidly, hence the name 'rapid eye movement' sleep. We breathe less deeply and more erratically. Our hearts beat faster, and our blood pressure rises. This is the time when we dream.

Are all the stages of sleep necessary?

The simple answer is yes. If we don't get enough deep sleep or REM sleep, then we can jeopardise our physical and mental health. If we don't get enough REM sleep then the brain will not wait 90 minutes before entering REM sleep, instead it can enter REM sleep as soon as we fall asleep.

Sleep cycles (continued)
- Scientific points
How much sleep do we need?

The first thing to address is how much do you need? Generally speaking, this varies throughout our lifetime. [395] Young children and teenagers need far more sleep than adults. Older people tend to have less sleep. The sleep cycle also alters during our lifetime. Teenager's brains move their bedtime and waking time back. Therefore, most teenagers will struggle to get to sleep before mid-night and struggle to wake before mid-morning. There is a debate about moving school hours to follow the science of sleep rather than impose on teenagers a sleep-wake regime that does not make sense scientifically. Older people's brains tend to move their sleep cycle forward. They retire to bed early, appear to sleep less and rise early. These findings are common but only you can decide what works for you and whether you are a night owl or a lark.

The amount of sleep varies from person to person and researchers debate the ideal amount of sleep required for an adult. 7 or 8 hours is often the quoted amount we need. However, there is great variance, and your own experience will dictate what you need.

How to approach sleep with a plan

We can form a plan for sleep by considering three significant biological systems that influence our sleep:
1. The light/dark system
2. The activating system
3. The tiredness system

> **Key Point**
>
> *When we have poor sleep patterns it is usually one or more of these three systems that we need to address.*

The light/dark system
- Scientific points

The light/dark system relies on our eyes detecting light and changes in it. When darkness falls, the pineal gland in our brain starts producing a substance called melatonin. Melatonin will help to send us to sleep.

When light hits the eyes, a message is sent to the brain advising it to stop producing melatonin, and we wake up.

The system is in reality much more complex because when light hits the retina it sets off a cascade of reactions. [396] First it activates an area of the brain called the suprachiasmatic nucleus (SCN). The SCN activates the paraventricular nucleus, which in turn activates an area called the T1T2 of the sympathetic nervous system. T1T2 then stops the superior cervical ganglion from working. If this ganglion is inactivated then the pineal gland can't produce melatonin and therefore we wake up! Let's just stick to knowing that we have a system that works with light and melatonin! [397]

Of interest, the SCN is also known as being the brain's clock. [398] This is important because it doesn't just work with light, **but it can also be set by habit**. Therefore, regular bedtimes will set the clock to go to sleep, with or without light.

The light/dark system (continued)
- Scientific points

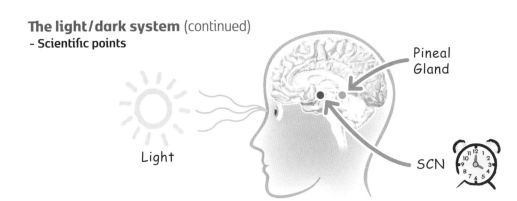

Practical points on how we can engage the light dark system to our advantage

- **Darkness** – We can simulate the hours of darkness by simply making sure that where we want to sleep is dark enough to allow melatonin to be produced. It helps to avoid short wavelength blue light from bright screens, such as smartphones and laptops, during the hour before sleep.
- **Early light** – Early morning bright light will wake our systems up. Interestingly, the retina at the back of the eye has rods and cones for vision but the light system does not use these. It uses different cells, which means that even those without sight or limited sight can still operate the light/dark system.
- **Regular sleep hours** – The biological clock within our brain will become set by sticking to regular sleeping times. If we vary our hours, particularly at weekends, then the clock will become confused and the sleep rhythm will be lost.
- **Beware catnaps!** – If your sleep is not good at night then catnapping could be the cause. The clock needs to be regulated and sleeping during the day can disrupt this. However, some people find a 'power nap' to be beneficial. Only you can work out what works for you.

The activating system
- Scientific points

This complex system within the brain is the main control for sleep.

The system is constantly monitoring sensory signals from our bodies and mind. It keeps us awake during the day and allows us to sleep at night, but only if it is not detecting a problem. It is called the reticular activating system and it will wake us up during sleep if it senses that something might be wrong. [399] For example, if we hear a noise, are too hot, or if our mind is troubled then it will wake us up. It's very important that we help this system to relax before we go to bed and whilst in bed. We can do this by addressing common things that could prevent us from getting to sleep or wake us up.

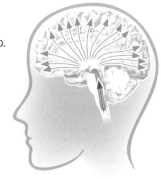

The Reticular Activating System

Practical points on how we can engage the activating system to our advantage:

Preparing for sleep

- **Restful lead in** – try to wind down for an hour or so before retiring. Too much exercise before bed, physically or mentally, can keep you awake.
- **Park your worries and any decision-making outside your head** – this is a lot easier said than done! So how could you do this? One method is to write down your thoughts and concerns. When you have done this, you need to remind your Chimp that you are now going to sleep. It needs to know that these things will be addressed in the morning and not before. Many people find this a useful practice.
- **Have a comfortable bed**
- **Keep still** – the activating system will be alerted if you keep changing position when you are in bed and trying to get to sleep. However, if you keep still, the mind loses the ability to appreciate where you are, and this will aid entry into sleep.
- **Active dreaming** – by this, I mean proactively thinking pleasant dream-like thoughts when you get into bed, rather than staying with reality.
- **Relaxation** – people use various methods and, as always, only you can work out what will work for you. Reading a book, a warm milky drink, listening to a relaxation tape, a hot bath and a gentle walk before retiring are all well recognised methods to aid sleep.

Total sensory deprivation
- Scientific points

If we deprive the brain of any form of outside stimulation, then we will enter into a deep relaxed state. Experiments show that when we subject people to sensory deprivation over a period of time, they can begin to hallucinate and experience anxiety. [400] [401] [402] However, if we subject ourselves to sensory deprivation, as we are falling asleep, this will help us to enter a state of relaxation, which will then aid sleep onset. Therefore, a silent darkened room is a real help to inducing sleep. [403] [404] [405]

Practical points to prevent or manage broken sleep because of the reticular activating system

Set the right temperature – being too hot or too cold in bed will cause the alerting system to wake you up. During REM sleep our bodies struggle to maintain a steady temperature. Therefore, if the temperature in the bedroom or bed is not suitable, we are likely to wake up.

Eliminate sound – if you know you are a light sleeper try to sound-proof your room as best you can or use ear plugs.

- **Relax if you do wake up during the night** – when we wake during the night, our Chimp is in full control. It is very likely that you will not be able to keep perspective on anything that you think about. This means worrying during the night could waste a lot of energy. The next day, when we are back into Human mode, we usually realise how unnecessary the worry was. Try to remind yourself that allowing your Chimp to think during the night is a waste of time! Broken sleep won't make much difference to your overall sleep requirements.
- **Avoid heavy meals** – or those that are more difficult to digest, such as cheese-based snacks. Difficulty with digesting food can result in restless sleep.

The tiredness system
- Scientific points

As we use energy in our body and mind, cells build up a product called adenosine. [406] This build-up of adenosine affects our mind and causes us to become tired. It does this by reducing the amount of a transmitter called acetylcholine in the anterior hypothalamus. This decreases our wakefulness. The build up of adenosine then pushes us into the deep non-dreaming sleep, in order to give us physical rest. During sleep, we clear this build-up of adenosine and so become refreshed and awake again. As we sleep, and the adenosine is removed, acetylcholine starts to build again, and this substance wakes us up. [407]

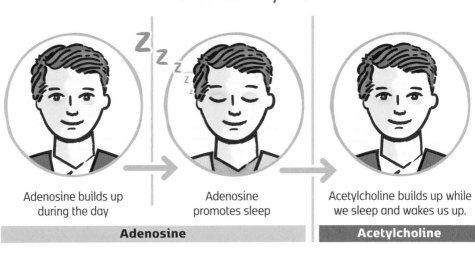

The tiredness system

Adenosine builds up during the day

Adenosine promotes sleep

Acetylcholine builds up while we sleep and wakes us up.

Adenosine | **Acetylcholine**

Practical points on how we can engage the tiredness system to our advantage

- **Avoid Caffeine** - if we block adenosine or promote more uptake of acetylcholine then we are likely to become alert and awake. [408] [409] This is exactly what caffeine does. Caffeine in coffee, tea or chocolate blocks the action of adenosine and promotes the action of acetylcholine, which keeps us awake. [410] Clearly, it's not a good idea to drink coffee before bed, even though a few people don't seem to be affected.
- **Regular hours** – if we keep regular sleep hours then the adenosine will not build up and we promote a sleep pattern that the adenosine follows. However, what happens to adenosine if we deprive ourselves of sleep and don't keep regular hours? The adenosine builds more receptors in our brain, so that over time we become chronically tired. [411] [408]
- **Don't engage emotions** – if you keep trying to get to sleep you are fighting this system. Counterintuitive as it might seem, don't try to get to sleep. The system will then relax and begin to put you to sleep.

A caution! – *Using alcohol to get to sleep*

Many people find having an alcoholic drink before bed helps them to sleep. It's not for me to judge, but rather to bring the facts to your attention. You can then decide. Alcohol will relax you and it can aid sleep initially. However, alcohol has three negative effects:

- It stops REM sleep
- It keeps us in the lighter stage of sleep
- It can wake us up during the night

If we don't get enough REM sleep then the next night, the mind tries to catch up and has more REM sleep. Catch-up REM sleep usually presents with vivid dreams and occasionally nightmares. The lack of depth of sleep leads to us feeling non-refreshed the next day.

The effects of the alcohol are short-lived and generally cause us to wake during the night and have disrupted sleep. [412] Consequently, alcohol can become the cause of insomnia. There is also the concern that for some, the use of alcohol can lead to a dependency problem.

Alcohol stops REM sleep

What about our Chimp and sleep?

The Chimp experiences sleep very differently to the Human. During sleep the Human rests and recuperates, whereas the Chimp becomes active and sorts itself out. [386] [413] [414] The problem is that the Chimp does not have very much influence from the Human or Computer once we begin sleeping. Some of the time during sleep, there is a lot of communication between the three systems, but this is more like dialogue than a struggle for control. When we put our heads down, the Chimp can come to life!

The Chimp takes over during sleep

There are two particular ways in which the Chimp can cause us sleep problems:
- The Chimp being unrealistic about sleep
- The Chimp exercising during sleep

Being unrealistic

To expect **consistent** sleeping hours is completely unrealistic. Yet many Chimps constantly expect to have a regular and reliable sleep pattern with unbroken hours. This expectation then becomes the source of stress.

When we struggle to get to sleep, or we have broken or non-refreshing sleep, then the Chimp becomes upset. Of course, having a poor sleep pattern is unpleasant, and a poor night's sleep might have some repercussions the next day. However, allowing your Chimp to become distressed, angry or anxious about not being able to sleep, won't help. It merely adds to your problems. Worrying about not sleeping is often a bigger problem than the lack of sleep itself.

Poor sleep could distress the Chimp

Therefore, some helpful facts might settle the Chimp down:

- Sleep disturbance is common and natural, and most people have disrupted sleep patterns at some point in their life
- Sleep patterns change and with effort we can influence those changes
- Sleep varies and some nights are better than others
- We all sleep far more than we think – try a sleep app to find out!
- Most people cope easily the next day, even with limited and poor sleep from the night before
- Allowing your Chimp to become emotional about sleep won't help it or you
- Half of sleep is about resting – so just rest if you can't sleep
- The deep sleep will happen once you do fall asleep, and this is what is needed; the hours are not as important

The Chimp exercises during sleep

The Chimp system comes to life during sleep. It uses sleep as an opportunity to sort out its thoughts and emotions. [393] [414] It does this alone much of the time, but also has periods when it engages with the Human. The brain effectively holds discussions between the systems and gets things in order. This helps the Chimp.

Having an active Chimp might be acceptable *while we sleep*, but the problem is that the Chimp starts to wake up as the Human tries to go to sleep.

This means that once we lie down the Chimp can begin to flood our mind with all of its concerns. It also means that if we wake during the night it will be the Chimp doing our thinking for us. These two specific problems need specific plans to manage them. It was mentioned earlier that one way to settle the Chimp down before sleep is to unwind the mind by writing down any specific problems or decisions that need to be made. This way we are letting the Chimp know that these can be put to one side for the night and will be picked up the next day.

The Chimp could also worry about not getting to sleep and the more it lies awake the more intense the anxiety can become.

The Chimp thinks catastrophically during the night

It is important to recognise that these racing or repetitive thoughts do not belong to you, but to your Chimp. By distancing yourself from these thoughts, the Chimp will be easier to settle and manage. This is especially true if you wake in the night. By distancing yourself from your Chimp, you might be able to accept that any thoughts during the night are likely to lack perspective and be exaggerated. The Chimp's thought processes will be irrational, and conclusions are usually catastrophic in nature. It is no wonder that when this happens to us, we feel terrible. The good news is that the following day when we have woken up fully, we will definitely see things differently.

It is worth programming your Computer to deal with the Chimp's night-time hijacks. If you wake in the night, the Computer will then remind your Chimp that the unhelpful worries and catastrophic thoughts it offers are not worth engaging with.

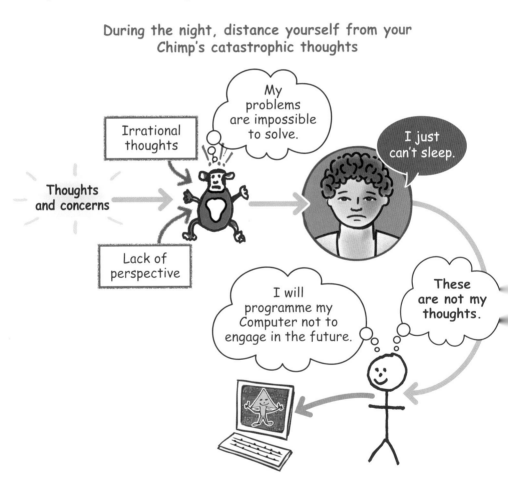

During the night, distance yourself from your Chimp's catastrophic thoughts

Keeping a sleep diary

If you do have some serious problems with sleep, then keeping a sleep dairy can help. It might reassure you or help you to identify what promotes a good night's sleep and what doesn't. It can also help you to spot patterns of behaviour that could be modified. There are lots of suggestions on the Internet and many sleep therapists who could help guide you through the problem. Disturbed sleep can be a sign of illness. If you are not managing to improve your sleep then you ought to seek help from your doctor.

Spotting patterns of behaviour

Summary of what can help or hinder sleep

The light/dark system
- Ensure darkness for sleep
- Use early morning light when you wake
- Set the brain's clock by keeping regular hours for sleep
- Don't cat nap if your night-time sleep is not good

The activating system

Getting to sleep:
- Restful lead into sleep
- Switch off electronic devices well before bedtime
- Park your worries outside your head
- Comfortable bed
- Keep still
- Active dreaming
- Relaxation methods

Staying asleep:
- Right temperature
- Eliminate sound
- Avoid heavy meals
- Relax if you do wake up

The tiredness system
- Avoid caffeine for several hours before sleep
- Keep regular hours
- Don't engage emotions – allow the system to work

Example: *Bob and Sarah provoke their Chimps*

Bob and Sarah are a couple who break the rules of sleep and then wonder why they struggle with tiredness. Bob's job involves making a lot of decisions. He doesn't wind down at night but often takes his problems to bed with him. As he can't sleep well, he has developed a habit of drinking alcohol to relax himself. Sarah has lots of family problems and avoids issues by ignoring them. This leaves her with a lot of unresolved matters and a lot of inner conflict. She finds she can get to sleep quickly but soon wakes up.

During the early hours of the morning, they both wake up and start a conversation. Bob's alcohol got him to sleep but then wakes him a few hours later. Bob is still ruminating about his decisions and Sarah talks about why her family are so dysfunctional. They are both in Chimp mode because their Humans are fast asleep, so the conversations are catastrophic in nature and never reach any conclusion but focus on the problems.

> **Key Point**
> *It doesn't take a genius to find a simple solution, but it does take insight to recognise that one is needed.*

This can be a useful reflection point for us all, when we are faced with any problem.

Unit 21 Reminders

- Recuperation is a proactive process
- Brief and longer rest periods are best planned
- Sleep can be improved by actively working on it
- The Chimp is generally unrealistic about expectations of sleep
- The Chimp exercises before and during sleep - this needs to be recognised and managed
- Programming the Computer and using lists can help manage the night-time Chimp

Unit 21 Exercises

Focus points and reflective exercises

1. **Recuperation**
2. **Advantages of time out and brief rest periods**
3. **Sleep review**

Focus 1: *Recuperation*

Exercise: *Giving advice*

Sharon is an Executive Director of a large company. She works long hours, from 8 a.m. to 6 p.m., and twice a week she goes to the gym to keep fit. She sets off for work at 7 a.m. and doesn't get home until 7 p.m., after commuting. When asked what she would like to improve on, she says:

- It would be good to get a cup of coffee before 11.30 a.m. and to get a lunch break but I am too busy
- I need a magic energy pill most days because I get so tired
- My tiredness makes me irritable and snappy with people, which I don't like
- It would be good to have chance to reflect on my work rather than just keep going

If Sharon gave lots of reasons why she couldn't change her lifestyle, what advice would you give her?

1. How much of your own advice might you need to take yourself?
2. Try the same exercise, of giving advice, for the example of Bob and Sarah in this unit.

Focus 2: *Advantages of time out and brief rest periods*

Exercise: *Listing advantages*

Take some time to list the advantages to yourself and others of you taking time out during your day or of timetabling in brief rest periods.

Our Chimps must consult the Computer constantly throughout the day. Therefore, by programming in the advantages of stopping for brief periods, it is likely that the Chimp will agree and push you to implement this. As a starter to help you to see possible personal advantages to temporarily and briefly stopping your day, here are some thoughts:

- It will help me to bring perspective back to my day

- It will help me to refocus when I start again
- It will help me to drop any stress levels that might be building
- It will probably make me more tolerant towards others
- It will give me chance to move into Human mode

Focus 3: *Sleep review*

The next exercise is applicable to those who struggle with sleep.

Exercise: *Reviewing your sleeping*

***There are two parts to this focus point:* The sleep diary** and **removing thoughts**.

Part 1 - *The sleep diary:* One way to check on sleep patterns is to complete a sleep diary. This is to monitor your sleep and to check on the hours that you are going to bed and rising, and the quality of sleep achieved. There are many apps now that can be used to feedback on your sleep, but a simple diary can also work well.

Sleep clinics will always try to establish patterns of sleep before jumping in with methods to deal with particular problems. By checking our hours and by finding potential reasons for poor sleep, we can often see where to focus our energy to bring about change.

Please don't forget that while you are monitoring your sleep patterns, one of the main reasons people have poor quality of sleep is because they worry about it. Accepting and working with your current pattern of sleep can remove that worry.

Part 2 - *Removing thoughts:* One of the commonest complaints around not being able to get to sleep is constant thoughts running through the mind.

Try writing down any thoughts and decisions that you want to remove from your head before you go to sleep. Remind your Chimp at this point that you will address these thoughts and decisions tomorrow but for tonight they will be parked up. Be firm with the Chimp!

Thoughts that occur during the night can be managed by programming your Computer. This programme needs to be practiced whilst you are awake, if it is to be fully implemented as an Autopilot in your mind. The programme could be:

1. On waking, immediately remind yourself that it is your Chimp that is in charge at this point and any thinking will be lacking in perspective and is likely to be ridiculously catastrophic
2. Remind the Chimp that you will discuss the concerns it has after you have woken up in the morning and come around from sleep
3. Also, remind the Chimp that everything changes with time
4. Disengage any emotion that you are experiencing and decide what emotion you would prefer to have
5. Distract yourself by deciding on something pleasant to focus on

Unit 22
Getting the best out of others

STAGE 7: *addresses our interactions with others.*

Unit 22: *offers reflections, ideas and encouragement on how we can understand, support and build relationships with others.*

Begin with yourself

If you want to get the best out of someone, there are several points you could consider. The most important is that you are in a good place yourself, before you interact with them. [415] [416] [417]

It's essential to meet your own needs and sort out your own emotions before you try to help others. Otherwise, it is likely that you will bring your own needs and feelings into any interaction. If your intention is to get the best out of someone then it's worth checking to see that this really is your agenda. That way you will only act in their best interests.

> **Key Point**
>
> *The probability of having successful relationships will rise significantly if you are in a good place yourself.*

Appreciating the mind of others

During early childhood, we naturally develop the ability to work out what others are thinking and how their mind is working. The ability to do this is called possessing a 'theory of mind'. [418] As children, we use this theory of mind to work out people's intentions and to predict their behaviour. We effectively get inside their head and see the world from their standpoint.

Virtual embodiment
- Scientific points

The self-image we hold can influence the way we behave. Surprisingly, this is not all self-image will do. An experiment involving 30 young men who were virtually embodied in the image of an older body of Einstein showed two remarkable changes during the experiment. They showed an increase in their thinking ability and they also decreased their bias towards older people. [419]

We can appreciate that *everyone has a Chimp* and that it might be talking to us. We can therefore look past their Chimp, to discover what their Human might want to say, and also how their Human might want the situation to be handled. [81] Most of us do not agree with the way that our Chimps handle many situations.

Example: *Stuart and his ten-year-old daughter*

Stuart believes that children ought to learn manners and not be rude or answer back: a very reasonable belief. He finds his daughter frequently answers him back and this pushes Stuart into Chimp mode. The Chimp doesn't usually operate with theory of mind and think about what is going on in someone else's head; it just reacts. If Stuart reacts to his daughter's Chimp then they both end up in Chimp mode, which is unlikely to help.

If Stuart recognises his daughter is in Chimp mode, then he can remain in Human mode, use theory of mind, and stop and ask, "what would my daughter's Human say about her own behaviour?" Unless he has produced a psychopath, she will say that *she would agree with him* and that her Chimp is hijacking her. She doesn't know how to manage her Chimp and she has no Computer programme to operate with, in order to behave as she wants to behave. If Stuart can see this, he could help his daughter to manage her Chimp and create an Autopilot for her to work with, should her Chimp begin to react. Therefore, he could calm the situation down, simply by asking if this is how she wants to be acting and how else would she want to respond to him. He could offer some automatic comments for when she becomes frustrated that would be acceptable to anyone she is interacting with.

> **Key Point**
>
> **Appreciating the mind of others is accepting that we all have Chimps and Gremlins that might be hijacking us.**

Helping the distressed person move from Chimp to Human

Probably the commonest question that I get asked about helping others, is: *"How do I help someone to move out of Chimp mode and into Human mode?"* In order to answer this in a practical way, we can turn to the neuroscience of how the brain works. The Chimp system will always have control of the brain if it feels there is a danger or something that needs addressing. [420] Therefore, if the Chimp is not happy it will hold on to the power and run the brain.

When the brain is operating in Chimp mode there are some potential drawbacks:
- It finds it very hard to listen or take in information
- It focuses on problems and not solutions
- It loses perspective

Therefore, when a person is in Chimp mode, they will find it very difficult to hold a productive conversation. They are likely to express emotion that is out of context or over emphasised and tend to repeat expression of feelings and not move forward.

From Chimp to Human: EUAR

There are some specific processes that can help someone to move out of Chimp mode and into Human mode. What follows is a suggested blueprint for managing someone's distressed Chimp and fulfilling its needs. The first thing to acknowledge is that the Chimp isn't necessarily going to be rational or work with logic. Therefore, while it remains distressed, offering solutions and rationalising are likely to irritate it further.

A way forward to help the Chimp is to use '**EUAR**':

Expression
Understanding
Acknowledgement
Recognition

Expression

Allowing someone to express their feelings, fears or beliefs is the starting point to settling a Chimp down. The Chimp needs to exercise. This means expressing emotion and feelings in order to be able to settle down and then be able to listen or discuss things in a reasonable way. While the Chimp is expressing its feelings, it's important to listen carefully and not judge or interrupt, otherwise the Chimp won't feel like you are taking it seriously. [421] [422]

Understanding

Understanding someone or seeing their point of view doesn't necessarily mean agreeing with them. As a general rule, the Chimp likes to know that someone understands their experience, opinion or feelings. At this point, the Chimp isn't looking for comment, answers or solutions. It is looking for understanding. It's helpful to ask questions if you are not clear exactly what they are saying to you, as this demonstrates that you do want to hear them.

Acknowledgment

Once we have understood someone's situation and how they feel, we usually feel some empathy. If you express an acknowledgement of what the person is going through or experiencing, then this will further settle their Chimp. Acknowledging suffering, distress or unfairness goes a long way to removing many feelings such as frustration, anger or emotional pain. Our Chimps have a need to know that someone understands their position.

Recognition

The final point for helping someone to settle their Chimp and move into Human mode is to demonstrate a recognition of the attempts that they have made to deal with their situation. Most people have tried to deal with a situation and also with their own feelings, before we get to listen to them. Giving someone recognition of what they have tried to do will help their Chimp to settle, even if it is just to recognise that they have made the effort to talk to you about the situation or problem.

Two potential pitfalls when using EUAR

I think it's important to mention two potential pitfalls at this point.

The first potential pitfall is following the above steps of EUAR without any feeling for the other person. In other words, just performing the steps without compassion. If you don't really care then don't try this suggested blueprint because this is a compassionate structured approach to help you to deal with someone that you care about, personally or professionally.

The second potential pitfall would be to dive in with solutions, facts or comments before the person is ready to take them in. When our Chimps are in a distressed state, they cannot receive information but need to express the feelings that are bottled up inside them.

The Chimp is likely to settle down and hand over to the Human, once it has:

- Expressed its feelings
- Been understood, by receiving acknowledgement for what it has, or is going through
- Had recognition of how it is trying to solve or manage the situation

It is at this point that their Human is able to listen and discuss things rationally with you. Recognising when the person has entered Human mode and when the time is right for facts, suggestions and advice, is a skill in itself.

The Human takes over

Assuming that someone wants to enter Human mode and has now managed this, here are some suggestions for helping them to address a difficult situation.

The facts

Begin by establishing the facts of whatever is distressing them. Often this brings in some additional information that can help to start clarifying a way forward. Also, look for truths rather than their impressions of what is happening.

The solutions and a plan

Ask how important it is for them to find a solution to their problem. By asking this question, it can help to focus someone's mind away from the problem and onto possible solutions. Encourage the person to talk through the possible solutions. Try to avoid offering your idea of solutions unless they are stuck, as this can be detrimental. If you do help with solution finding then remember that your role is to act as the Human in the situation by bringing perspective, reality and a set of values into the picture. Once possible solutions are offered, a plan of action can be made to achieve them.

Moving from Chimp to Human

First help their Chimp:
- Expression
- Understanding
- Acknowledgement
- Recognition

Then help their Human:
- Establish the facts
- Search for the solutions
- Form a plan

Helping people to see the difference between their 'wants' and their 'needs'

Our Chimp and Human have needs and wants. Sometimes, it is difficult for someone to differentiate between what their Chimp needs and wants and what their Human needs and wants. Computers don't have a need or want, as they are programmed with an agenda.

Example: *The territorial dispute*

Imagine an office situation where there are two people working. They have distinct roles but often there is a grey area of who should do the work. Barney has taken on some work that probably should have been taken by Chris. Chris has just realised this and is about to confront Barney.

Dividing up the mind of Chris, we can see the three systems preparing to engage. I will take typical stances but clearly these will vary depending on the person involved.

The Chimp's stance on 'needs' and 'wants'

I have a need for a defined role and this represents my territory. I believe my territory has been invaded and therefore needs to be reclaimed. This is about winning. In the Computer I have placed the underpinning belief that if my territory is invaded then I am vulnerable and might be seen as weak. Alternatively, the underpinning belief could be that people who invade my territory are rude and need to be confronted. There could be many alternative beliefs.

I want Barney to be reprimanded for this invasion and made to recognise his wrongdoing and apologise.

The Human's stance on 'needs' and 'wants'

I have a need to feel fulfilled and to be respected. I will want to check the facts of the situation and see if Barney has realised he has over-stepped the mark. In the Computer I have placed the underpinning belief that people make genuine mistakes and it's best to point them out tactfully and supportively. He might or might not respond constructively, or how I would like, but this isn't about winning it's about solution finding.

I want to find a solution that helps both of us and prevents this from happening again.

The Computer's stance

The Computer relies on being programmed **BEFORE** the event happens, otherwise, it will simply hand back to the Chimp for a reaction. If it is programmed by the Chimp, and not the Human, then the Computer will encourage the Chimp and fuel its indignation.

So, when you start considering helping others to get the best out of themselves, it's worth looking at how to help them to:

- Manage their Chimp
- Programme their Computer
- Establish what their Human wants

If Chris can gain insight into how his mind is working, he can choose to go with his Human or with his Chimp. He can also check how his Computer is programmed to manage the situation and alter this, if it's not what he wants.

Building a relationship with an uncooperative person

Don't lose your dignity! Please recall, that to get the best out of others, you need to be in a good place yourself before going into any interaction.

In this section we will cover several points:

- Thoughts you might have about someone
- Judging or using your judgement
- Accepting someone as they are
- Developing compassion
- Stopping your Chimp from being prodded
- When someone lacks emotional skills

Thoughts you might have about someone

As you meet an uncooperative person, what thoughts have you got? When we approach another person, we will have thoughts about them.

Fundamentally, our feelings usually follow our thoughts. If we focus on good things, we feel good, and if we focus on negative things, then we feel negative. We always have a choice about what we want to focus on and think about, even though at times it might not be easy. Therefore, if you want to have a constructive relationship with someone, it's wise to see if you can have some good thoughts about them, and focus on these. The question to ask yourself is: how you are viewing this person? There is a big difference between judging someone and using your judgement when assessing someone.

Judging or using your judgement

Example: *Marie and her best friend in conflict*

Marie had organised a trip with several friends. Her best friend, April, was keen to go on the trip and helped with some of the organisation. Marie was really excited about the trip until April called her the night before. April called to apologise but said that her boyfriend was not in a good place and she wanted to be with him the following day. Marie has got two choices:

- Judge her friend
- Use her judgement

Chimps tend to judge and Humans tend to use judgement.

If Marie's Chimp is angry with her friend *and judges her*, she might say that April was not a good friend and was lacking in her sense of commitment. In doing this, Marie will generate emotions in her own Chimp that will likely leave her feeling negative. It is also likely to sour the relationship.

If Marie's Human *uses her judgement*, then she can see the dilemma that April is in and also appreciates April didn't intend to hurt Marie. Marie can also see that it is not the end of the world. If she takes this approach, Marie will retain perspective. Although she might be disappointed, her friendship is likely to remain intact.

Using your judgement means operating in Human mode. For example, seeing a person not as 'difficult' but as 'reacting in an unhelpful way' is more constructive and not judging them. The person is probably in Chimp mode, and they themselves might not agree with nor want what is happening. Labelling people as 'difficult', 'awkward' or any other negative label isn't very helpful. Once you have negatively labelled someone, it is likely that you will feel negatively towards them. The reality is, that some individuals do appear to have an unhelpful approach to others, and to life in general, but often this is a behaviour rather than a personality trait.

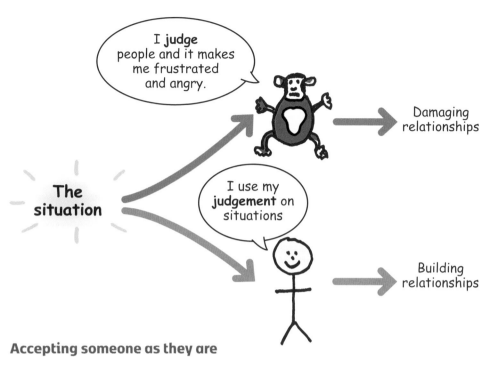

Accepting someone as they are

Example 1: *The wayward daughter*

I worked with a man who was struggling to build a relationship with his daughter. The daughter was in her thirties and had a long history of dishonest and devious behaviour. The father would keep on trying to help his daughter to mend her ways. About every six weeks the father's Chimp would get things off its chest, only for the tension to rebuild again. He would become upset and frustrated by his daughter's actions. The problem lies with the fact that the father's Chimp had expectations that were reasonable but unrealistic. His daughter is very likely to continue with these behaviours until she sees the need for change. If the father can switch to Human mode, he can then accept his daughter, as she presents, and be *realistic* about what is probably going to occur and then plan how to manage this. If he does this, his upset and frustration can be diminished or removed.

Example 2: *The 'challenging' colleague*

A professional businessman worked alongside a colleague, whom he found to be very irritating. He believed that the colleague was lazy but clever enough to get away with a shirking attitude towards work. Every time the man went to work and saw the colleague, he would become agitated. The man felt an injustice was happening. He wanted this colleague to change their ways or at least be found out and suffer the consequences. Getting the man to accept that this colleague would be highly unlikely to change, and also unlikely to be found out, was a struggle. The result was that the colleague was fine, but the man was not.

Clearly the man had options of confronting the colleague, reporting the colleague's work ethics to a line manager or focussing on his own work.

Some situations we can't change, and most people don't wish to change themselves. Therefore, a starting point is to accept this reality and decide what to do, *after* you have accepted the reality.

Example 3: *The imperfect partner*

Many relationships suffer because one of the partners doesn't accept the other as they are. Instead, they build up hopes or dreams of how that person will be, which are unlikely to happen. Repeatedly demanding something from a partner that they can't or don't want to offer isn't helpful and won't make you feel good. We sometimes forget that relationships are built on accepting people as they are and working with them, rather than imposing what we think they should be like. Demanding change is not an option, it's an imposition; but deciding on whether you want to stay or leave is an option.

Developing compassion

Chimps can often turn negative emotions into compassion by seeing pain in others. If you take the time to understand someone and their situation, it is likely that you will become more compassionate towards them. This is one of the foundations of team building. A common experience, or an insight into the other person, can help you to form a bond with them. [423] Compassion frequently develops when you are able to see the world from someone else's position. Encouraging your Human to do this can help your Chimp to have emotions that are more constructive.

> **Key Point**
>
> *Accepting someone, as they are, is the basis of a sound relationship.*

Working with your Chimp can turn negative emotions into compassion

Getting the best out of others

Some commonplace scenarios involving others

Stopping your Chimp from being prodded

The Computer in our minds frequently works by either linking items together or by an item triggering a reaction or response in either Chimp or Human.

Forming links in the mind
- Scientific points

One area, where we see links taking place, is in the hippocampus of the brain. There are three different cells that work together. One stores factual information (place cell), one stores a sensory experience such as touch, smell or emotion (sensory cell), and the third cell (conjunction cell) connects the first two cells, when an event occurs. Consider a friend who grows roses. When we smell roses, we think of the friend's house, and when we think of the friend's house, we often think of roses and sometimes appear to smell them.

This important principle of linking experiences and trigger points is used a lot by the brain and utilised in psychological treatments.

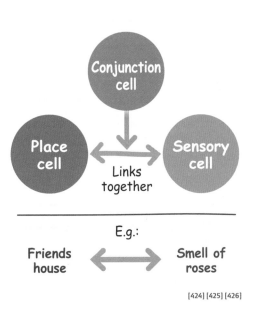

[424] [425] [426]

How can we recognise links and turn them to advantage in day-to-day life?

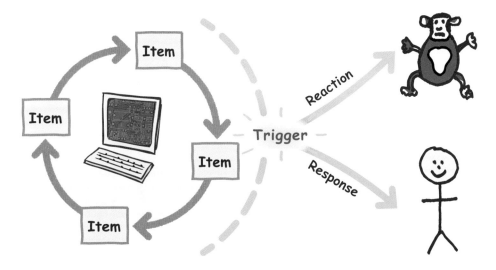

The Computer links items together to trigger a reaction or response.

Example: *Claire and the trigger*

Claire works with someone she finds difficult to get along with. Other people also find this individual difficult to work with. The result is that conversations often focus on the individual and these conversations then trigger negative emotions in Claire. When Claire's Computer hears the individual's name being mentioned it evokes a negative feeling. This then prods her Chimp into action.

If Claire recognises this trigger point, then it's not difficult to avoid it, by being politely assertive and asking to change the topic of conversation. What Claire can do is to link something constructive to the person whom she finds it difficult to get along with. For example, whenever their name crops up Claire can think of how she wants to come across to others. This trigger point will then help her to avoid making caustic comments or negative remarks about them. Alternatively, whenever their name comes up Claire can think of how she could avoid any conflict and also how she can get the best out of them. This link would help Claire to have a constructive, if not pleasant, relationship with the person.

> **Key Point**
>
> *We can use trigger points to advantage by knowing how to link positive actions to negative emotions.*

When someone lacks emotional skills

We all have different talents and gifts. Emotional skill is one of these. We can all work on improving our emotional skills, but some people struggle to make progress or lack the insight to see that there is a problem with their skill-base.

If you believe that someone's emotional skill-base is poor, then why get upset when they display this? Ask yourself are you working in Human or Chimp mode? Are you accepting the reality of your belief or are you judging and then expecting to see change? Accepting that some people will remain in Chimp mode throughout their lives and not display any emotional skills is not easy, but certainly easier than expecting them to change.

Common themes involving others

The next section will look at specific areas within relationships that can create problems and how to solve them. These areas include:
- Whose perspective: yours or theirs?
- Speaker thinkers and thinker speakers
- Imposing our needs onto others - Crossing boundaries
- Trying to win against others
- Empowering others
- When a relationship isn't working

Whose perspective: Yours or theirs?

Clashes with others in our life can often be avoided by seeing the world through their eyes.

Whenever we search for facts and information, our Chimp is very selective in what it chooses to see and hear. When it has gathered its own selected 'facts', it interprets them as best it can, but this inevitably has its own bias and previous experience mixed into the interpretation. This means that the Chimp can only ever have its own perspective on what is happening. Someone who possesses emotional insights can grasp this and will therefore move into Human mode and try to understand the perspective that others have on situations. The natural working of the Human is to gather facts, and in order to do this the Human listens first and sees another person's perspective. This is the essence of good teamwork.

We can programme our Computers to learn to switch from Chimp to Human simply by constantly reminding ourselves to see things through another person's eyes. This gives a 3D perspective to life rather than the Chimp's 2D perspective.

The Card Game

The Chimp loves to play card games. The game is often subtle and has to be recognised. The card game happens when someone expresses an opinion as if it were a fact, imposes this 'fact' upon you and then expects you to respond. In other words, they deal the pack of cards, have a look at what is in their hand and then decide on the rules of the game: you effectively can't win.

The solution is to recognise what has happened and take the cards back and not engage with the game. Unless you do this, the relationship with the person will become frustrating, especially if you can't see what they are doing. Your role can be to recognise what is happening and to disengage from the card game. It will help the other person to see what they could be unintentionally doing.

Example: *Joanne and her therapist*

Joanne has told her therapist that since she has lost her husband, she has changed. She says: "I know that nothing in my life can be the same again and I cannot ever be as happy". The therapist cannot move. Joanne has given the therapist an opinion but has presented it as a fact. If Joanne is unable to move from this 'fact', then there is little the therapist can do, except to work with this opinion and its consequences. Alternatively, what the therapist could do, is to collect the 'fact' cards in and deal again with some truths that can be worked with. The truths could suggest that with time, Joanne might find these feelings change, but until then she will have to work with them. A different approach could be that the therapist accepts Joanne's opinion as fact and asks how Joanne intends to work with this fact. Another route could be to ignore the opinion and instead work through the grieving process of Joanne losing her husband. This way the 'facts' might change as the process progresses. There are various ways then that the card game can be played.

Example: *Louise enters the card game*

Louise has a poor memory for anniversaries and her husband has become upset because she has forgotten their wedding anniversary. He confronts Louise with the statement "You really have to sort this out. What are you going to do about it?". Louise now feels bad and doesn't really have an answer. She has tried with a calendar and a phone reminder but forgot to check both.

What her husband has done is given an opinion as a fact. He has dealt the cards and called the rules. He has stated that **Louise has a problem** and that **Louise must sort it out**. The reality is that Louise might not think she has a problem; she merely has a poor memory. If she doesn't feel it is that important then there is nothing to sort out. What she could do is to collect the cards in and disengage. Alternatively, she could offer her opinion on the situation. If Louise wants to help her husband, then it would help to bring some insight into what he is doing, so that he has a chance to change his approach.

Example: *Duncan's card game*

Duncan has become exasperated by his teenage son, Peter, playing games on a computer for hours at a time. Duncan eventually challenges his son and says, "If you had any respect for me, you would limit the time for your games to an hour a day". Peter becomes frustrated by this remark and doesn't know how to respond. This is because Duncan has given an opinion as a fact and left his son perplexed in how to respond. The reality is that Peter doesn't agree with his father. He feels that he gives respect towards his father and the time he spends on his computer has nothing to do with respect. Duncan might hold this opinion, but he is playing a card game with his son by stating his opinion and expecting his son to accept the opinion as a fact. For Duncan to get the best out of his son it would help to recognise what he is doing and to address his real concern, which is about the time his son is spending on the Computer.

Speaker thinkers and thinker speakers

Sometimes our expectations of others are unrealistic because we don't have all the scientific facts. A good example of this is the way that we think and speak. It seems that about half of us like to speak and think as we speak, whereas the other half like to think before they speak. [427]

If this isn't recognised, then the speaker thinkers tend to dominate meetings with their ideas and opinions. The thinker speakers remain silent, or even worse, they are pressed to give an idea or opinion before they have had chance to think. By simply understanding how different people can operate, we can allow for this and become more realistic in our working with them. Thinker speakers tend to need a fair amount of time to do their thinking but will then optimise their performance within teams or relationships.

When we operate in Chimp mode, we impose our beliefs and expectations, without taking into account the facts. The Human will always search out facts and then where necessary use these facts to modify our behaviour.

Imposing our needs onto others - *Crossing boundaries*

Sometimes, we don't recognise that we are imposing our needs onto others. The result is that we see a series of relationships going wrong, as this imposition continues.

This is demonstrated when we cross professional boundaries and move into an inappropriate personal relationship. Our need for security, a partner or a friend overrides our professional and ethical stance.

Example: *Perry's confusion*

Perry is a social man who enjoys the company of others. His Chimp has a need for companionship but doesn't recognise the difference between 'personal' and 'personable'. He regularly ends up in affairs, which he regrets, because he crosses boundaries in order to satisfy his Chimp. Being personable means being approachable and friendly. Being personal means going beyond personable and sharing feelings with a measure of intimacy. If Perry doesn't want to have affairs, then he needs to recognise that Chimps frequently try to use others to satisfy their own needs, regardless of the consequences.

Example: *Tina and her substitute mother*

Tina's Chimp constantly feels vulnerable and has learnt that finding a substitute mother is one way to overcome this. This might be fine, provided that person wants to be a substitute mother. However, if Tina imposes her need for a mother onto someone who doesn't want this, she is likely to damage that relationship. It would be better if Tina looked at why her Chimp was insecure and what other options there are to make it feel secure.

Trying to win against others

The Chimp typically relates to others by wanting to impress them and gain approval. One way that it thinks it can do this, is to see any interaction as a battle to win and demonstrate superiority or esteem. Any battles that are lost are usually seen by the Chimp as failures, humiliation or weakness. The Chimp therefore lives in a world of battles that are all about proving its self-esteem.

The Human typically doesn't see battles but sees peace-making and resolution during interactions. This means the Human generally thinks about the other person before themselves. The Human mostly relies on values and self-assessment for esteem and therefore doesn't need to 'win'.

Example: *Audrey's Chimp in action*

Colleagues in her office perceive Audrey as being difficult and combative in her approach to life. Sadly, Audrey is in the grip of her Chimp's way of seeing the world and others. Her Chimp won't back down and rarely apologises. If apologies do take place, then they are always linked to reasons why others or circumstances have caused the situation. The only way forward for Audrey to stop her Chimp's behaviour is to move into Human mode by challenging the Gremlins her Chimp has placed into her Computer. Gremlins, such as:

- I must be seen to 'win' in all interactions
- I need to demonstrate that I am strong
- Being strong is shown by winning
- I cannot display any fault or weakness
- I can't see interactions in any other way than win or lose

These Gremlins need replacing with Autopilots, such as:
- My opinion of myself isn't based on 'winning' but on values
- If my Chimp wants strength, then being pleasant demonstrates strength
- I will feel much better if I am not so defensive
- The Human way of interacting is better than the Chimp for building relationships
- I have a choice in how I want to interact

How could you help Audrey?

If her behaviour is Chimp driven then what the Chimp is looking for is approval and praise. By approving and praising the traits that you feel are worthy, her Chimp can change and demonstrate the behaviours that you approve of. If Audrey is willing to engage in a heart-to-heart conversation then compassion and support, coupled with some insights, can make a big difference.

Empowering others

Nobody likes to be disempowered: so why do we do this to people? Our Chimp's nature is to dominate and not to consider the effects on others. Our Human will rationally accept that everyone likes to be empowered. Empowerment doesn't necessarily mean that we allow people to do exactly what they like. It means that we have listened to and considered the opinion or wishes of others. In other words, given them a voice. The opinions or wishes then need to be discussed and managed. If we disempower people then we will provoke their Chimps.

When a relationship isn't working

Sometimes, when we are interacting or forming more in-depth relationships, we have to accept that it won't always work out. If we have common ground or experience this can help our Chimps to bond with others. If we share similar values, then this will help the Human to bond with others. If either Chimp or Human become unhappy then the relationship is likely to be unproductive. We have to learn to live with the other person's Chimp, as well as their Human.

When things are not going to work out it's probably best to see this as a mismatch rather than a failure. The Chimp will see things in terms of success and failure whereas the Human will see things in terms of experience and learning.

Unit 22 Reminders

- Getting yourself into a good place is the best basis for any relationship
- Humans employ theory of mind; Chimps usually don't
- To help the Chimp we can use 'EUAR'
- The Human and Chimp have different needs and wants
- Feelings usually follow thoughts
- Accepting someone as they are is the basis of a sound relationship
- We can use trigger points to advantage by knowing which ones make us feel better and which ones to avoid

Unit 22 Exercises

Focus points and reflective exercises

1. **Expectations of others**
2. **Assertiveness**
3. **Judging or using your judgement**
4. **The ideal...?**

Focus 1: *Expectations of others*

Exercise: *Reviewing expectations of others*

Select a relationship: a parent, partner, friend or colleague and perform your own reality check on your expectations of them.

Start with a list of their good points and traits. Then, if the person has traits or behaviours that irritate or disturb you, list them and try to form a plan of how you will manage your attitude and approach to each one of these behaviours.

If their behaviours or attitudes cannot change or they are unwilling to change then consider your own options. Whatever happens, try to be constructive and remain in a good place. Remember, ***it is your change in approach that will make the difference*** not your requests or demands that they change.

Focus 2: *Assertiveness*

Lacking assertiveness in a relationship can lead to frustration on both sides. Assertiveness has three simple steps to it. Let the person know:

1. What you don't want them to do
2. How the person's actions are making you feel
3. What you do want them to do

For example, if someone were raising their voice to you, then to be assertive would be to say:

1. That you do not want them to raise their voice to you
2. It is making you feel uncomfortable
3. You would like them to speak in a quieter non-aggressive way

Exercise: *Being assertive*

Try to recognise throughout the day, any times when you experience uncomfortable feelings, and ask if being assertive would help.

Practicing assertiveness when you are alone, and speaking the words that you want to say out aloud, can help to put it into practice when the occasion arises.

Focus 3: *Judging or using your judgement*

Exercise: *Moving from the judging Chimp into Human using judgment*

The next few occasions when you recognise your Chimp is judging someone, try to move into Human mode by using your judgement. To do this, you will need to remove your feelings and consider the situation as an onlooker. Try to see why a person might be acting the way they are and to consider if they are in Chimp mode and might regret their behaviour, with time.

If you can do this, try to assess the difference it makes to your own emotions and hopefully you will find yourself in a better place. Using your judgement doesn't mean excusing unacceptable behaviour, it just means approaching it in Human mode and with understanding.

Focus 4: *The ideal....?*

This focus point is to consider what we think of as being the ideal person in a given relationship. Having worked a lot with teenagers, its often very sobering for them to consider what the ideal teenager would be like if they were the parent. This often stops them in their tracks, as they compare this ideal to the reality of what they are really like.

Exercise: *The ideal....*

In our lives we take on many different roles: parent, sibling, friend, colleague, daughter, son, partner, line manager and you could add many more. Write down some significant roles that you play within your life. Try to define your idea of the ideal person, who would fit that role and then compare yourself to see how you are doing. It can be quite sobering and provoke changes to your thinking and behaviour that will reap rewards within your relationship.

Sometimes it helps to ask the other person how they would like you to be.

Unit 23
A basis for relationships

Unit 23: *will look at the way the Chimp and Human enter relationships. They have a very significantly different basis. We will consider the relationship we have with a parent as an example.*

The relationship we have with our parents

Not everyone has a problematic relationship with their parents, but many do. Several of the principles that we will consider apply to all relationships, so even if you have a great relationship with your parents, there will be aspects that you might find useful.

Why do problems commonly occur with parents?

We both need to consider how we view important relationships

Our childhood years and the basis of the problem

Throughout this unit I will use the term 'parent' to represent any significant adult who is acting in this capacity.

Here is the problem: we are in Chimp mode during early childhood most of the time and our Human is poorly developed. [428] [429] [430] [431] When we are very young, we are genetically pre-programmed to look to our parents for security. [432] [170] [171] [172] Relying on our parents is clearly a critical in-built survival drive that compels us to turn to them for help. When we turn to our parents, our Chimp necessarily has to perceive them with an unwavering belief that they are all knowing, invincible, completely dedicated to us and perfect. This belief makes our Chimp feel secure. [300]

Therefore, what we have is a young child, led by their Chimp, believing that their parents are perfect and that *this perfect relationship is the normal situation*. Therefore, the child's Chimp *expects* their parents to demonstrate this every day all day.

Many young children idolise their parents and can be heard boasting about how good they are and what they can do. Along with these beliefs, we have the expectation that our parents will protect us and do what is best for us. There might be variances on this theme, individually, but most people experience this unrealistic scenario. It is helpful for a child's survival to see their parents as perfect and it is built into our genes to do this.

The childhood myth

As we grow older and become more aware, we start to have experiences where the above just doesn't seem to ring true. Our parents appear not to be superman or superwoman. We continue to compare our parents to the expected mythical ideal parent and sadly they fall short. We might now become judgemental towards our parents, as in our eyes, they have not lived up to the parents that they should have been. They are not all knowing, they are not always right, they do not always protect us, they sometimes put themselves first and at times they seem to be unpleasant!

Childhood reality dawns

So far in this dialogue, what we have done is to see the world through the Chimp's eyes. It's a world that it must create in our early years, in order to get us through them safely. However, it's not useful to enter adulthood still holding on to this Chimp's view of our parents.

If we now shift to Human mode, we can see a very different picture. As we shift from Chimp to Human, we will be able to see things rationally and with perspective and the reality of the way it is.

Our adult years

Our parents might or might not have opted to be parents. Either way, parenthood is placed upon them and there are usually no lessons beforehand. They are still human beings trying to cope with life themselves. They have their own internal struggles and now they have the additional responsibilities of parenthood. Most parents make an effort to be the best parent that they can be, but even in their own eyes, they often fall short. If we don't accept that this is the way it is, but we hold on to the mythical parent-figure that our Chimps expect, it will inevitably evoke negative emotions within us. We either direct these negative emotions and unrealistic expectations at our parents or we torment ourselves with them. Often we do both.

A basis for relationships

As you can imagine, our Chimps tend to do this with every relationship we have. [301] They impose what 'should be' and then become upset when they meet reality. On the other hand, our Humans accept reality and work with what people can offer us within relationships.

> **Key Point**
>
> *It helps to see our parents as people with their own struggles and not as people who should be super beings.*

The Chimp expects ideal parents

The Human accepts real parents

Dealing with past experiences

How do we deal with any negative past experiences with our parents, partner or friend?

It's helpful to follow the same procedure that we have previously been using – three steps.

Step 1: *Exercising the Chimp*

First allow yourself to express any emotions you have from any negative experiences. Find a willing listener but not your parents if it's about them! Don't hold back because you are just expressing what the Chimp might feel, and you might or might not agree with your Chimp. The exercise is to get the emotions and thinking out, no matter how irrational, depressing or hostile these emotions or thoughts might be. It's helpful to go through specific experiences that might support the Chimp's feelings.

However, this is only the first step to moving on and not one that is helpful if it is done without steps two and three. If you just vent your feelings, with examples, it might make the situation worse because all you are doing is becoming frustrated or angry and not doing anything about it.

Key Point

You must follow through with steps two and three to effect change.

Step 2: *Identifying and addressing each concern*

Form a list of all your grievances and specific examples, rather than just keep jumping from one to another. Some people tell me that their list is too long! If so, form specific areas with titles that your examples can go under, that way you only need to list a few examples. Here is an illustration of what I mean:

The list

- They never encouraged me
- I just got criticised all the time
- They broke their promises
- They put other people before me
- They put themselves first
- They belittled me
- They were cruel to me
- They abused me
- They neglected me
- They wrongly punished me
- I just didn't feel loved
- I didn't feel I got the amount of attention I needed

It is very important to do this exercise properly. In order to process our emotions, we almost always need to express them first. [433] [434] It is sometimes necessary to repeat the exercise because our Chimps like to make sure that they have been given permission to keep going until they are ready to move on. You can see that step 2 (forming a list) and step 1 (expressing emotion) are merged together in this example. We are exercising our Chimp, but at the same time identifying specific problems and then exercising again.

Now, having identified the problems, we need to find some answers. The situation with parents does not lend itself easily to having specific answers for specific grievances because part or most of the problem is within us. This is because we are not accepting that it is our interpretation and the way that we are dealing with the situation, which is causing us a problem. So let's look at some examples of home truths that could be used.

Key Point

Remember you are not trying to correct your parents; you are trying to understand them.

Home truths

- **Parents have Chimps too:** this means they will be hijacked and at times act unhelpfully. This doesn't give them an excuse but it does help us to understand that they might not even agree with how they have behaved. If you continue to judge them, then expect more grief. It is no different to not forgiving yourself for something you have done, that you might now regret. It is damaging and very unhelpful.
- **Seeing the world through your parent's eyes:** it can help to try and imagine being your own parent and trying to appreciate what they might have been going through, while trying to raise you. Seeing the world through somebody else's eyes and with their perspective might help you to understand them.
- **Understanding and approval are different:** just because you can see things differently with understanding doesn't mean that you approve of what happened or that you can now see it as acceptable. It merely means that you understand what happened. In cases of neglect there is reason to feel hurt, but I assume that people want to move on from this position.
- **Parents can struggle to manage:** some parents have very little parenting skills and try to compensate in destructive ways. For example, if they become stressed by not being able to deal with a situation, they can act inappropriately. Worry or inability to manage a child or adolescent can be converted into shouting and unacceptable behaviour from the parent.
- **Values can be an issue:** some parents have very different values to their children. Your parents might have very different values to you. Having to accept this situation is not always easy but we have little choice. By recognising and accepting that this is the case, we can begin to understand and manage to find a way forward.
- **Most parents have tried:** try to get a balance and see when and what *they did provide for you.* Recognise when they clearly made an effort to please you. If possible, form a list of when you can appreciate that your parents did put you first.

- **Accept that some parents might appear as not particularly nice people:** there could be a myriad of reasons why they are coming across like this. They might be damaged from their own past. They might be holding their own grievances. They might be feeling vulnerable and unwanted themselves, or of course they might just be unpleasant. For whatever reason they are acting in an unpleasant way, it doesn't help to keep judging them.

- **Try seeing your parents as fellow adults who are just as frail as everyone else:** most adults struggle at times with life and responsibilities. Our parents will certainly have flaws and struggles of their own. If we can adjust and see our parents as people in their own right, then we can relate to them more appropriately when we become adults. It also means we are less harsh on ourselves as adults.
- **Parents are parented:** Don't forget that your parents have their own experience of being parented and this might be influencing how they are interacting with you.

When you have formed a list that identifies the areas or experiences that you want to address, try to match them up with the statements above. It helps to have some understanding and compassion from either someone who is a willing listener or by showing yourself some kindness.

We have to accept that some parents do neglect or abuse their children and compassionate understanding would be appropriate. To emphasise again, what we are not doing is condoning or justifying any form of neglect or abuse. These are terrible circumstances. What we are doing is trying to understand why they might have happened. By understanding what happened, we have the first step for moving forward from the experiences.

Step 3. The plan

A difference in approach

Before you begin, it is important that you understand what you are going to do. You are going to revisit all of your experiences and look at them in a different way; otherwise nothing will change within you. You can't change what has happened but you can change the way you see it and deal with it. Effectively we are going to see parents through the eyes of a Human and not a Chimp. We will bring some truths into the picture about parents.

It might take a few attempts to go over some of the items on your list and to see them from your Human's viewpoint and not your Chimp. You will need to decide when you think you are ready to move on and accept things or when you feel you need to be a bit firmer with yourself and get over what can't be changed and then move on.

Key Point

Forming a plan for going forward will mean making a decision to change your stance.

> **Key Point**
>
> *Decide what you want the relationship with your parents to be like now; check that you are being realistic!*

So where do we go from here?

If you can accept the differences that you might have with your parent, you can then determine what you want now. Try not to fall back into an unrealistic stance with expectations that are unlikely to happen. Also, try not to fall into the trap of believing that they will suddenly change. Try to work with 'what is' rather than 'what could be'; hopefully as you change your stance with them, they might change their stance with you.

Sometimes finding a 'substitute' parent can work very well for some people who still feel the need for a parent figure in their lives. The word 'parent' could mean biological or it could mean something quite different. Some people accept that a parent is someone who fulfils a loving and caring role towards them, irrespective of age or bloodlines. Common sense warns you that it would be unwise to impose being a parent onto someone who does not wish to be a 'parent'. Be careful!

> **Key Point**
>
> *At some point in life, we have to grow up, take responsibility and behave like an adult!*

Often problems arise not from the past but in the present because we have not moved on fully into an adult life. Adults usually find a partner or friends that supplement or take over the parenting needs that we have. [301] [302] [435] However, having a parent figure around can be reassuring; go with what works for you.

Deceased or absent parents and grievances

Occasionally a parent has died or is absent and their child has been left with some conflict or grievance with the deceased or absent parent. This situation can cause grief and a feeling of unfinished business or a need for resolution. Many therapists have helped countless people to come to terms with this situation and to find resolution. One of the main methods used is to write a letter to the deceased or absent parent explaining what your feelings are. By doing this we can often process the situation and it can be resolved. [436] [247]

Finally, if you are still struggling with any aspect of your relationship with a parent then seek out suitable professional help.

Unit 23 Reminders

- In childhood our Chimp has an unrealistic view and expectation of a parent
- The Chimp can carry this unrealistic view and expectation through into our adult life unless we intervene
- Changing to a Human perspective on any relationship removes unrealistic views and expectations

Unit 23 Exercises

Focus points and reflective exercises

1. *Seeing relationships from two standpoints*
2. *Giving advice*

Focus 1: *Seeing relationships from two standpoints*

It can be useful to view your relationship with a parent, or anyone else, from the Chimp's and then the Human's standpoint. The Chimp's standpoint will be an expectation of virtual perfection. The Human's standpoint will be to accept the reality of how people are.

Exercise: *Two standpoints*

Choose a couple of people in your life and see if you can work out the two different standpoints that your Human and your Chimp have. The purpose of the exercise is to recognise when your Chimp might be hijacking you into seeing their unrealistic and usually perfectionist standpoint. If you allow this to happen it can be damaging to the relationship.

Focus 2: *Giving advice*

Imagine Leroy is a friend of yours and he has come to you with the following problem. He says he feels that his father has never shown an interest in him, having walked out on his mother when Leroy was just two years old. Leroy is now 25 and has just become a father. He is in a steady relationship and delighted at becoming a Dad. However, his own father has now got in touch saying that he wants to be part of his grandson's life.

Exercise: *Managing the situation*

If you wanted to help Leroy to sort this out in his mind, how could you take this forward, without offering an opinion, but allowing Leroy to have a structured approach to getting beyond the problem?

Unit 24
Optimising relationships

Unit 24: *considers the foundations for forming and maintaining successful relationships. If we wish to strengthen relationships, it can take quite a bit of reflection and some practical work. This unit offers ideas and suggestions for you.*

The foundations for successful relationships

If we now look at adult-to-adult relationships, what are the principles in the previous unit about the parent-to-child relationship that still apply?

First, and most importantly, the Human and Chimp approach to relationships is very different.

Differences between the Human and the Chimp when someone approaches a relationship

- Expectations of who they are
- Expectations of what they should deliver
- Dependency
- Self-serving bias
- Constant wariness
- Conditional support

- Acceptance and understanding of them as they are and of who they are – no comparison
- Realistic expectation of what they can deliver
- Mutual benefits to both
- Acceptance of faults and failures
- Trust
- Unconditional support

Our Chimps are by nature quite needy. Therefore, many of the Chimp's traits can be brought into a relationship without us recognising this is happening. Sometimes it is quite clear and this is demonstrated in very dysfunctional relationships. For example, when someone is very insecure and constantly checks to see where their partner is, what they are doing and who they are speaking to. Most of us are not so insecure but can still subtly bring our Chimp's fears or needs into a relationship and these subtleties can be detrimental. With this in mind, we will look at the severe end of the Chimp's spectrum and then consider how this could still be seen in a milder form within our own relationships. The examples will all demonstrate the milder presentations of the Chimp.

Expectations

Just as our Chimps defined the perfect parent in the last unit, they will also define what a perfect friend, partner or colleague should be like, and then use this as a standard. Nobody can live up to these unrealistic standards. There is little attempt by the Chimp to understand someone as a person or to allow for differences. The Human approaches someone *by accepting them as they are* and tries to *understand them* and their behaviours.

Example: *Joyce and her moody partner*

Joyce loves Harry but finds it difficult to tolerate the days he becomes moody.

Joyce decides

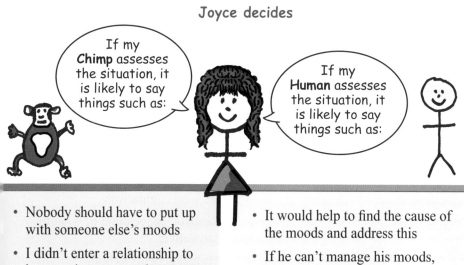

If my **Chimp** assesses the situation, it is likely to say things such as:

- Nobody should have to put up with someone else's moods
- I didn't enter a relationship to have to tip toe around a partner
- He needs to deal with his problem

If my **Human** assesses the situation, it is likely to say things such as:

- It would help to find the cause of the moods and address this
- If he can't manage his moods, then we need a plan for those days
- Talking through feelings can often help, so I will let him express them

I am sure that you can think of many other possible comments. These example comments are there for you to recognise the difference between the Chimp's approach and the Human's approach. The Chimp is problem centred and reactive; the Human is accepting but solution focussed.

Dependency

Our Chimps will often transfer dependency from the parent or peer group onto the person they meet. This can then lead to a belief that this friend or partner is responsible for looking after them in some way, such as being responsible for their happiness or security. The Chimp will often expect that a friend or partner will be available to them on request. The Human is independent but welcomes the company of others, often forming a mutually beneficial relationship that is supportive.

Example: *Daniel's insecurity*

Daniel has been through a number of relationships and for various reasons they have not worked out. His Chimp is now insecure and is trying to deal with this. He has formed a new relationship with Kevin. Kevin is easy going and supportive but he is finding Daniel's clinginess difficult to deal with.

Daniel's Choice

- When I'm not with Kevin I cannot cope and need to know when he will be back
- I constantly focus on things I am doing wrong that will cause this relationship to end
- I keep calling him to be reassured that he still feels something for me

- I can be happy when I am alone and in my own company, I need to work on being more independent
- It's important to be myself if this relationship is going to work
- It's great to be in a relationship with someone, but the relationship I hold with myself is as important

Again, the Chimp has focussed on problems and fears with reactive measures, whereas the Human is focussed on building foundations and forming plans for emotional stability.

Self-serving bias

The Chimp will not tolerate errors, flaws or mistakes in others and often not in itself. In order to prevent self-criticism, it's not unusual for our Chimps to justify our mistakes and criticise the same mistakes made by others; this is known as a self-serving bias.

Self-serving bias
- Scientific points

When teaching this concept to medical students, I used a simple scenario of a car accident. I asked six students to leave the group, so that they did not know what we were looking for. Three then entered the room. The first three were told that there was a traffic queue and they were stuck in their car in that queue. A car came from behind, overtook the queue, skidded and finally crashed into a wall. They were then asked, "Why do you think the driver crashed?" Their answers were always along the lines of "They were driving too fast" or "They were not paying attention".

The next three students were then brought into the room. This time, the scenario was that you were driving along the road and came upon a queue of traffic. You skidded, overtook the queue and crashed into a wall. They were then asked, "Why do you think you crashed?" Their answers were always along the lines of, "I must have had a brake failure" or "There might have been oil on the road".

It's not unusual, when we are in Chimp mode, to be self-serving and condemn others but justify our own actions.

When I discussed the results of the experiment with the students, and they moved into Human mode, they were far more objective and were able to correct self-serving bias.
One way, to avoid self-serving bias, is to always put yourself into someone else's position and think of reasons why you might have acted in the way that they did, before you make a judgement call. [437] [438]

Self-serving bias can damage relationships. The Chimp's lack of tolerance with errors or mistakes is often more pronounced in friendships or partners compared to strangers. Our Chimps ask so much more of those close to us. If we find our Chimps are being unreasonable and unforgiving to those we are close to, then it can be easy to move into Human mode by asking yourself the question, "If I had made this error, how would I like my partner or friend to help me through it?". This can remove self-serving bias and also give us a way forward.

Constant wariness

The nature of the Chimp is one of vulnerability with accompanying wariness.

In order to make it feel secure the Chimp might put demands on friends or partners. Alongside this, the Chimp might also demonstrate a lack of trust by constantly checking motives. The Human recognises that all relationships have to be based on trust because we can never know, for example, where our partners are all of the time, or what our friends might be thinking. The Chimp's lack of trust often leads to checking behaviour and the destruction of the relationship. If we move into Human mode, then we can put trust in people and not allow ourselves to jump to conclusions. We can recognise that feeling secure is something we need to build internally for ourselves. Creating and living by your Stone of Life is the foundation for this stability.

Example: *Gerry and his colleagues*

Gerry is the foreman on a building site. He constantly checks to see if all the men are working and challenges anyone who he thinks might be slacking.

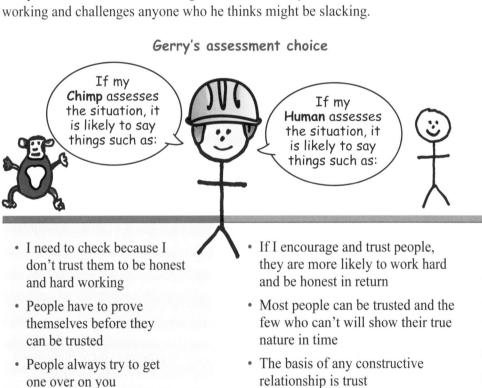

Gerry's assessment choice

If my **Chimp** assesses the situation, it is likely to say things such as:

- I need to check because I don't trust them to be honest and hard working
- People have to prove themselves before they can be trusted
- People always try to get one over on you

If my **Human** assesses the situation, it is likely to say things such as:

- If I encourage and trust people, they are more likely to work hard and be honest in return
- Most people can be trusted and the few who can't will show their true nature in time
- The basis of any constructive relationship is trust

Clearly, there needs to be a balance between naive trust and constructive trust. Pause and think of the consequences to the work output and the general mood on the building site of employing either the Chimp's approach or the Human's approach.

Conditional support

The Chimp typically likes to check on who is putting the most into a relationship. It thinks in terms of, 'Who owes who?' If the Chimp has done something for someone then it feels that they now owe a favour back. Any acts of kindness by the Chimp are frequently attached to conditions or expectations of a return. When we are in Human mode, we accept that when we do something pleasant for someone, then the reward is our feeling good about it; there are no strings attached. In Human mode we give unconditionally.

Example: *Ty and the loan*

Ginny has fallen behind with her rent but is due some money in a few months time. Ty, a friend of hers, has offered to lend her the money for her rent. After a few weeks have passed, Ginny receives her money and pays Ty back. Here is how the Chimp and Human would typically view this:

Ty and his stance on giving

If my **Chimp** assesses the situation, it is likely to say things such as:

If my **Human** assesses the situation, it is likely to say things such as:

- You owe me for when I helped you out
- If I need any type of help you should give me that help
- You need to remember and be indebted to me forever

- It was my choice to lend you the money, therefore you owe me nothing
- I did this to help you, not for any other reason
- Relationships are not about measuring who gives the most

What I am hoping to do, by giving these examples, is to promote you to think about your own relationships and how you could improve them and feel better about yourself. You might not agree with a lot of the statements given; that is not a concern. My concern is that I promote us all to reflect on whether we are allowing our Chimps to quietly hijack us without us realising it. Please consider your own Human and Chimp responses and reactions, and reflect on your own answers.

Building and maintaining relationships

Any relationship, whether professional or personal, is healthy if it builds us up. Check on whether you are being built up and also if you are building up the other person in the relationship. 'Building up' can mean improving in areas such as:

- Better self-confidence
- Better self-esteem
- Feeling more secure
- Feeling appreciated
- Feeling understood
- Feeling a sense of belonging

If we step back from this list, we can see that most of the items are about how we feel. In other words, the person is calming our Chimp down and bringing the best out of us.

A danger can occur when our Chimp is fooled. For example, by someone who settles the Chimp using superficial flattery or gifts, which are given to make the giver more popular with the recipient, rather than given to please them.

It isn't just the Chimp that can be built up. Our Humans are built up with many of the same things that build up our Chimps. However, our Humans are specifically strengthened by sharing values and with caring, selfless, constructive behaviours. Being selfless and putting the other person first in any interaction is a clear way to improve any relationship. Being a doormat, and trying to please someone by always putting them first, isn't helpful! It's a fine balance but one worth giving some time to think about, if you wish to build a sound relationship with someone.

How do we work practically to build and maintain a relationship?

Stating the obvious: most relationships manage to develop regardless of any effort put into them. However, if we put effort into a relationship then it stands to reason that if that effort is effective, we will develop and maintain good friendships and intimate relationships.

So, exactly what do we do to be effective? One way of clarifying what we need to do, is to ask ourselves what we would like from a partner of friend, and then to see if this is what we are offering others.

The list of what we might want, would probably include some of the following. Someone who:

- Listens to us
- Takes a genuine interest in how we feel
- Understands us
- Puts our interests first
- Encourages us
- Is warm and approachable at all times
- Doesn't take offence at our mistakes
- Is trying to make us happy
- Knows what we like and makes this happen

Matching expectations of others against self

The one in five rule

A caution when trying to form personal or professional relationships:

- Four out of five people are on our side and wanting to befriend us
- One of those will be very supportive of us at all times
- There are a lot of people who are reasonable, have morals and are generally pleasant individuals
- However, the reality is that *one in five* people won't be pleasant, won't be reasonable and will criticise us, whoever we are and whatever we do

There is therefore no point in taking their criticisms and comments seriously. There are a lot of them out there and they are by nature or choice unpleasant, especially on social media. Their comments and actions merely reflect an unpleasant and often cowardly attitude.

The one in five rule

Much of my work involves supporting people, who are being criticised, usually unfairly, by unpleasant people. The answer is always the same; we cannot stop those one in five people from being unpleasant, caustic and unkind. We can, however, choose to ignore them and listen to our Troop and other reasonable people. Even better, we can rely on ourselves to gain strength.

Key Point

There are some unpleasant people, who you will never please – don't hand them your time, energy or happiness.

Dealing with past relationships

Sadly, some personal relationships don't work out and can leave us feeling upset. Initially the end of a relationship can devastate us as our Chimps feel that the end of the world has come. Our Humans, on the other hand, know that life will go on and it is likely that another relationship will develop in the future.

How do we manage to look back on a relationship, that meant so much to us, but has come to an end? We have previously looked at how we are likely to go through a grief reaction but what happens once we have gone through the grief? Is it really just ghost emotions we are experiencing? We have previously covered the idea of ghost emotions returning to remind us of what we once suffered, but what about the rational thoughts that appear?

For example, 'What if I had been different?' or 'I really believe this was the right person for me', or 'We had a great relationship and now I feel I have failed because it's gone'.

One way to come to terms with these thoughts can be to put a relationship into time context. All relationships have a time period where they operate. Some of these times are short and some long but most are time limited. If we can move into Human mode, and remove the emotions that we are experiencing, then we can accept that when we were in a relationship, we did what we did and we were as we were, and these are fixed in time. Of course, things could have been different but they weren't. We can only change the way we act in the future.

Similarly, every relationship changes on a day-by-day basis. Sometimes the relationship changes and doesn't work anymore. It is helpful not to see this as a failure but rather as a natural change. Learning to recognise that a relationship has run its course, and knowing when to move on, is a success. Whatever time period the relationship ran for, it was a success for that period. Relationships are only seen as being a failure if we allow the relationship to go on too long, instead of letting go and celebrating the success we had for that time the relationship did work. Of course, it's always good to try and revitalise a failing relationship. The difficult part is recognising when to stop.

Losing relationships that meant a lot to us, will mean we need time for grieving. We will also need plans to manage ghost emotions and a rationalisation of our beliefs in order to move forward.

Unit 24 Reminders

- The Chimp and Human have very different approaches to relationships
- The Chimp can have unrealistic expectations of others
- The Chimp can form an unhelpful dependency on others
- Building relationships is a skill that we can acquire
- The one in five rule can help us to come to terms with an unreasonable person

Unit 24 Exercises

Focus points and reflective exercises

1. *Removing any self-serving bias*
2. *What are you offering others?*
3. *The one in five rule*
4. *The emotional criminal*

Focus 1: *Removing any self-serving bias*

It is always very easy for us to justify to ourselves any actions we take, even if they are not correct. In our minds, we appreciate all of the possible influences that led us to make an error or a poor judgement call. For example, if we have just had upsetting news or if we are feeling unwell, then we understand why we might have acted out of turn or out of character. We then usually put our error into context and although we might feel unhappy with our actions, we feel at ease that they made sense to us. When someone else makes an error or acts under the influence of their Chimp, we need to have a plan if we don't want this to spoil our relationship with them. One way to increase our emotional skills and to build our relationships, is to ask ourselves the question:

"If I had made this error, how would I like my partner or friend to help me through it?"

The answer might be that I want them to:

- Understand the influences on me at that time
- Stand by me
- Allow me to express how I feel
- Help me to get over it

Exercise: *What would you want your friend or colleague to do?*

Draw up a list of points that you would like someone to follow when you have made a mistake or acted out of turn.

Imagine that you are now in their place and practice these points with someone who has made a similar mistake. By practicing, your Computer will have a chance to take over when the situation arises.

Focus 2: *What are you offering others?*

Relationships are a two-way street. The most productive relationships are also built on an unconditional giving to others. One way to improve relationships is to work on yourself and see your reward as being the best friend or partner that you can be and being proud of this.

Exercise: *What you would like from others*

Draw up a list of the features that you would want a partner or friend to display towards you. It would be helpful to also add in things such as, their general outlook on life, their sense of humour, their reliability and stability of mood. Then check on whether you are offering these yourself. If you are not doing this, then form plans to turn this around. Address the things that are stopping you from being the ideal friend or partner that you would clearly want to be. Please remember, the reason you are unlikely to be presenting to the world the person that you want to be, is because there is interference from the Chimp and Computer. The work is about managing the Chimp and tidying up the Computer.

Focus 3: *The one in five rule*

We have already covered the need for a Troop and how to use this to gain strength. The one in five rule strongly supports this. If we can gain the skill of dismissing those unkind people's comments and seeing them for what they are, then emotional stability can be achieved. I cannot emphasise enough that for every unkind person, there are many more kind people who would willingly help us. Many of us watch terrible emotional attacks on others and would willingly help the person, if they reached out. We are never alone. Support for anyone is out there if they reach out. There are a lot of people with kind and big hearts only too willing to help.

Exercise: *Recognising and dealing with the one in five person*

If we first accept that we will meet many people who don't like us, then it won't come as a surprise when we meet them. If they choose to be unkind or are unpleasant towards you, then try to recognise they fall into the one in five category. Most of them are like it with all of us – it isn't you!

Ask yourself how much time you want to spend trying to please those who are never going to be pleased or listening, and giving any importance to what they have to say about you. Then consider how much time you want to spend pleasing those who love you, and listening and giving time to what they have to say about you. In Human mode, we all have a choice on the importance we place on things; in Chimp mode, we don't!

Have a plan to deal with an unpleasant person when the situation arises. The plan could include:

- Recognising a one in five person
- Dismissing their opinions or actions by looking to the opinions of my Troop
- Reinforcing that you are not alone and reaching out for help
- Talk through the experience and your feelings
- Celebrate the good people in your life

Focus 4: *The emotional criminal*

It's often said that people are either energisers or energy sappers. Whenever we meet someone, they inevitably have some measure of emotional affect on us. If the interaction is a negative one, then negative emotions can occur. Most of this negative emotional affect is unintentional but it is still negative.

Exercise: *Seeing yourself as the emotional criminal*

Reflect on whether your interactions with others are energising or energy sapping and if you are evoking welcome or unwelcome emotions in them. It's surprising how few people stop to think about the effect they are having on others. A simple energising approach can make a big difference to others in your life and ultimately to you yourself. Decide if you are being an emotional criminal and if so, what you can do about changing it!

Unit 25
Communication

Unit 25: *covers effective communication by considering: emotive words, Chimp and Human approaches to communication and a check on our communication skills.*

One of the most important factors for forming a sound relationship is good communication. This unit will look at the differences in the way that the Chimp and Human communicate. The Chimp communicates by using feelings and the Human communicates by using facts.

The two key concepts covered will be, the way that the Human and Chimp:

1. Perceive information and express themselves
2. Operate during verbal interactions

Key concept 1: *How the Chimp and Human perceive information and express themselves*

The Human hears the words; the Chimp hears the music

The way the Human and Chimp perceive and express communication is very different. Whenever a communication is received by the mind it evokes an emotional reaction in the Chimp and a rational response in the Human. [154] [155] [156] [260] [204] [439] Most of the time, we are not aware that there has been an emotional reaction. These emotional reactions vary in strength and generally we are only aware of the stronger ones. [164] [440] When words are involved, some words evoke strong emotions, for example, death, anxiety and love. These are termed emotive words. Other words don't usually have any significant emotional content attached to them, such as desk, cloud or apple. However, for one individual, apple might bring back memories of an experience that did have an emotional content and therefore will evoke an emotional response. The important point is that most words will have some form of emotional attachment and therefore evoke reactions in us. [440]

STAGE 7

Psychopaths and emotive words
- Scientific points

Individuals termed 'psychopathic' have measurably different reactions to emotive words compared with the rest of us. One experiment showed words or jumbled letters on a screen. The words were of two types: emotive words and non-emotive words. When subjects saw a word, they were to press a button and the time taken to recognise the word was recorded. The non-psychopathic group recorded faster reaction times to emotive words compared with non-emotive words. The psychopathic group showed no difference in reaction times to emotive and non-emotive words. This is because their brains are wired differently and they do not experience emotion as intensely as the typical brain. Their amygdala also shows less activation with emotionally based moral situations. This is one reason why the psychopath doesn't feel empathy in the same way as everybody else. [441] [442]

We know how important it is to choose our words carefully in difficult conversations because the wrong word, especially one with a strong emotional attachment, can change everything!

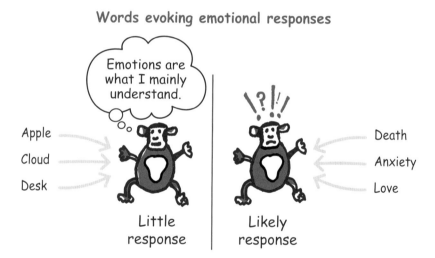

Words evoking emotional responses

Practical application

Before you engage in a significant conversation with someone, it helps to think about the kind of emotional reaction that you would like from them. Consider which words would either help or not help you to get the reaction that you want.

For example, here are two sentences, at the extreme ends of communication, when trying to get someone to stop tapping their pen on a desk while you are trying to concentrate:

"Incessantly tapping your pen is very irritating and it annoys me, can you stop immediately"

"I'm sorry to interrupt you but I am finding it difficult to concentrate when you tap your pen"

The first sentence has a number of emotionally evocative words: 'incessantly; irritating; annoys; immediately' – these are more likely to evoke emotionally negative reactions in the other person. The second sentence has a few emotionally evocative words: 'Interrupt; difficult' – and these are unlikely to evoke emotionally negative reactions. The word 'sorry' can evoke positive emotions. You can decide for yourself which of the two sentences might be better for achieving the outcome that you want.

Try this exercise: think about someone who you find it difficult to get on with and use some words to describe them. Now try using different words, that might give a similar description, but that don't evoke negative reactions within yourself. The situation might not change, but you are much less likely to cause yourself to have distressing feelings.

For example, when describing the person that you dislike, you might say: "I find him selfish, ego-driven, thoughtless, devious and undermining". Describing the person using these words, even though they might be true, is very likely to make you feel disturbed: you have evoked your own negative feelings. You could change your description to: "He appears to have an absence of emotional intelligence and builds his world around himself". This sentence removes your emotions, looks at him objectively and also doesn't have the strong emotive words. Both sentences say similar things, but the second sentence is much more likely to leave you feeling more at ease. It is a trainable skill to learn to be non-judgemental and objective when making comments. [443]

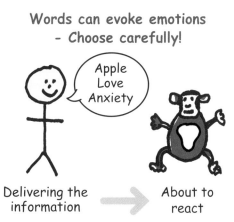

Words can evoke emotions - Choose carefully!

Delivering the information → About to react

Communicating with words and emojis
- Scientific points

Try reading the following text and then see what research has shown us.

'It's been a really great day. In fact, perfect, but we are expecting some rain soon.'

Research shows that when we send a text like this, with a strong positive content, but add a negative emoji, the person receiving the text is likely to perceive the sender as not being in a good mood. This is because our Human interprets the facts of the message but our Chimp interprets the feelings of the message. In many people, the negative emoji can override the positive message. What is happening is that our rational Human interpretation of the facts is being overridden by our Chimp's feelings that have been induced by the negative image of the emoji. [444]

When we communicate with someone, their Chimp hears the feeling of the message, whereas the Human hears the words.

For example, here are two responses to the question 'What do you think about the rise in child neglect?'

Response 1 – "I think it's disgusting and unacceptable. No child should be subjected to abuse or neglect. It leaves me angry at times to see such lack of care."

Response 2 – "I think every child should have a great childhood with support and encouragement. Every child deserves a great environment with security and happiness."

Although each respondent might have similar thoughts, someone who hears these responses can find their Chimp will receive the messages very differently. The Human hears the words and will rationally understand the position of both respondents. However, the Chimp only hears the feelings that the words evoke. Therefore, respondent 1 will evoke a negative feeling in the Chimp and the Chimp might well not like the message or the respondent! Respondent 2 will evoke a positive feeling in the Chimp and the Chimp is liable to like both the message and the respondent.

Key Point

When message are sent: the Human hears the words but the Chimp only listens to the music.

Make sure the music is good!

A word of caution regarding the Stone of Life and negative statements! Can we turn some of them around?

Our Stone of life might have some negative but helpful truths on it. For example: it might have that 'Life is not fair', and this might be helpful in accepting some injustices. However, if we have a lot of negative statements on our Stone of Life then it can have a negative effect on our Chimps. It is possible to turn some negative truths around so that both Human and Chimp find them pleasant and helpful.

For example, 'There will always be setbacks in life', could be turned round into 'Setbacks are opportunities for learning'.

Key concept 2: *How the Chimp and Human operate during verbal interactions*

Effective communication begins with listening

Why is listening so important? Unless we establish the facts of what we are discussing and both start from the same page, then it is unlikely that we will get very far. It's important to make sure we enter into a significant conversation in Human mode because the Human will start by establishing the facts.

Example: *Owen and the cream cakes*

Owen has two children, aged 5 and 6. He had bought two cream cakes for them to eat after their evening meal. He specifically told them not to eat the cakes before this. Later in the afternoon he stepped outside to do some gardening. On returning he saw that the cream cakes had gone and both children had cream on their faces. Owen's Chimp went into melt down. It raged and brutally told the children that he was not only disappointed in them but they were a great disappointment to him. He left the room with both children in tears.

Let's repeat the scenario, but this time Owen's Human will respond. When Owen's Human saw the cream on the children's faces, he asked what had happened. The children then told him that their mother had come home and brought some great desserts for after the evening meal. She had said that they had been so good this week, that they could eat the cream cakes now, and have desserts tonight as well. Just by fact finding, scenarios can change in front of us and stop our Chimps from humiliating us!

> **Key Point**
> *Effective communication begins with listening.*

Who is speaking?

Conversations take place in one of four scenarios:

Human to Human
Human to Chimp
Chimp to Human
Chimp to Chimp

Recognising which interaction is occurring can be helpful for moving the conversation into a Human-to-Human interaction. We have previously covered how to move yourself from Chimp into Human mode and also how to move somebody else from Chimp into Human mode.

If we look at the difference between the Human and Chimp approaching a conversation then it becomes obvious why Human-led conversations are usually productive, whereas, Chimp-led conversations are usually unproductive.

Differences in approach when communicating

	Chimp	Human
Agenda	• Intention is to get you on to their page • To win • Express emotion • Attack and defend • Go on feelings and impressions • Don't give way or change stance • Look good and innocent • Have excuses ready • Be a victim, if all else fails	• To help both of you to be on the same joint page • Understand the other person • Allow expression from the other person • Look for a solution • Establish the facts • Be willing to change or compromise • Accept responsibility
Method	• Speak first • Shout and be emotional • Interrupt • Use emotive words • Dominate with speed and volume • Intimidate with body language • Focus on the problem • Be devious if necessary • Offer only the truths that serve your purpose	• Listen first • Remain calm • Speak slowly and softly • Recognise opinions are not facts • Reason and discuss • Find common ground • Be reconciliatory • Focus on solutions • Accept differences
Beliefs	• I have spoken therefore you have heard • You have heard therefore you have understood • You have understood therefore you now know why I said it • If this isn't true then I will get frustrated and you are at fault	• I have spoken but I need to check if you have heard • You have heard but I need to check that you have understood • You have understood but I need to check you know why I said it • We are trying our best but I might have to try again

Getting into Human mode might mean that you have to establish what your Chimp wants to say, and its agenda, and make sure that you agree with this, and then represent it.

You can always keep a check on yourself to ensure that you remain in Human mode throughout the conversation.

Checking your communication skills

Sometimes we have to give instructions to someone to make sure that they get important things done and done correctly. I have worked with many team leaders and parents who complain about their team or children not listening. By observing their communication styles, some important factors can be seen to be missing. Here is a simple checklist that could be used to measure how you are doing, and how you might improve, when imparting instructions; one particular step is critical!

Checklist for auditing important instructions

1. Are the instructions clear in your own mind before you begin?
2. Did you stick to the main points when explaining?
3. Can the person repeat the instructions?
4. Do they know why they need to follow the instructions?
5. ***Do they understand the consequences of following or not following the instructions?***
6. Have you checked that they agree, and nothing is preventing them from following the instructions?
7. Have you thanked them?

Point 1 is very important. If we don't have a clear idea of what we are about to say then it might become a very ineffective message. One of my favourite amusing newspaper headlines appeared in Utah in the USA. It read:

Utah Poison Control Centre reminds everyone not to take poison'.

It helps to think about the message we are putting across!

On point 2, when we try to take in information there is obviously a limit to what we can retain. This is very variable and depends on the person and the information. However, sticking to the points and not wandering into extra detail will help. The critical step is highlighted in blue because this step hands responsibility to the person who is listening. Once we appreciate the consequences of what we are doing or not doing, we are much more likely to take things seriously.

Example: Bella and her sister

Bella has asked her sister to call in at the Chemist to get some medication for their mother. Bella didn't emphasise the request, even though it is important that the mother gets her medication. Her sister heard the request but forgot to collect the medication. Her sister felt bad about this and went back into town to get the medication. This made a mess of her plans for the day.

Had Bella added step 5, it would have been much more likely that her sister would have remembered. In this example, step 5 would have been to ask her sister what the consequences would be to her and her mother if she forgets. When the sister realises that forgetting would mean her day will be messed up, then she is much more likely to remember what she needs to do.

Unit 25 Reminders

- The Human hears the words; the Chimp hears the music
- Effective communication begins with listening
- Conversations can have four ways to operate between two Chimps and two Humans
- Chimps and Humans have different agendas, methods and beliefs when communicating
- We can use an audit list to check on how we are communicating

Unit 25 Exercises

Focus points and reflective exercises

1. Making good music
2. Reviewing the Truths on your Stone of Life
3. Checking on who is speaking

Focus 1: *Making good music*

Exercise: *Phrases and expressions reviewed*

When communicating with someone, reflect on the expressions that you are using and ensure that the general feel of the communication is a positive one. It can help if you reflect on phrases that you might use commonly. Make sure these are positive in nature and also consider introducing some new positive ones that suit your style of communication.

Focus 2: *Reviewing the Truths on your Stone of life*

Exercise: *Ensuring your Stone of Life has some inspiring Truths*

Some Truths can be rephrased so that they become inspiring rather than something difficult to accept. For example: 'Sometimes things won't go your way' can be rephrased as; 'When things don't go your way, there is always an opportunity, if you look for it".

Focus 3: *Checking on who is speaking*

Exercise: *Listening or expressing?*

In a conversation that is important or filled with emotion, check whether you are in Human mode or Chimp mode. One way to do this is to ask if you are listening and considering what is being said, or whether you are listening and waiting to speak, in order to challenge anything that is said. Checking on who is speaking might not be as easy as you think! It is a good exercise to do because it can make for very good communication skills.

Unit 26
Robustness and resilience

STAGE 8: *will focus on becoming robust and resilient, and cover troubleshooting.*

Unit 26: *will revise some key points and pull the previous stages together.*
By doing this, we will be able to understand how to achieve robustness and resilience

At the start of this course a question was posed: "How do you want to be?" Probable answers might include:

- Be confident
- Be happy
- Have peace of mind
- Be successful
- Understand myself
- Possess a positive self-image
- Be able to manage emotions, thinking and behaviour

When you consider these aspects, it becomes apparent that the characteristics that most people would like to have, rely on the person being both robust and resilient. We can look at how the previous stages have led us to the position where we can put into place processes and skills that will make us both robust and resilient. I will start with some definitions of how we could view the terms, 'robust' and 'resilient'.

Definitions of robustness and resilience

Robust: Robust is *being prepared and ready*. This means that you are in a great place, each day, **BEFORE** you engage with the world.
Becoming robust means having plans in place to manage your own mind and also whatever situations you meet in life.

Forming plans is something anyone can do by just making the effort. How robust you are will depend on how good your plans are.

Resilient: Resilient is being able to bounce back and manage the challenges of life. In other words, *remaining* in a great place or *returning* to a great place throughout the day, and ending the day in a great place. Resilience is a skill.

Two common questions

Before we begin to form plans for robustness, it might be worth answering two questions that commonly get asked.

What if I make plans and they are not working? The plans will always need revising because life and circumstances change. Many aspects of our plans will hold true throughout our life and those aspects that are not working, can be revised

What if my plans are really good but I just can't implement them? What has been emphasised throughout this course is that resilience is a skill. Therefore, there will be days when we struggle to apply the skill and we have to see this not as a failure but as a learning point. What is true is, that the more we practice, the better we will become at using the skills of mind management. Therefore, becoming resilient, by acquiring and maintaining skills, can be a difficult thing to do and is something to work on throughout life.

Example of robustness and resilience: *A toy for a child*

If we want to create a toy suitable for a child to play with, then we make plans. When we think the plans are ready, we create the toy and have made it robust. In other words it is prepared and fit for purpose.

However, once the child then plays with the toy, we want to see if the toy survives the rough and tumble it will receive. If it does, then it is resilient. If it is not resilient then we either, go back to our plans and make them more robust, or we give instructions not to subject the toy to certain conditions, like throwing it against a wall or putting it under water.

Similarly, as an individual you can become robust with your plan to face the world but accept that there might be limitations as to what you can subject yourself to. For example, there will be certain conditions or situations that your Chimp will not tolerate, such as an unacceptable job, incompatible relationship or unfulfilling surroundings.

Key Points

- **To become robust, we form a plan**
- **To become resilient, we learn a skill.**

An understanding of how robustness can be achieved

We have now covered the major factors that will help you to become robust. These major factors will be given without detail, as these details can be found and revised by rereading the previous relevant units.

Major Factors

STAGE 1

We began in stage 1 by gaining an understanding of our mind and the rules by which it works. In order to do this, we applied the Chimp model for accessing and working with the mind.

The components of the mind

For robustness, we need to have an understanding of:
- Exactly who we are and be able to clearly separate ourselves from the machine
- The nature and functioning of our unique Chimp
- The nature and functioning of our Computer

The rules of the mind:

Power	>		The **Chimp** can overpower the Human
Speed	> >		The **Computer** will take over if it is programmed
Advice			The **Computer** can advise decisions

For robustness, we need to have an understanding of:
- How we can manage the machine but cannot control it
- How and why the Computer plays *a major role* in mind management

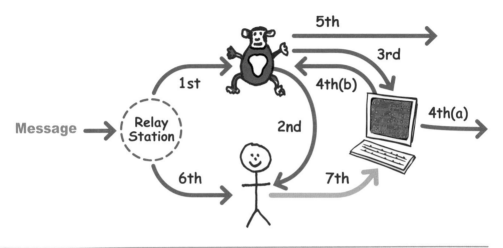

The order that information is dealt with:

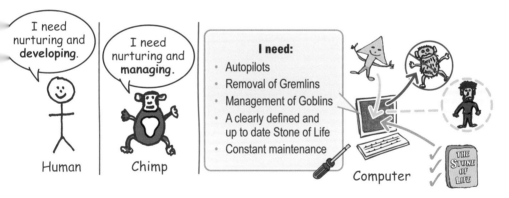

Looking after self and each part of the machine

For robustness, we need to have an understanding of:
- The need to nurture, manage and develop all parts of the machine and ourselves
- How to use development time to maintain and improve our mind management

STAGES 2 and 3

Stages 2 and 3 were focused on the Chimp. Unless we intervene, the Chimp system will run our mind. Therefore, a lot of work needs to be done to understand and work with the Chimp system.

> There could be trouble if you don't understand me and manage me.

For robustness, we need to have an understanding of:
- How to manage our Chimp in a structured way (three stage process)
- How to use distractions and rewards to manage the Chimp
- How to recognise, and manage, positive and negative aspects of our Chimp
- How to nurture and befriend our Chimp
- How to express emotion effectively and constructively
- The importance of making changes after expressing emotion
- How to recognise, and manage, emotional messages from the mind
- How to manage our biological drives

STAGE 4

Stage 4 focused specifically on the Computer. In this stage, we looked at how the Computer forms habits and processes information. This helps us to programme the responses that we want to have to both trivial and significant life events. This stage also gave insights into how the Computer can help the Chimp to process information, which helps us to come to terms with losses, changes and injustices.

> Programme me for helpful habits and managing responses.

For robustness, we need to have an understanding of:
- Why and how habits are formed
- The main factors that will help promote a change of behaviour or life style
- How to change habits
- How the Chimp, Human and Computer process life events
- How to programme the Computer to help process events and emotion
- How to manage grief, loss and other significant life events

STAGE 5

Stage 5 covered some critical elements, when it comes to achieving emotional stability, peace of mind, happiness and confidence. The two major stabilisers of the mind were covered: the Stone of Life and the Troop.

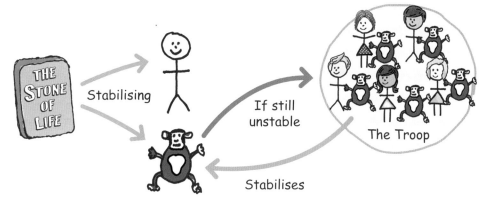

For robustness, we need to have an understanding of:
- The importance of the Stone of Life and the Troop
- How to work with reality and Truths and how to live with them
- How to establish values and live by them
- How to gain and retain perspective
- How to form, manage and maintain your Troop

STAGE 6

Stage 6 covered managing and preventing stress, managing your environment and giving attention to recuperation and sleep, in order to improve our physical well being.

The Chimp takes over during sleep

For robustness, we need to have an understanding of:
- How to manage stress by working with the stress reaction
- How to recognise and remove chronic stress
- How to prevent stress from occurring
- How to create and manage our environment
- The importance of recuperation
- The nature and management of sleep

STAGE 7

Stage 7 focused on our relationships with others and how to get the best out of these.

For robustness, we need to have an understanding of:
- The way that other people's minds are structured and function
- How to help move someone from Chimp to Human
- Accepting and working with someone as they are
- What the foundations of a successful relationship are
- How to communicate effectively with others

How to form a plan to establish robustness

The major factors listed under each stage can now be turned into a checklist of practical actions. It is impossible to check on all of the following points, all of the time. I would suggest that you choose a few at a time to reflect on and ensure that you know how to put them into action. Some points will resonate much more than others, so be selective! These points could be ideal for reflecting on during development time.

Suggested menu checklist for a robustness plan

STAGE 1

- Always separate yourself from your machine
- Don't take responsibility for the nature of your machine but do take responsibility for managing it
- Remind yourself that you always have a choice to operate by one of three systems: Human, Chimp or Computer
- Respond to situations; don't react to them
- Remind yourself that the role of the Chimp is to alert you to a problem and your role is to come up with a solution
- Remind yourself to start from where you are and what you have got and NOT from where you want to be and what you have not got
- Apply the NEAT exercise when you can
- Look after yourself by having an excellent relationship with yourself
- Be your biggest fan and not your worst critic

STAGE 2 and 3

- Use the three-step process and bananas to manage your Chimp
- Use some grade A hits that you know will settle your Chimp
- Fully accept the nature of your Chimp and nurture it
- Know how to commit to tasks if motivation is missing
- Don't allow your Chimp to compare itself to others
- Don't allow your Chimp to do a juggling act
- Prepare for the day ahead by programming your Computer
- Express emotion constructively
- Follow up exercising the Chimp with a plan for the future
- Interpret emotional messages correctly and then plan
- Recall: Chimps use emotions to dwell on; Humans use emotions to act on
- Recognise and manage emotional scars and ghosts
- Address any emotions that lie under the surface of your mind
- Fulfil your drives constructively and draw a line where needed

STAGE 4

- Use the Triangle of Change where appropriate
- Hold a positive and correct self-image
- Define success via your Human
- See acceptance as a skill
- Use a Computer programme to process simple events
- Address any grief, loss and other significant life events

STAGE 5

- Be in possession of an up-to-date Stone of Life
- Remind yourself regularly of your Truths of Life
- Work with reality and not expectation of what should be
- Have clear values and live by them
- Regularly praise yourself when you live by your values
- Constantly remind yourself about keeping perspective
- Be in regular contact with your Troop
- Maintain your Troop

STAGE 6

- Prevent stress from occurring, whenever possible, by avoiding or managing your trigger points
- Remove stress, as it appears
- Check for any chronic stress and remove the cause
- Look after your recuperation
- Look after your sleep
- Create and maintain the right environment for you
- Make sensible lifestyle choices

STAGE 7

- Appreciate that others have Chimps and Gremlins
- Accept people as they are
- Be able to help someone to move from Chimp to Human by using EUAR
- Use your judgment, if appropriate, rather than judging others
- Communicate in Human mode
- Listen carefully with a view to understanding others
- Hold your own happiness and don't give it to others
- Recognise and apply the one in five rule

Example of forming a robustness plan:

The first thing to do is to select the main points from the suggested menu that resonate with you.

For example, these could be:

1. Respond to situations; don't react to them
2. Be your biggest fan and not your worst critic
3. Know how to commit to tasks if motivation is missing
4. Don't allow your Chimp to compare itself to others
5. Remind yourself regularly of your Truths of Life
6. Maintain your Troop

These six points could now be focused on daily. By focussing regularly on these points, they will become second nature as you programme your Computer to bring them to the forefront of your mind. By putting the points into practice, the skill of applying them will be acquired, thus leading to resilience.

Clearly, the checklist can be changed from time to time to develop more skills. For most people, the Stone of Life will be the most crucial point to work on and keep active every day in their mind.

Becoming resilient

Now that you have formed a checklist of actions, you have a robustness plan to work with that is specific to you and you are ready to engage with the world outside.

To become resilient you need to practice using your robustness plan on a day-to-day basis and develop the skill of managing your mind.

Example: *Egan and his poetry*

Egan writes and publishes poetry. He has a small following of people who really enjoy his work and push him for more poems. He was persuaded to publish some of the poems on his website. As soon as he did this, he saw several very caustic comments appear. He recognised that these comments were unkind and unnecessary. He also accepted that you can't please everybody all of the time. However, his Chimp refused to settle and he found himself in a bad place and wishing that he had never published his work.

Egan did have a robustness plan, which included some Truths on his Stone of Life. Some of these Truths were:
- The one in five rule is something I strongly believe in and therefore I can accept some people will be unfriendly
- I will thrive if I live by my own values
- My friends are the people whose opinion matters

He also has a plan that under stress he will turn to his Troop for help.

The robustness plan seems to be a good one. However, Egan found it didn't stop his Chimp from eating him alive. What does he do?

First, his plan needs to be looked at. Did he follow it? Did he reflect and really spend time thinking about his truths and allow them to settle his Chimp? If he did this, did he turn to his Troop and gain their support and perspective. Often, people's plans don't work because they don't really put them into action; but what if he did put them into action? Then the plans need to be reviewed.

Could he add further Truths that will settle his Chimp? Does the plan need adjusting? For example, does the Truth 'Everybody has critics and everybody has fans' help him? If so, he can add this to his list of Truths. Perhaps the Truth 'Any criticism is going to upset my Chimp but it doesn't have to upset me, and I can manage my Chimp' might help, or 'I won't give my happiness to someone else'. A different Truth could be, 'If I choose to listen to the criticisms then I must accept the consequences'.

If new Truths don't seem to help, perhaps Egan can lean more heavily on his Troop and ask them to say why his poetry does help them and what they find good about it. Their statements will mean much more to him and will counter the criticisms.

Example: *Jana and her robustness loophole*

Jana presented a really good plan for robustness. She had thought of every eventuality and every situation and had plans and answers to manage them all. There was just one snag. Jana didn't like herself and had low self-esteem. It's really important that she looks after herself and this includes having a good self-image. Without a good relationship with herself, none of her plans are going to be very strong. Jana's task is to build her self-esteem by working with her Human and basing her self-esteem on the values that she demonstrates, rather than listening to her Chimp and basing her self-esteem on more superficial things.

> **Key Point**
>
> **Robustness and resilience can only be achieved if the Chimp is managed and the Computer is tidied up: All Gremlins need to be found and removed!**

Building resilience in children
- Scientific points

Often when parents raise a child they want to develop resilience in the child but are not sure how to do this.

Research shows that parents who constantly criticise their child can create self-doubt and lack of self-confidence in the child. [445] Parents who constantly praise their child can produce a child with narcissistic tendencies. [446] [447] [448] The parent who gives appropriate measures of praise and criticism would appear to be developing a resilient child, but this doesn't appear to be the case. Research indicates that a child who is taught how to self-assess with appropriate measures of self-praise and self-criticism is more likely to develop into a resilient adult. Research on four-year-old children shows that if they are **taught how to self-assess** they usually develop into resilient teenagers many years later. [449] [450] [451]

Very young children necessarily look to their parent for approval of what they are doing, as the child is operating in Chimp mode. The Chimp must look externally and turns to the parent and relies on the parent's judging of their worth. The child sees this as a measure of their success or failure. If the child continues to look externally then when they reach their teenage years, they replace their parent's judgement with their peer's judgement. [452] [453] In other words, they remain in the hands of external sources to judge them. I call this 'The Fridge Door Syndrome'. (See the next page).

However, if we teach children self-assessment, we are promoting the Human circuits to operate and these will look internally to decide on how they are doing. Instead of seeing *a judging statement* by an external person as being valid, they will only see *an opinion* from another person and will be able to **weigh up and accept or reject that opinion**. Furthermore, they will look to their own Human to decide on how they are doing and this will result in resilience.

The Fridge Door Syndrome and its cure
- Scientific points

The fridge door syndrome typically occurs when a child goes to school and is asked to paint a picture for their parent. On returning home, the child gives the parent the picture and the parent encourages the child by praising the picture. The parent then adds statements such as: I am really proud of you, you are so clever and I love you. To show how proud they are, the parent puts the picture onto the fridge door. This then lets the world know just how clever the child is. The problem here is that the child is being programmed to look externally to what it produces, in order to establish its worth. Not only this, the child is very likely to be in Chimp mode when receiving this praise and will add on that with this praise come the statements that the parent loves them and is proud of them. The child cannot rationalise and distinguish the subtlety of what the parent really means. Further, the child sees the world looking at its painting on the fridge door and accepts that their worth is dependent on what they produce and is decided by everyone else just how good they are.

The alternative scenario is to establish internal self-worth that is unconditional. In this case, when the child comes home with the painting, the parent asks the child to put the painting to one side for a moment. The parent then internalises the child's worth by letting them know that they are loved unconditionally and the parent is proud of the child just for being who they are. At this point the child can join in when looking at the painting and with encouragement can decide whether they think the painting is good or not and whether it should be put onto the fridge door. Clearly there might be help needed to encourage the child to praise or constructively criticise their own work and also clearly, the parent would compliment and encourage the child's efforts as well as their achievement. They key points are; that the child first establishes that they are validated without having to achieve anything, and that they are encouraged to praise and constructively criticise their own work by a parent who guides them with this.

This move to self-praise and self-validation has its roots in object-relations theory. Applying this becomes the cure for someone suffering with the fridge door syndrome by believing that, what they achieve and what others think, is more important than their own opinions. The cure is to replace the approving parent or others by entering Human mode and place yourself in the assessment and approving position.
[454] [455]

When we consider how children build resilience, we can apply this principle to ourselves even as adults. To bolster resilience we can learn how to constructively assess how we are doing by moving into Human mode. We can learn to praise and constructively criticise ourselves in a supportive and helpful manner, that will promote self-development. We can perceive other people's comments as opinions and not as judgements that must be valid. In Human mode we can easily reject any unhelpful comments from others, if we don't agree with their opinion about us, or what we do. As adults, we can ask others for their opinion not seek their approval and thereby build resilience.

Unit 26: Reminders

Robustness means having your mind ready and prepared to face the world

Robustness is achieved by:
- Nurturing and developing your Human
- Nurturing and managing your Chimp
- Keeping your Computer up to date
- Understanding and working within the rules of the mind

Resilience means acquiring the skill to manage your mind and keep it robust, whatever is happening in your life.

Resilience:
- Is a skill
- Is unlikely to be achieved at all times
- Will improve the more we practice
- Occurs, as we work within the rules of the mind

Unit 26

Focus points and reflective exercises

1. *Being robust*
2. *Warming up and warming down the Computer daily*
3. *Resilience by losing self*

Focus 1: *Being robust*

Recall that, **robustness = a plan: resilience = a skill.**

It's worth stopping and spending time to reflect on what becoming robust means to you. How will you be prepared to face the world and to manage yourself?

The question is; "Have you put in the work and formed a plan for becoming robust?"

Exercise: *Forming your personal robustness plan*

Use the 'menu checklist' to select the areas that you would like to give attention to. Work through each area one by one and ensure that you have a management plan for each one.

For example: if one area is 'to use grade A hits to settle your Chimp', then the plan could be to have five grade A hits ready to remind yourself of these, whenever something unsettles your Chimp. You can then return to Human mode.

Focus 2: *Warming up and warming down the Computer daily*

It doesn't take very much time to reset your mind into Human mode. Try to establish a fixed routine that allows you just a few minutes to prepare yourself for the day ahead. If you can also allocate time at the end of the day to warm your Computer down this will also be very helpful for keeping you in a good place.

Exercise: *Warming the Computer down*

At the end of the day, go back over events and ensure that your Chimp has not put into the Computer any distortions or misinterpretations of what has happened. These are new Gremlins that could grow with time. Go into Human mode and if necessary reinterpret events with the facts and the truth. Where appropriate, add perspective to the events of the day by asking how important they will be in a few months time. Finally, look to your values and how you stood by them, if you want to increase your chances of a good night's sleep!

Focus 3: *Resilience by losing self*

People who see themselves as being important will become more sensitive to criticism, more self-conscious and preoccupied with what others might think about them. Those who see themselves as not being so important gain a perspective about their significance in life itself. Therefore, what others might think becomes unimportant and self-esteem becomes based on self-assessment.

Exercise: *Questioning how others see you*

It is important to have good self-esteem but this can be balanced with perspective on our own self-importance. Try to enhance your resilience by being able to lose yourself and not take things so seriously. Laughing at yourself or your situations can be a great way to gain resilience.

Unit 27
Troubleshooting

If we can learn to manage our mind then resilience will be established. The question is; "What could stop us from becoming resilient?"

Unit 27: *will cover -*
- *Five reasons why the Stone of Life might not be working*
- *Five common causes that prevent resilience*

Troubleshooting

We will begin by looking at the Stone of Life and why it might not be functioning correctly, and then move on to common problems that prevent resilience.

In an ideal situation your Stone of Life would be up and running and able to manage your Chimp. This means that whenever the Chimp tried to react to anything distressing or unhelpful, it would look into the Computer and immediately be reminded about working with reality and truths, living by your values and keeping perspective. So, why is it that our Chimps can still get loose?

In order to answer this question, we will consider five common reasons why the Stone of Life might not be working. Methods will be offered for correcting any deficits and we will see how to keep the Stone of Life prepared and active.

Five common reasons why the Stone of Life might not be working:

1. The Computer is in sleep mode
2. The components are not all accurate
3. The Chimp or Human has unmet needs
4. There is a health problem or lack of self-care
5. The Computer needs to offer a grade A hit

1. The Computer is in sleep mode

An example might shed light on the commonest reason why the Stone of Life isn't working.

Example: *The puzzled team*

Everyone in the office was excited to have a new all singing and dancing printer and photocopier. According to the manual, it would be able to do everything. The problem was that it wouldn't feed the paper through. After some time of enduring frustration and cursing, it was decided to send the machine back or get the maintenance team out. One of the staff phoned for help and explained the problem. The reply was simple: 'Have you switched it on?'

I think most people have experienced this embarrassing situation, when trying to operate some electrical goods. We forget to switch the machine on and of course when we do, it helps! This leads us to the first and most common problem why the Stone of Life just doesn't seem to be working.

The Computer in your head is very similar to a real computer. If you wanted to retrieve some information from your real computer then you would wake it up by switching it on and pressing the right keys. This might be quick to do, but it does take a few moments. The rules of the mind dictate that once the Chimp has received any information it will quickly check with the Computer to see if there is any advice or if the Computer can take over with a response. [7] [3] The Chimp is impatient and if your Computer doesn't respond immediately, then the Chimp will act. [456] [457] In other words, if your Computer isn't switched on, then by the time it has begun to display information, it is too late. The Chimp won't wait, but instead it will act and hijack the system. Once the Chimp has asked the Computer for advice, it won't do it a second time. Therefore, it is vital that the Computer is constantly ready for action.

Why has the Computer gone to sleep and how do we switch the Computer on and keep it running?

It is vital that the Computer is constantly ready for action

It is quite simple. The Computer will go to sleep, if we don't continually use the programmes that we have input. To keep the Computer active, we must keep using the programmes.

Example: *Joseph and the Defibrillator*

Joseph works in a factory and is the lead for first aid. He has been shown how to operate the defibrillator, which will be used if someone collapses from a heart problem. However, he has not practiced the resuscitation procedure for many months. One day at work, a man collapsed and fell into unconsciousness. Joseph was called to the scene and the defibrillator was ready for him to use. Joseph realised that, although he did know what to do with the device, he wasn't confident and checked the instructions. At this point, his Chimp panicked and took over. Joseph became agitated and fumbled with the equipment.

What happened, demonstrated two pathways in Joseph's mind that were not warmed up: the use of the defibrillator and also the management of his Chimp in an emergency. Let's repeat the scenario but this time Joseph has his Computer warmed up.

Joseph has been regularly practicing using the defibrillator and revising the procedures. The process has been practiced enough to make the Computer ready to act at any point without hesitation. It is warmed up and is kept switched on by this regular practice. The pathways in the brain are being used regularly. When he attends the scene of the collapse, he knows exactly what to do. His Computer is programmed to act and takes over from his Chimp. However, during the process, Joseph's Chimp panics. It starts questioning if he is really doing things right and asks, "What if this man doesn't make it?". These are all natural Chimp reactions. However, Joseph has also practiced how to manage his Chimp, ready for any emergencies. Joseph's Computer has regularly gone through a rehearsal of emergency situations and is prepared with answers for the Chimp. Many of these answers are from Joseph's Truths on his Stone of Life. The Computer's answers to the Chimp include:

- All I can do is my best, and that I will do
- Anything I can do is better than doing nothing at all
- I can deal with any outcome
- The best chance of success is to put all my focus onto the process

The above Truths resonate with Joseph. When his Chimp attempts to hijack him by panicking, Joseph's programmed and rehearsed (warmed up) Computer immediately answers the Chimp and Joseph remains in Computer mode.

For your Stone of Life to settle your Chimp down, you need to revise and reflect on your Stone of Life regularly and keep it up to date.

Just a few minutes every morning can set the scene for the rest of the day. A few minutes of reflection at intervals during the day, can also help immensely.

This time of reflecting and reminding yourself of the details on your Stone of Life will keep them very much awake. They will then be ready to respond immediately if a situation arises where the Chimp needs these details. Without regularly reminding yourself of the details on your Stone of Life, the Chimp will not be able to find them and will act on its own initiative from an emotional basis. [457] Without reminders, your Computer will effectively go into sleep mode. [458] [459]

Revise and reflect for a few minutes every day

2. The components are not all accurate

If your Stone of Life doesn't appear to be stabilising your emotions, then I suggest you check and revise all of the details on it. It is important that you make sure that they are all very relevant to you and your daily life. If some of the details are not resonating with you then they won't work. The Chimp is not fooled by anything that does not have a ring of truth to it.

Example: *Yvette and a false truth!*

Yvette has put a Truth onto her Stone of Life: 'Everything works out in the end'. For some people this rings true, as they believe that even when things go wrong, life will continue and they will move on.

For others, the phrase 'Everything works out in the end' is just not true. Sometimes life takes a turn for the worse that could affect us permanently. Therefore, this is not a truth that resonates with everyone. When Yvette challenges this truth, she accepts that it would be a great belief to have, but she doesn't believe it herself. Now that she has recognised this, she needs to find a statement that is true for her and will help her in difficult times. For example, she might resonate with 'Even when things go wrong, I will still have some good days'.

Example: *Missing elements*

Seamus is using the Truths on his Stone of Life to manage his Chimp. He has noticed that every time he meets a setback his Chimp overreacts and the Truths on his Stone of Life don't seem to hold up. It could be that some Truths are missing and the ones he has are not relevant for setback situations. One possibility for Seamus, is that a missing Truth could be, 'When I enter Human mode, I am an adult and can manage anything that life brings to my door'. If he believes this and yet it is missing from the Truths on his Stone of Life, then it is likely that his Chimp will continue to panic and react to many situations that it doesn't need to react to.

Alternatively, it could be about gaining perspective in the moment that the setback happens. Seamus tries to gain perspective by laughing at himself or at the situation, but this fails. He tries to think about how short life is, and not to waste time worrying, but this also fails. The problem he faces is that there is nothing wrong with what is currently on his Stone of Life. It just needs an addition. This time he discovers a different way to keep perspective. He asks himself, "Will I still have these emotions in one week's time, or will they have changed?". Alternatively, he could ask a similar question, "Are these emotions permanent or will they change with time and with the changing situation?". For Seamus, this could work. It is important to makes sure that we have sufficient information on our Stone of Life.

Example: *Changing truths and values*

It helps to keep on adding any new truths or values that you may uncover during your journey through life. As we mature, it is very likely that some of the truths we once believed might no longer hold true and therefore need removing or adjusting.

For example, if you believe that everyone could be pleasant, if they are given the opportunity, then you will live your life according to this belief. If someone is unpleasant then you are likely to try to give them the opportunity to be pleasant or to look for a reason that is stopping them from being pleasant. This is a reasonable belief and your actions will feel correct, even if they do not achieve what you would like to happen.

However, if you then change your belief and now believe that not everyone is pleasant, regardless of what you do for them, then you will approach life very differently. It's important that you work out what you believe, and keep your Stone of Life up to date, because this will then affect your behaviours and interpretations of situations.

> **Key Point**
>
> *Revising and refining your Stone of Life regularly, ensures it is always accurate for that time in your life.*

3. The Chimp or Human has unmet needs

The Stone of Life is there as a guide and reassurance for the Chimp and Human. What it can't do is to satisfy their needs. Therefore, it is important to ensure that the Chimp's needs, which are based on drives, are being met. If they are not being met then the Chimp might not listen to the Stone of Life. It's also important to meet your own Human needs.

Key Point

The Stone of Life will help prevent a Chimp reaction but won't necessarily help to fulfil a Chimp's needs.

Management of a Chimp reaction is different to fulfilling the Chimp's needs.

Example: *Lorna and her Chimp's Troop*

Virtually all Chimps need a Troop to support them. [460] [461] This Troop will usually consist of some family and some friends who will always stand by the Chimp, whatever the circumstances or despite what the Chimp might have done! If the Chimp doesn't have a Troop then it will continually search for one and will not settle down. Without a Troop, it might well express inappropriate emotion in many areas of its life. [301]

Lorna has developed a perfect Stone of Life, but she can't understand why she still feels unhappy. She recognises that she doesn't have a solid Troop for her Chimp. To counter the concern about the lack of a Troop, she has put a Truth on her Stone of Life, which she firmly believes, that 'Given time, everyone will have a Troop'. For many people this truth will help to settle their Chimp, but for Lorna her Chimp's need for a Troop is so strong that it isn't going to settle until it finds one.

The reality is that our Chimps have fundamental needs to keep us healthy. There are certain things individual Chimps can and can't tolerate.

The Chimp won't settle down unless its needs are met

The difference between what a Chimp **wants** to make it settle and what a Chimp **needs** is a fine line, and only you can work out what this is for your Chimp.

Example: *Oscar and the office job*

Our Chimps need to feel secure and in the right part of the 'jungle' to meet its territory drive. Making sure that you are living in the right environment, both professionally and personally, is essential to meeting the needs of the Chimp. The Stone of Life is unlikely to settle down a Chimp that is unhappy in an environment that it can't tolerate.

Oscar works in an office but knows that his Chimp wants to work outdoors. He loves the team in the office and doesn't want to leave them. He keeps revising and going over his Stone of Life, but his Chimp continues to feel unsettled. A Chimp that knows it needs to move on but also feels that it can't, and likes to stay with familiarity, is a very common situation. Oscar could try to sort out this dilemma by using logic and putting down the pros and cons for staying in the office or getting an outside job. However, he might have to accept that his Chimp will keep on telling him that he is in the wrong part of the jungle.

Example: *The broken relationship*

Harry has been in a relationship with Pauline for ten years but it isn't working. They have tried to make things work, but feel even more distant from each other. Harry wants to have a relationship that is warm with a good home life, but Pauline wants a relationship that is energising and adventurous. The Stone of Life won't necessarily work here because they are trying to address both the wants and the needs of their Chimps. A practical approach to solve the problem is required.

4. There is a health problem or lack of self-care

Health problems or a lack of self-care can cause your mind to malfunction. Lack of self-care includes inattention to sleep, [462] poor nutrition, [463] [464] or a lack of relaxation, [465] [466] which can all prevent your mind from working correctly. [467] Falling in love also has a dramatic impact on the way our mind functions.

There is a health problem or lack of self-care
- Scientific points

The mind is very difficult to manage under certain circumstances. Inadequate rest, poor sleep, the influence of alcohol, some medications and some recreational drugs can dramatically affect our ability to make sound decisions. Falling in love can also severely affect our judgement. When we fall in love major changes in hormones and transmitters occur in our brain. The key effects are that the amygdala and dorsolateral areas of the brain become less active, which impairs our warning system and also our rational decision-making. As well as having impaired judgement, these brain changes also affect our risk taking ability and we therefore become impaired in our judgement and take risks that normally we would not take. Falling in love, therefore, could almost be seen as our mind malfunctioning! [468]

It's obvious that you have to look after yourself; otherwise both the physical and emotional machines will not be able to function correctly. If you are in any doubt about your physical or mental wellbeing then please see your doctor and get a check-up. Try to be practical and ensure that you are looking after your physical health, particularly in relation to rest and sleep. A tired machine will default to Chimp mode, regardless of how good the Stone of Life is. [469] [369] Physical and mental health problems need addressing with the appropriate professional, but self-care is our own responsibility.

Example: *Cheryl and tiredness*

Cheryl has a great Stone of Life and regularly reviews it. She can't work out what is wrong because everything seems in place and she is in agreement with it. On closer inspection of her lifestyle, four common themes become apparent:

1. Cheryl doesn't know how to say no to requests and overloads herself with work and deadlines
2. She doesn't look after her sleep and has an inconsistent and unplanned sleep pattern
3. Her nutrition leaves a lot to be desired and she eats erratically
4. She doesn't allow for any planned downtime during the day

It doesn't take a genius to work out why she feels tired all the time and what the solution might be. A tired or stressed physical body will play havoc with the mind. Tiredness usually results in the person entering Chimp mode, with irritability often being the sign of a tired Chimp. [470] [471]

Key Point

Don't neglect your health and wellbeing if you want your mind to be in a good place

5. **The Computer needs to offer a grade A hit**

A Truth of Life is a 'fact' that resonates with you and is a general rule of life. Sometimes, we need to add a grade A hit to manage specific situations.

Example: *Dennis and the spider*

Dennis' Chimp is terrified of spiders. A fear of spiders and snakes is specifically inbuilt from birth, in both humans and chimpanzees but most people overcome this. [472]

If Dennis turns to the Truths on his Stone of Life, he is unlikely to have a mention of spiders on this. This is because his fear of spiders is a specific problem to him. Therefore, Truths such as, 'I can deal with anything that life brings to my door' and 'Not all experiences in life will be pleasant', don't help!

Dennis needs to create a grade A hit specific to the spiders he fears. For example, he might believe, 'A fear of spiders is natural, but spiders don't attack people if unprovoked' or 'Spiders in my country, can't harm me', (if that is true!). In other words, there are some specific situations that will be helped by forming grade A hits for them, to supplement the Truths on the Stone of Life.

Summary for addressing the five common reasons why the Stone of Life might not be working.

The Stone of Life:
1. Can only work effectively, if the Computer is turned on and ready
2. Needs to be reviewed regularly and must have accurate details on it
3. Cannot address unmet needs
4. Is greatly dependent on good physical and mental health
5. Grade A hits are needed for some specific situations

Five common problems that prevent resilience

We will now consider five common problems that can prevent resilience:
1. Muddling yourself up with the machine
2. Constant maintenance is being neglected
3. Forgetting that resilience is a skill and not a fixed process
4. Not using the Troop to settle the Chimp
5. Confidence is not being achieved

1. Muddling yourself up with the machine

This is a very common problem. It is easy to see why we often confuse our actions with those of the Chimp or Computer. The essence of my work is to explain that we share our mind with a machine and not to muddle ourselves up with it.

I often hear despairing phrases, such as:
- Why do I eat so much?
- What is wrong with me?

- I am one of life's worriers
- I always get so nervous
- Why do I hate myself at times?
- I can't seem to stop overreacting

The list is clearly endless!

None of these are accurate statements. In all of these cases the person is describing a machine. The most helpful way forward, is to answer these questions or statements by approaching them with an understanding of the machine. Therefore, for example, the responses could be changed to:

- Why does my Chimp have such as a strong eating drive and how do I manage this?
- How can I understand my machine better and work with it?
- My Chimp is a worrier, so how do I reassure it and suitably programme my Computer?
- My Chimp is nervous by nature; I need to accept this and learn how to reassure it
- If I don't accept the machine, **as it is**, then I will experience frustration at times
- Learning how to present to the world the real me begins with knowing who the real me is and then learning how to prevent the machine from interfering and start enhancing me
- My Chimp is doing a great job by overreacting and drawing my attention to things. I need to do a great job by listening to it and then offering perspective

2. Constant maintenance is being neglected

Just as physical fitness needs constant work and time allocating to it, psychological fitness also needs constant work and time allocating to it.

Maintenance means, checking that you are looking after yourself and your Chimp, and constantly tidying up your Computer.

Tidying your Computer means reflecting on your beliefs and processing any events that are going on in your life. Processing events on a day-to-day basis means checking on your perception and interpretation of what is happening, and then where necessary, forming the appropriate, accurate and helpful beliefs.

3. Forgetting that resilience is a skill and not a fixed process

Example: *Marilyn and progress*

Marilyn resonated with the Chimp model and felt she had been doing well for several years. There had been lots of improvements to her self-image, her confidence and her peace of mind.

Following an incident at work, Marilyn found herself losing confidence, becoming unhappy and unable to stop the feelings her Chimp was giving her. She came to see me in quite a distressed state. Her opening words were: "After all these years, I have learnt nothing".

It's very important to make sure that you interpret things correctly. This was not about her learning anything. It was about accepting that the nature of the Chimp doesn't change and it is a skill to constantly manage this. The Chimp might be trained but its nature will always remain the same. It's true that the more we practice, the better we get at managing the Chimp. However, it is a skill, and realistically, sometimes our skill can let us down. Therefore, it isn't that we haven't learnt anything; we just need to practice day by day. We often need to remind ourselves that the Chimp is built to be much stronger than we are, in order for it to protect us. Therefore, expect days when the Chimp will flex its muscles. The skill is to know what process or what action to take in a situation where the Chimp needs managing. It is also a skill to clean up the Computer and find any hidden Gremlins and to implement helpful Autopilots that will manage the Chimp. Resilience is based on recognising and accepting that it is a skill and not a process.

4. Not using the Troop to settle the Chimp

The ideal situation is when we manage the Chimp by using the Human's programming of the Computer. It's not surprising, therefore, that it fights back! If the Computer fails to settle the Chimp, there is another way forward. Rather than repeatedly trying to use the Computer to settle the Chimp, we can change our approach and work with what the Chimp is wanting.

If we recall that the Human always looks internally to gain stability but the Chimp looks externally for stability, then this is our starting point. Instead of turning inwardly to the Computer, we can look externally to the Troop. Investing in your Troop and turning to them whenever you cannot settle your Chimp is a very constructive and sensible way of managing the situation. Gaining emotional stability is not about being completely independent. It is about managing the mind effectively. It is not a failure to turn outwardly for help but rather an alternative method to manage emotions. Most people would prefer to be able to manage their emotions without turning to others but to try and do this on every occasion can lead to unnecessary stress.

The Troop will offer reassurance and a sense of belonging and this can offer the Chimp security. Talking through difficulties and problems with a friend can help the Chimp immensely. [88] [81] [143] This is because they usually offer what the Chimp wants: expression of feelings, understanding, acknowledgement and recognition.

The Chimp will look for support and security from its troop

Don't forget to avoid being bitten by roaming Chimps, who might or might not be friendly, but are not part of your Troop. By this, I mean people who don't necessarily mean you any harm but are not reliably on your side. Therefore, they can distance themselves from you or even be critical, without being concerned about your feelings. Remember that not all Chimps are friendly!

5. Confidence is not being achieved

We have looked at confidence as being based on doing your best and being able to deal with any consequences of not achieving your best or failing.

Despite knowing this, many people still find it difficult to build confidence. What we need to answer is: "Why do people struggle to build confidence and how do we address this?"

The first important point to remember is that all of the work that we do is based on developing a skill. This means it will take time and effort to achieve a state of confidence and that there will be times when we don't manage it. If you struggle with confidence, I hope this won't discourage you from persevering. When we look at the force of nature within us, which is basing confidence on achievement, it's not surprising that we often default back to this stance. The Chimp is determined to look at what it feels it can achieve and be concerned with what others might think. It is not just the force of the Chimp looking at achievement; it is also being reinforced by our society and what we experience right from childhood. We praise achievement much more than effort. This is the basis of the Fridge Door Syndrome, which most of us are subjected to as children.

As in the previous unit, we can build confidence by taking assessment out of the hands of other people and placing it firmly back into our own hands. If we decide that what matters is whether we do our best, live by our values and do the right thing then we move away from the Chimp stance that all that matters is achievement. It is important to recognise that this doesn't mean we are not trying to achieve; what it does mean is that we accept that all we can ever do is to give everything and hope we can achieve.

Unit 27 Reminders

Five common reasons why the Stone of Life might not be working:
1. The Computer is in sleep mode
2. The components are not all accurate
3. The Chimp or Human has unmet needs
4. There is a health problem or lack of self-care
5. The Computer needs to offer a grade A hit

Five common problems that can prevent resilience:
1. Muddling yourself up with the machine
2. Constant maintenance is being neglected
3. Forgetting that resilience is a skill and not a fixed process
4. Not using the Troop to settle the Chimp
5. Confidence is not being achieved

Unit 27

Exercises

Focus points and reflective exercises

1. **Distancing yourself from your machine**
2. **Maintenance work**
3. **Being forgiving of self**
4. **Keeping the Computer running**

Focus 1: *Distancing yourself from your machine*

Try to check on whether you are falling into the trap of muddling yourself up with your machine, with phrases such as:

- I am angry
- I am upset
- I am sorry I shouted
- I get over worried

Remind yourself that these statements are untrue, as it is your machine that is responsible for these experiences; *your role is to take responsibility and prevent them or manage them.* The most important part of the Chimp model is the ability to separate who you are from your machine. It is worth reminding yourself on a regular basis of who you are and what your machine is. All too often we can forget and this can lead us to some despondency and low self-esteem.

The starting point for resilience is to know who you are.

Exercise: *Reminding yourself of who you are*

The blank piece of paper exercise from Unit 1 can be very helpful for finding yourself. We can also find ourselves in other ways. Here are some examples:

1. Look back on events that didn't go too well and ask yourself: how you would have liked to have acted, what you would have liked to have thought and what emotions you would have liked to have experienced. This will demonstrate how the Chimp or Computer had hijacked you. Now look forward, and imagine similar events in your life, and ask: "how will I prevent myself from being hijacked" and "how will I programme my Computer with beliefs or behaviours that will help me to achieve what I want".

2. Imagine going forward one year in a time machine and looking back over that one year and ask how you would like to have been. What values would you have wanted to demonstrate? What emotions and behaviours would you have wanted to have, when setbacks or unwelcome things happen? By doing this, you will find the real you and not the muddled up picture that the world might see of you, after interference from either Chimp or Computer.
3. Imagine that you have entered a paradise. The people in this paradise are really great people, who have great personalities. Try and describe the great personality traits that they would have. Think about this; you are very likely to be describing the real you. **Once you have found yourself:** *be pleased!*

Focus 2: *Maintenance work*

Exercise: *Checking for Gremlins*

At the end of each day, try and set aside a few minutes to go through the day and check for any Gremlins that have entered the machine. Reconsider how you have interpreted events that have happened during the day. Make sure that you are working with the facts and reality and not allowing your Chimp to make interpretations based on feelings or suspicions. Checking each day, will keep your Computer Gremlin free.

Focus 3: *Being forgiving of self*

Recall: are you your biggest fan or your worst critic? Resilience and managing the Chimp are skills. It is very helpful to be forgiving and understanding of yourself. None of us can always get things right. We all forget lessons we have learnt. If it were your best friend, how would you address them if they got things wrong?

Don't allow your Chimp to make destructive, critical comments about you.

Exercise: *Being your biggest fan*

Try for a brief time to be your biggest fan and encourage yourself with understanding and some compassion.

Focus 4: *Keeping the Computer running*

Exercise: *Regular reminders*

Throughout the day, every so often, try spending a few minutes, to remind yourself of the key truths on your Stone of Life. This will keep your truths at the forefront of your mind. Your Computer will be up and running and will stop the Chimp if it tries to go against the truths.

By reminding yourself of truths, you will reset the mind into Human mode, manage your Chimp and have a powerful ready to act Computer.

Going Forward

– a few lines to encourage you and wish you well

Thank you for going through the programme. I hope that you will be able to spend just a few minutes each day to remind yourself of things that resonate with you and put them into practice. By reflecting on how you are doing, you will bring about changes. Every day that we get up, we change. We are a different person from the one who went to bed the night before. By making small changes to your thinking and behaviours, over time, you will move yourself in the direction that you want to go.

Sometimes, life seems more like a circus than a jungle. If you see life as a circus then stay in your seat and laugh at the clowns: don't enter the ring and risk becoming one.

We would all like to live in a pleasant and safe world, some people become disillusioned as they hold on to this expectation. I often say to these people that we don't live in a world made of chocolate, sit around the camp fire at night singing Kumbaya and holding hands; I only wish we did. However, we can create our own world and invite in like-minded people. I hope some of the themes in this book will help you to do this.

I wish you every success, peace of mind and happiness going forward.

Further support

If you would welcome support and meet with other like-minded people while developing your skills, then Chimp Management is a company that is dedicated to helping people to get themselves into a good place. We are here to support you and welcome all enquiries.

Please see our website, *chimpmanagement.com* for information and details.

References

– for those who want to follow up on points

1. Kringelbach, M. L., & Berridge, K. C. (2016). Drive and Motivation in the Brain. In Miller, H. (Ed.), *SAGE Encyclopedia of Theory in Psychology* (pp. 240244). SAGE.
2. Saper, C.B., & Lowell, B.B. (2014). The Hypothalamus. *Current Biology, 24*(23), 1111-1116.
3. Rolls, E., & Grabenhorst, F. (2008). The Orbitofrontal Cortex and Beyond: From Affect to Decision-Making. *Progress in Neurobiology, 86*(3), 216–244.
4. Godoy, L.D., Rossignol, M.T., Delfino-Pereira, P., Garcia-Cairasco, N., & de Lima Umeoka, E.H. (2018). A Comprehensive Overview on Stress Neurobiology: Basic Concepts and Clinical Implications. *Frontiers in Behavioural Neuroscience, 12*(127), 1-23.
5. Hwang, K., Bertolero, M., Liu, W., & D'Esposito, M. (2017). The Human Thalamus Is an Integrative Hub for Functional Brain Networks. *Journal of Neuroscience, 37*(23), 5594-5607.
6. Sherman, S. M. (2016). Thalamus Plays a Central Role in Ongoing Cortical Functioning. *Nature Neuroscience, 19*(4), 533-541.
7. Rolls, E. T. (2019). The Cingulate Cortex and Limbic Systems for Emotion, Action, and Memory. *Brain Structure and Function, 224*(9), 3001-3018.
8. Wittkuhn, L., Eppinger, B., Bartsch, L., Thurm, F., Korb, F., & Li, S. (2018). Repetitive Transcranial Magnetic Stimulation Over Dorsolateral Prefrontal Cortex Modulates Value-Based Learning During Sequential Decision-Making. *NeuroImage, 167*, 384-395.
9. Passingham, D., & Sakai, K. (2004). The Prefrontal Cortex and Working Memory: Physiology and Brain Imaging. *Current Opinion in Neurobiology, 14*(2), 163-168.
10. Barbey, A.K., Colom, R., & Grafman, J. (2013). Dorsolateral Prefrontal Contributions to Human Intelligence. *Neuropsychologia, 51*(7), 1361-1369.
11. Cochrane, T. (2020). A Case of Shared Consciousness. *Synthese, 197*(1), 1-19.
12. Squair, J. (2012). Craniopagus: Overview and the Implications of Sharing a Brain, *University of British Columbia's Undergraduate Journal of Psychology, 1*(1), 1-8.
13. Dominus, S. (2011, May 25). Could Conjoined Twins Share a Mind?. *The New York Times Magazine*. https://www.nytimes.com/2011/05/29/magazine/could-conjoined-twins-share-a-mind.html.
14. Phelps, E. A. (2004). Human Emotion and Memory: Interactions of The Amygdala and Hippocampal Complex. *Current Opinion in Neurobiology, 14*(2), 198-202.
15. Zander, T., Horr, N., Bolte, A., & Volz, K. (2016). Intuitive Decision Making as a Gradual Process: Investigating Semantic Intuition–Based and Priming–Based Decisions with fMRI. *Brain and Behavior, 6*(1), 1-22.
16. Hiser, J., & Koenigs, M. (2018). The Multifaceted Role of the Ventromedial Prefrontal Cortex in Emotion, Decision Making, Social Cognition, and Psychopathology. *Biological Psychiatry, 83*(8), 638-647.
17. Tyng, C., Amin, H., Saad, M., & Malik, A. (2017). The Influences of Emotion on Learning and Memory. *Frontiers in Psychology, 8*(1454), 1-22.
18. Dolcos, F., Katsumi, Y., Weymar, M., Moore, M., Tsukiura, T., & Dolcos, S. (2017). Emerging Directions in Emotional Episodic Memory. *Frontiers in Psychology, 8*(1867), 1-25.
19. Schiller, D., Monfils, M., Raio, C., Johnson, D., Ledoux, J., & Phelps. E. (2010). Preventing the Return of Fear in Humans Using Reconsolidation Update Mechanisms. *Nature, 463*(7277), 49-53.
20. Schiller, D., Raio, C., & Phelps, E. (2012). Extinction Training During the Reconsolidation Window Prevents Recovery of Fear. *Journal of Visualized Experiments, 66*, 3893-3893.
21. Bechara, A., Tranel, D., Damasio, H., Adolphs, R., Rockland, C., & Damasio, A. R. (1995). Double Dissociation of Conditioning and Declarative Knowledge Relative to the Amygdala and Hippocampus in Humans. *Science, 269*(5227), 1115–1118.
22. LaBar, K., LeDoux, J., Spencer, D., & Phelps, E. (1995). Impaired Fear Conditioning Following Unilateral Temporal Lobectomy in Humans. *Journal of Neuroscience, 15*(10), 6846-6855.
23. Peters, S. (2012). *The Chimp Paradox: The Mind Management Program to Help You Achieve Success, Confidence, and Happiness*. Vermilion.
24. Peters, S. (2018). *The Silent Guides*. Bonnier Books UK.
25. Mora-Bermudez, F., Badsha, F., Kanton, S., Camp, J., Vernot, B., Köhler, K., Voigt, B., Okita, K., Maricic, T., He, Z., Lachmann, R., Pääbo, S., Treutlein, B., & Huttner, W. (2016). Differences and Similarities Between Human and Chimpanzee Neural Progenitors During Cerebral Cortex Development. *ELife, 5*, 1-24.
26. Freeman, H., Cantalupo, C., & Hopkins, W. (2004). Asymmetries in the Hippocampus and Amygdala of Chimpanzees (Pan troglodytes). *Behavioral Neuroscience, 118*(6), 1460-1465.
27. Houston, S. M., Herting, M. M., & Sowell, E. R. (2014). The Neurobiology of Childhood Structural Brain Development: Conception Through Adulthood. *Current Topics in Behavioral Neurosciences, 16*, 3–17.
28. Dubois, J., Dehaene-Lambertz, G., Kulikova, S., Poupon, C., Hüppi, P.S, & Hertz-Pannier, L. (2014). The Early Development of Brain White Matter: A Review of Imaging Studies in Fetuses, Newborns and Infants. *Neuroscience, 276*, 48-71.
29. Blakemore, S. (2008). The Social Brain in Adolescence. *Nature Reviews. Neuroscience, 9*(4), 267-277.
30. Goel, V. (2019). Hemispheric Asymmetry in the Prefrontal Cortex for Complex Cognition. *Handbook of Clinical Neurology, 163*(3), 179-196.
31. Tsujimoto, S. (2008). The Prefrontal Cortex: Functional Neural Development During Early Childhood. *The Neuroscientist, 14*(4), 345-358.
32. Michaud, A. (2019). The Mechanics of Conceptual Thinking. *Creative Education, 10*(2), 353-406.
33. Hebscher, M., & Gilboa, A. (2016). A Boost of Confidence: The Role of The Ventromedial Prefrontal Cortex in Memory, Decision-Making, and Schemas. *Neuropsychologia, 90*, 46-58.
34. Spalding, K., Schlichting, M., Zeithamova, D., Preston, A., Tranel, D., Duff, M., & Warren, D. (2018). Ventromedial Prefrontal Cortex is Necessary for Normal

Associative Inference and Memory Integration. *The Journal of Neuroscience, 38*(15), 3767-3775.
35. Xi, C., Zhu, Y., Niu, C., Zhu, C., Lee, T.M.C., Tian, Y., & Wang, K. (2011). Contributions of Subregions of the Prefrontal Cortex to the Theory of Mind and Decision Making. *Behavioural Brain Research, 221*(2), 587-593.
36. Avery, R.D., Zhang, Z. Avolio, B. & Kreuger, R.F. (2007). Developmental and Genetic Determinants of Leadership Role Occupancy Among Women. *Journal of Applied Psychology, 92*(3), 693-706
37. Hiser, J., & Koenigs, M. (2018). The Multifaceted Role of the Ventromedial Prefrontal Cortex in Emotion, Decision Making, Social Cognition, and Psychopathology. *Biological psychiatry, 83*(8), 638–647.
38. Dougherty, D., Chou, T., Buhlmann, U., Rauch, S., & Deckersbach, T. (2020). Early Amygdala Activation and Later Ventromedial Prefrontal Cortex Activation During Anger Induction and Imagery. *Journal of Medical Psychology, 1*, 1-8.
39. Moreno-López, L., Ioannidis, K., Askelund, A.D., Smith, A.J., Schueler, K., & van Harmelen, A.L. (2020). The Resilient Emotional Brain: A Scoping Review of the Medial Prefrontal Cortex and Limbic Structure and Function in Resilient Adults with a History of Childhood Maltreatment. *Biological Psychiatry: Cognitive Neuroscience and Neuroimaging, 5*(4), 392-402.
40. Roos, L., Knight, E., Beauchamp, K., Berkman, E., Faraday, K., Hyslop, K., & Fisher, P. (2017). Acute Stress Impairs Inhibitory Control Based on Individual Differences in Parasympathetic Nervous System Activity. *Biological Psychology, 125*, 58-63.
41. Diano, M., Celeghin, A., Bagnis, A., & Tamietto, M. (2016). Amygdala Response to Emotional Stimuli without Awareness: Facts and Interpretations. *Frontiers in Psychology, 7*, 2029.
42. Wolf, R.C., Pujara, M.S., Motzkin, J.C., Newman, J.P., Kiehl, K.A., Decety, J., Kosson, D.S., Koenigs, M. (2015). Interpersonal Traits of Psychopathy Linked to Reduced Integrity of the Uncinate Fasciculus. *Human Brain Mapping, 36*(10), 4202-4209.
43. Stevens, F. L., Hurley, R.A., Taber, K.H., Hayman, L.A., Hurley, R.A., & Taber, K.H. (2011). Anterior Cingulate Cortex: Unique Role in Cognition and Emotion. *The Journal of Neuropsychiatry and Clinical Neurosciences, 23*(2), 121-125.
44. Maurer, J.M., Paul, S., Anderson, N.E., Nyalakanti, P.K., & Kiehl, K.A. (2020). Youth With Elevated Psychopathic Traits Exhibit Structural Integrity Deficits in the Uncinate Fasciculus. *NeuroImage Clinical, 26*, 102236.
45. Oishi, K., Faria, A.V., Hsu, J., Tippett, D., Mori, S., & Hillis, A.E. (2015). Critical Role of the Right Uncinate Fasciculus in Emotional Empathy. *Annals of Neurology, 77*(1), 68-74.
46. Fuster, J. M. (2008). *The Prefrontal Cortex.* Academic Press.
47. Bechara, A., Damasio, A.R., Damasio, H., & Anderson, S.W. (1994). Insensitivity to Future Consequences Following Damage to Human Prefrontal Cortex. *Cognition, 50*(1), 7-15.
48. LeDoux, J. E. (2000). Emotion Circuits in the Brain. *Annual Review of Neuroscience. 23*(1), 155–184.
49. Gratton, G. (2018). Brain Reflections: A Circuit-Based Framework for Understanding Information Processing and Cognitive Control. *Psychophysiology, 55*(3), 1-26.
50. Tau, G.Z., & Peterson, B.S. (2010). Normal Development of Brain Circuits. *Neuropsychopharmacology, 35*(1), 147-168.
51. Lodish, H., Berk, A., Zipursky S.L., Matsudaira, P., Baltimore, D., & Darnell, J. (2000). *Molecular Cell Biology* (4th ed.). W. H. Freeman.

52. Morrell, P. & Quarles, R.H. (1999). Myelin Formation, Structure and Biochemistry. In Siegel, G.J., Agranoff, B.W., Albers, R.W., Fisher, S.K., & Uhler, M.D. *Basic Neurochemistry: Molecular, Cellular and Medical Aspects* (6th ed.). Lippincott-Raven.
53. Arnsten, A. F. (2009). Stress Signalling Pathways That Impair Prefrontal Cortex Structure and Function. *Nature Reviews. Neuroscience, 10*(6), 410–422.
54. Arnsten, A., Mazure, C. M., & Sinha, R. (2012). This is Your Brain in Meltdown. *Scientific American, 306*(4), 48–53.
55. Tovote, P., Fadok, J.P., & Lüthi, A. (2015). Neuronal Circuits for Fear and Anxiety. *Nature Reviews. Neuroscience. 16*(6), 317-331.
56. McLeod, P. (1987). Visual Reaction Time and High-Speed Ball Games. *Perception, 16*(1), 49–59.
57. Woods, D.L., Wyma, J. M., Yund, E. W., Herron, T. J., & Reed, B. (2015). Factors Influencing the Latency of Simple Reaction Time. *Frontiers in Human Neuroscience, 9*(131), 1-12.
58. Ng, T.W.H., Eby, L.T., Sorensen, K.L., & Feldman, D.C. (2005). Predictors of Objective and Subjective Career Success: A Meta-Analysis. *Personnel Psychology, 58*(2), 367-408.
59. Miao, C., Humphrey, R.H., & Qian, S. (2017). A Meta-Analysis of Emotional Intelligence Effects on Job Satisfaction Mediated by Job Resources, and a Test of Moderators. *Personality and Individual Differences, 116*, 281–288.
60. Urquijo, I., Extremera, N., & Azanza, G. (2019). The Contribution of Emotional Intelligence to Career Success: Beyond Personality Traits. *International Journal of Rnvironmental Research and Public Health, 16*(23), 4809.
61. Libbrecht, N., Lievens, F., Carette, B., & Côté, S. (2014). Emotional Intelligence Predicts Success in Medical School. *Emotion, 14*(1), 64–73.
62. Hills, P., & Argyle, M. (2001). Emotional Stability as a Major Dimension of Happiness. *Personality and Individual Differences, 31*(8), 1357-1364.
63. Bajaj, B., Gupta, R., & Sengupta, S. (2019). Emotional Stability and Self-Esteem as Mediators Between Mindfulness and Happiness. *Journal of Happiness Studies, 20*(7), 2211–2226.
64. Cohrdes, C., & Mauz, E. (2020). Self-Efficacy and Emotional Stability Buffer Negative Effects of Adverse Childhood Experiences on Young Adult Health-Related Quality of Life. *Journal of Adolescent Health, 67*(1), 93-100.
65. Campo, M., Champely, S., Louvet, B., Rosnet, E. Ferrand, C., Pauketat, J.V.T., & Mackie, D.M. (2019) Group-Based Emotions: Evidence for Emotion Performance Relationships in Team Sports. *Research Quarterly for Exercise and Sport, 90*(1), 54-63.
66. Lazarus, R. S. (2000). How Emotions Influence Performance in Competitive Sports. *The Sport Psychologist, 14*, 229–252.
67. Chiang, Y.T., Fang, W.T., Kaplan, U., & Ng, E. (2019). Locus of Control: The Mediation Effect Between Emotional Stability and Pro-Environmental Behavior *Sustainability, 11*(3), 820.
68. Michely, J., Rigoli, F., Rutledge, R.B., Hauser, T.U. & Dolan, R.J. (2020). Distinct Processing of Aversive Experience in Amygdala Subregions. *Biological Psychiatry: Cognitive Neuroscience and Neuroimaging 5*(3), 291-300.
69. Phan, K., Fitzgerald, D., Gao, K., Moore, G., Tancer M., & Posse, S. (2004). Real-Time fMRI of Cortico Limbic Brain Activity During Emotional Processing *Neuroreport, 15*(3), 527-532.

70. Koch, S.B.J., Mars, R.B., Toni, I., & Roelofs, K. (2018). Emotional Control, Reappraised. *Neuroscience & Biobehavioral Reviews, 95*, 528-534.
71. Hare, B., Melis, A.P., Woods, V., Hastings, S., & Wrangham, R. (2007). Tolerance Allows Bonobos to Outperform Chimpanzees on a Cooperative Task. *Current Biology, 17*(7), 619-623.
72. Hare B. (2017). Survival of the Friendliest: Homo Sapiens Evolved via Selection for Prosociality. *Annual Review of Psychology, 68*(1), 155-186.
73. Aldao, A., & Nolen-Hoeksema, S. (2013). One Versus Many: Capturing the Use of Multiple Emotion Regulation Strategies in Response to an Emotion-Eliciting Stimulus. *Cognition and Emotion, 27*(4), 753-760.
74. Horr, N., Braun, K., & Volz, C. (2014). Feeling Before Knowing Why: The Role of the Orbitofrontal Cortex in Intuitive Judgments - an MEG Study. *Cognitive, Affective, & Behavioral Neuroscience, 14*(4), 1271-1285.
75. Darke, P. R., Chattopadhyay, A., & Ashworth, L. (2006). Going With Your "Gut Feeling": The Importance and Functional Significance of Discrete Affect in Consumer Choice. *Journal of Consumer Research, 33*, 322–328.
76. Isen, A. M., Daubman, K. A., & Nowicki, G.P. (1987). Positive Affect Facilitates Creative Problem-Solving. *Journal of Personality and Social Psychology, 52*(6), 1122–1131.
77. Slovic, P., Peters, E., Finucane, M. L., & MacGregor, D. G. (2005). Affect, Risk, and Decision Making. *Health Psychology, 24*(4), 35–40.
78. Nicolle, A., & Goel, V. (2013). What Is the Role of Ventromedial Prefrontal Cortex in Emotional Influences on Reason?. In Blanchette, I. (Ed.), *Emotion and Reasoning*. Psychology Press.
79. Konu, D., Turnbull, O., Karapanagiotidis, T., Wang, H.T., Brown, L.R., Jefferies, E., & Smallwood, J. (2020). A Role for the Ventromedial Prefrontal Cortex in Self-Generated Episodic Social Cognition. *NeuroImage, 218*, 116977.
80. Gross, J. (2002). Emotion Regulation: Affective, Cognitive, and Social Consequences. *Psychophysiology, 39*(3), 281-291.
81. Ochsner, K., Bunge, S., Gross, J., & Gabrieli, J. (2002). Rethinking Feelings: An fMRI Study of the Cognitive Regulation of Emotion. *Journal of Cognitive Neuroscience, 14*(8), 1215-1229.
82. John, O., & Gross, J. (2004). Healthy and Unhealthy Emotion Regulation: Personality Processes, Individual Differences, and Life Span Development. *Journal of Personality, 72*(6), 1301-1334.
83. Goldin, P., Mcrae, K., Ramel, W., & Gross, J. (2008). The Neural Bases of Emotion Regulation: Reappraisal and Suppression of Negative Emotion. *Biological Psychiatry, 63*(6), 577-586.
84. Zilverstand, A., Parvaz, M., & Goldstein, R. (2017). Neuroimaging Cognitive Reappraisal in Clinical Populations to Define Neural Targets for Enhancing Emotion Regulation. A Systematic Review. *NeuroImage, 151*, 105-116.
85. Stephens, R., & Umland, C. (2011). Swearing as a Response to Pain-Effect of Daily Swearing Frequency. *The Journal of Pain, 12*(12), 1274-1281.
86. Gross, J., & Levenson, R. (1993). Emotional Suppression: Physiology, Self-Report, and Expressive Behavior. *Journal of Personality and Social Psychology, 64*(6), 970-986.
87. Larsen, J., Vermulst, K., Eisinga, A., English, A., Gross, R., Hofman, T., & Engels, J. (2012). Social Coping by Masking? Parental Support and Peer Victimization as Mediators of the Relationship Between Depressive Symptoms and Expressive Suppression in Adolescents. *Journal of Youth and Adolescence, 41*(12), 1628-1642.
88. Torre, J.B., & Lieberman, M.D. (2018). Putting Feelings Into Words: Affect Labeling as Implicit Emotion Regulation. *Emotion Review, 10*(2), 116-124.
89. Reeve, J., & Lee, W. (2012). Neuroscience and Human Motivation. In Ryan, R. M. (Ed.), *The Oxford Handbook of Human Motivation*. (pp. 365–380). Oxford University Press.
90. Berkman, E. (2018). The Neuroscience of Goals and Behavior Change. *Consulting Psychology Journal: Practice and Research, 70*(1), 28-44.
91. Cope, J. (2011). Entrepreneurial Learning from Failure: An Interpretative Phenomenological Analysis. *Journal of Business Venturing, 26*(6), 604-623.
92. Fischer, M., Mazor, K., Baril, J., Alper, E., DeMarco, D., & Pugnaire, M. (2006). Learning from Mistakes. Factors That Influence How Students and Residents Learn from Medical Errors. *Journal of General Internal Medicine, 21*(5), 419-423.
93. Metcalfe, J. (2017). Learning from Errors. *Annual Review of Psychology, 68*(1), 465-489.
94. Mason, L., Peters, E., Williams, S.C., & Kumari, V. (2017). Brain Connectivity Changes Occurring Following Cognitive Behavioural Therapy for Psychosis Predict Long-Term Recovery. *Translational Psychiatry, 7*, 1-7.
95. Barsaglini, A., Sartori, G., Benetti, S., Pettersson-Yeo, W., & Mechelli, A. (2014). The Effects of Psychotherapy on Brain Function: A Systematic and Critical Review. *Progress in Neurobiology, 114*, 1–14.
96. Moore, M., Jackson, E., & Tschannen-Moran, B. (2016). *Coaching Psychology Manual* (2nd ed.). Wolters Kluwer.
97. Koller, D., & Goldman, R.D. (2012). Distraction Techniques for Children Undergoing Procedures: A Critical Review of Pediatric Research. *Journal of Pediatric Nursing, 27*(6), 652-681.
98. Sahiner, N. C., & Bal, M. D. (2016). The Effects of Three Different Distraction Methods on Pain and Anxiety in Children. *Journal of Child Health Care, 20*(3), 277–285.
99. Sadeghi, T., Mohammadi, N., Shamshiri, M., Bagherzadeh, R., & Hossinkhani, N. (2013). Effect of Distraction on Children's Pain During Intravenous Catheter Insertion. *Journal for Specialists in Pediatric Nursing, 18*(2), 109-114.
100. Baumeister, R. F., Bratslavsky, E., Finkenauer, C., & Vohs, K. D. (2001). Bad is Stronger Than Good. *Review of General Psychology, 5*(4), 323–370.
101. Vaish, A., Grossmann, T., & Woodward, A. (2008). Not All Emotions Are Created Equal: The Negativity Bias in Social-Emotional Development. *Psychological Bulletin, 134*(3), 383–403.
102. Tugade, M., Fredrickson, B.L., & Feldman Barrett, L. (2004). Psychological Resilience and Positive Emotional Granularity: Examining the Benefits of Positive Emotions on Coping and Health. *Journal of Personality, 72*(6), 1161-1190.
103. Ford, B. Q., Lam, P., John, O. P., & Mauss, I. B. (2018). The Psychological Health Benefits of Accepting Negative Emotions and Thoughts: Laboratory, Diary, and Longitudinal Evidence. *Journal of Personality and Social Psychology, 115*(6), 1075–1092.
104. Catalino, L.I., Arenander, J., Epel, E., & Puterman, E. (2017). Trait Acceptance Predicts Fewer Daily Negative Emotions Through Less Stressor-Related Rumination. *Emotion. 17*(8), 1181-1186.
105. Öhman, A., Carlsson, K., Lundqvist, D., & Ingvar, M. (2007). On the Unconscious Subcortical Origin of Human Fear. *Physiology & Behavior, 92*(1), 180-185.
106. Öhman, A. (2005). The Role of the Amygdala in Human Fear: Automatic Detection of Threat. *Psychoneuroendocrinology, 30*(10), 953–958.
107. Williams, L.M., Phillips, M.L, Brammer, M.J, Skerrett,

D., Lagopoulos, J., Rennie, C., Bahramali, H., Olivieri, G., David, A.S., Peduto, A., & Gordon, E. (2001). Arousal Dissociates Amygdala and Hippocampal Fear Responses: Evidence from Simultaneous FMRI and Skin Conductance Recording. *NeuroImage, 14*(5), 1070-1079.
108. LeDoux, J.E., & Pine, D.S. (2016). Using Neuroscience to Help Understand Fear and Anxiety: A Two-System Framework. *The American Journal of Psychiatry, 173*(11), 1083-1093.
109. Fullana, M.A., Albajes-Eizagirre, A., Soriano-Mas, C., Vervliet, B., Cardoner, N., Benet, O., Radua, J., & Harrison, B.J. (2018). Fear Extinction in the Human Brain: A Meta-Analysis of fMRI Studies in Healthy Participants. *Neuroscience & Biobehavioral Reviews, 88*, 16-25.
110. Knowlton, B.J., & Yin, H. H. (2006). The Role of the Basal Ganglia in Habit Formation. *Nature Reviews Neuroscience, 7*(6), 464-476.
111. Heatherton, T.F. (2011). Neuroscience of Self and Self-Regulation. *Annual Review of Psychology, 62*(1), 363-390.
112. Kawamoto, T., Ura, M., & Nittono, H. (2015). Intrapersonal and Interpersonal Processes of Social Exclusion. *Frontiers in Neuroscience, 9*(62), 1-11.
113. Kringelbach, M.L. (2005). The Human Orbitofrontal Cortex: Linking Reward to Hedonic Experience. *Nature Reviews Neuroscience, 6*(9), 691-702.
114. Menzies, J.R., Skibicka, K.P., Dickson, S.L., & Leng, G. (2012). Neural Substrates Underlying Interactions Between Appetite Stress and Reward. *Obesity Facts, 5*(2), 208-20.
115. Bechara, A., Damasio, H., & Damasio, A.R. (2000). Emotion, Decision Making, and the Orbitofrontal Cortex. *Cerebral Cortex, 10*(3), 295–307.
116. Harari, D., Swider, B. W., Steed, L. B., & Breidenthal, A. P. (2018). Is Perfect Good? A Meta-Analysis of Perfectionism in the Workplace. *Journal of Applied Psychology, 103*(10), 1121–1144.
117. Bollini, A.M., Walker, E.F., Hamann, S., & Kestler, L. (2004). The Influence of Perceived Control and Locus of Control on the Cortisol and Subjective Responses to Stress. *Biological Psychology, 67*(3), 245-260.
118. Flett, G.L., & Hewitt, P.L. (2006). Perfectionism as a Detrimental Factor in Leadership: A Multi-Dimensional Analysis. In Burke, R.J. & Cooper, C (Eds.). *Inspiring Leaders*. Routledge.
119. Tannenbaum, M. B., Hepler, J., Zimmerman, R. S., Saul, L., Jacobs, S., Wilson, K., & Albarracín, D. (2015). Appealing to Fear: A Meta-Analysis of Fear Appeal Effectiveness and Theories. *Psychological Bulletin, 141*(6), 1178–1204.
120. Bloodgood, D.W., Sugam, J.A., Holmes, A., & Kash, T.L. (2018). Fear Extinction Requires Infralimbic Cortex Projections to the Basolateral Amygdala. *Translational Psychiatry, 8*(1), 1-11.
121. Unal, C. T., Paré, D., & Amano, T. (2010). Synaptic Correlates of Fear Extinction in the Amygdala. *Nature Neuroscience, 13*(4), 489–494.
122. Boksem, M.A.S., Kostermans, E., & De Cremer, D. (2011). Failing Where Others Have Succeeded: Medial Frontal Negativity Tracks Failure in a Social Context. *Psychophysiology, 48*(7), 973-979.
123. Zink, C.F., Tong, Y., Chen, Q., Bassett, D., Stein, J.L., & Meyer-Lindenberg, A. (2008). Know Your Place: Neural Processing of Social Hierarchy in Humans. *Neuron, 58*(2), 273-283.
124. Kedia, G., Mussweiler, T., & Linden, D.E.J. (2014). Brain Mechanisms of Social Comparison and Their Influence on the Reward System. *Neuroreport, 25*(16), 1255–1265.
125. Diwas, S. K. C., Staats, B. R., Kouchaki, M., & Gino, F. (2020). Task Selection and Workload: A Focus on Completing Easy Tasks Hurts Performance. *Management Science, 66*(10), 4397-4416.
126. Masicampo, E.J., & Baumeister, R.F. (2011). Consider It Done! Plan Making Can Eliminate the Cognitive Effects of Unfulfilled Goals. *Journal of Personality and Social Psychology, 101*(4), 667-83.
127. Ames, D.R., & Wazlawek, A.S. (2014). Pushing in the Dark: Causes and Consequences of Limited Self-Awareness for Interpersonal Assertiveness. *Personality and Social Psychology Bulletin, 40*(6), 775-790.
128. Speed, B.C., Goldstein, B.L., & Goldfried, M.R. (2018). Assertiveness Training: A Forgotten Evidence–Based Treatment. *Clinical Psychology, 25*(1), 1-20.
129. Sargunaraj, M., Kashyap, H., & Chandra, P.S. (2020). Writing Your Way Through Feelings: Therapeutic Writing for Emotion Regulation. *Journal of Psychosocial Rehabilitation and Mental Health, 7*, 1-7.
130. Malenka, R.C., Nestler, E.J., & Hyman, S.E. (2009). *Molecular Neuropharmacology: A Foundation for Clinical Neuroscience* (2nd ed.). McGraw-Hill Medical.
131. Frank, M. J., & Claus, E. D. (2006). Anatomy of a Decision: Striato-Orbitofrontal Interactions in Reinforcement Learning, Decision Making, and Reversal. *Psychological Review, 113*(2), 300–326.
132. Kerns, J.G., Cohen, J.D., MacDonald, A.W., Cho, R.Y., Stenger, V.A., & Carter, C.S. (2004). Anterior Cingulate Conflict Monitoring and Adjustments in Control. *Science, 303*(5660), 1023-1026.
133. Adcock, R.A., Thangavel, A., Whitfield-Gabrieli, S., Knutson, B., & Gabrieli, J.D.E. (2006). Reward-Motivated Learning: Mesolimbic Activation Precedes Memory Formation. *Neuron, 50*(3), 507-517.
134. Maslow, A.H. (1954). *Motivation and Personality*. Harper & Row.
135. Berridge, K.C. (2004). Motivation Concepts in Behavioral Neuroscience. *Physiology & Behavior, 81*(2), 179-209.
136. Ryan, R.M., & Deci, E.L. (2017). *Self-Determination Theory: Basic Psychological Needs in Motivation, Development & Wellness*. The Guildford Press.
137. Diener, E., Lucas, R.E., & Scollon, C.N. (2006). Beyond the Hedonic Treadmill: Revising the Adaptation Theory of Well-Being. *American Psychologist, 61*(4), 305-314.
138. Joranby, L., Frost Pineda, K., & Gold, M.S. (2005). Addiction to Food and Brain Reward Systems. *Sexual Addiction & Compulsivity, 12*(2), 201-217.
139. Volkow, N.D., Wang, G.J., & Baler, R.D. (2010). Reward, Dopamine and the Control of Food Intake: Implications for Obesity. *Trends in Cognitive Sciences, 15*(1), 37-46.
140. Ogden, J. (2010). *The Psychology of Eating: From Healthy to Disordered Behavior* (2nd ed.). Wiley-Blackwell.
141. Gibson, E. L. (2012). The Psychobiology of Comfort Eating: Implications for Neuropharmacological Interventions. *Behavioural Pharmacology, 23*(5), 442-460.
142. Pennebaker, J. W., Zech, E., & Rime, B. (2001). Disclosing and Sharing Emotion: Psychological, Social, and Health Consequences. In Stroebe, M. S., Hansson, R. O., Stroebe, W., & Schut, H. (Eds). *Handbook of Bereavement Research* (pp. 517–543). APA.
143. Lepore, S.J., Ragan, J.D., & Jones, S. (2000). Talking Facilitates Cognitive-Emotional Processes of Adaptation to an Acute Stressor. *Journal of Personality and Social Psychology. 78*(3), 499–508.
144. Fedurek, P., Slocombe, K.E., Hartel, J.A., & Zuberbühler, K. (2015). Chimpanzee Lip-Smacking Facilitates Cooperative Behaviour. *Scientific Reports, 5*, 13460.
145. Beer, J. S., John, O.P., Scabini, D., & Knight, R.T. (2006). Orbitofrontal Cortex and Social Behavior: Integrating Self-monitoring and Emotion-Cognition

Interactions. *Journal of Cognitive Neuroscience, 18*(6), 871-879.
146. Berthoz, S., Armony, J.L., Blair, R.J.R., & Dolan, R.J. (2002). An fMRI Study of Intentional and Unintentional (Embarrassing) Violations of Social Norms. *Brain, 125*(8), 1696-1708.
147. Kong, F., Yang, K., Sajjad, S., Yan, W., Li, X., & Zhao, J. (2019). Neural Correlates of Social Well-Being: Gray Matter Density in the Orbitofrontal Cortex Predicts Social Well-Being in Emerging Adulthood. *Social Cognitive and Affective Neuroscience, 14*(3), 319-327.
148. Howell, R.T., & Hill, G. (2009). The Mediators of Experiential Purchases: Determining the Impact of Psychological Needs Satisfaction and Social Comparison. *The Journal of Positive Psychology, 4*(6), 511-522.
149. Fujita, F., & Diener, E. (2005). Life Satisfaction Set Point: Stability and Change. *Journal of Personality and Social Psychology, 88*(1), 158–164.
150. Mruk, C.J. (2006). *Self-Esteem Research, Theory and Practice: Toward a Positive Psychology of Self-Esteem.* Springer Publishing Company.
151. Dogan V. (2019). Why Do People Experience the Fear of Missing Out (FoMO)? Exposing the Link Between the Self and the FoMO Through Self-Construal. *Journal of Cross-Cultural Psychology, 50*(4), 524-538.
152. Golkar, A., Lonsdorf, T.B., Olsson, A., Lindstrom, K.M., Berrebi, J., Fransson, P., Schalling, M., Ingvar, M., & Öhman, A. (2012). Distinct Contributions of the Dorsolateral Prefrontal and Orbitofrontal Cortex During Emotion Regulation. *PloS One, 7*(11), 48107.
153. Perlstein, W.M., Elbert, T. & Stenger, V.A. (2002). Dissociation in Human Prefrontal Cortex of Affective Influences on Working Memory-Related Activity. *Proceedings of the National Academy of Sciences, 99*(3), 1736-1741.
154. Purves, D., Augustine, G., Fitzpatrick, D., Katz, L., LaMantia, A.S., McNamara, J.O., & Williams, S.M. (2001). *Neuroscience* (3rd ed.). Sinauer Associates.
155. Barrett, L., Mesquita, B., Ochsner, K., & Gross, J. (2007). The Experience of Emotion. *Annual Review of Psychology, 58*(1), 373-403.
156. Nummenmaa, L., & Saarimäki, H. (2019). Emotions as Discrete Patterns of Systemic Activity. *Neuroscience Letters,* 693, 3-8.
157. Berrios, R., Totterdell, P., & Kellett, S. (2015). Eliciting Mixed Emotions: A Meta-Analysis Comparing Models, Types, and Measures. *Frontiers in Psychology, 6*(428), 1-15.
158. Keltner, D., Sauter, D., Tracy, J., & Cowen, A. (2019). Emotional Expression: Advances in Basic Emotion Theory. *Journal of Nonverbal Behavior, 43*(2), 133-160.
159. Brooks, A. (2014). Get Excited: Reappraising Pre-Performance Anxiety as Excitement. *Journal of Experimental Psychology, 143*(3), 1144-1158.
160. Levitt, E. E. (2015). *The Psychology of Anxiety.* (2nd ed.). Routledge.
161. Grupe, D., & Nitschke, J. (2013). Uncertainty and Anticipation in Anxiety: An Integrated Neurobiological and Psychological Perspective. *Nature Reviews Neuroscience, 14*(7), 488-501.
162. Troy, A., Shallcross, A., Brunner, A., Friedman, R., & Jones, M. (2018). Cognitive Reappraisal and Acceptance: Effects on Emotion, Physiology, and Perceived Cognitive Costs. *Emotion, 18*(1), 58-74.
163. Barrett, L., Gross, J., Tamlin, C., & Benvenuto, M. (2001). Knowing What You're Feeling and Knowing What to do About it: Mapping the Relation Between Emotion Differentiation and Emotion Regulation. *Cognition & Emotion, 15*(6), 713-324.
164. Barrett, L.F., Lewis, M., & Haviland-Jones, J.M. (2016). *Handbook of Emotions* (4th ed.). Guilford Press.
165. Goleman, D. (1998). *Working with Emotional Intelligence.* Bantam Books.
166. Moseley, R. (2018). The Limbic System. In Braaten, E. (Ed.), *The SAGE Encyclopedia of Intellectual and Developmental Disorders.* (pp. 964 – 966). SAGE Publications Inc.
167. Gross, J. (2014). *The Handbook of Emotional Regulation* (2nd ed.). Guildford Press.
168. Hentschel, U., Smith, G., Ehlers, W., & Draguns, J.D. (1993). *The Concept of Defense Mechanisms in Contemporary Psychology: Theoretical, Research, and Clinical Perspectives.* Springer-Verlag.
169. Paulhus, D.L., Fridhandler, B., & Hayes, S. (1997). Psychological Defense: Contemporary Theory and Research. In Hogan, R., Briggs, S. & Johnson, J. (Eds.), *Handbook of Personality Psychology.* (pp. 543-579). Academic Press.
170. Bowlby, J. (1969). *Attachment. (Attachment and Loss, Vol. 1).* Basic Books.
171. Bowlby, J. (1973). *Separation: Anxiety and Anger. (Attachment and loss, Vol. 2).* Basic Books.
172. Bowlby, J. (1980). *Attachment and Loss: Vol 3: Loss: Sadness and Depression.* Hogarth Press.
173. Maciejewski, P., Zhang, B., Block, S., & Prigerson, H. (2007). An Empirical Examination of the Stage Theory of Grief. *JAMA, 297*(7), 716-723.
174. Schaefer, S. M., Jackson, D.C., Davidson, R.J., Aguirre, G. K., Kimberg, D.Y., & Thompson-Schill, S. L. (2002). Modulation of Amygdalar Activity by the Conscious Regulation of Negative Emotion. *Journal of Cognitive Neuroscience, 14*(6), 913-921.
175. Tamir, M. (2016). Why Do People Regulate Their Emotions? A Taxonomy of Motives in Emotion Regulation. *Personality and Social Psychology Review, 20*(3), 199–222.
176. Urry, H. L., Van Reekum, C. M., Johnstone, T., Kalin, N. H., Thurow, M. E., Schaefer, H. S., Jackson, C. A., Frye, C. J., Greischar, L. L., Alexander, A. L., & Davidson, R. J. (2006). Amygdala and Ventromedial Prefrontal Cortex are Inversely Coupled During Regulation of Negative Affect and Predict the Diurnal Pattern of Cortisol Secretion Among Older Adults. *The Journal of Neuroscience. 26*(16), 4415-4425.
177. McLeod, S.A. (2019, April 10). Defense Mechanisms. Simply Psychology. https://www.simplypsychology.org/defense-mechanisms.html
178. Hariri, A.R., Bookheimer, S.Y., & Mazziotta, J.C. (2000). Modulating Emotional Responses: Effects of a Neocortical Network on the Limbic System. *NeuroReport, 11*(1), 43-48.
179. Anderson, M. C., & Levy, B. J. (2011). On the Relationship Between Interference and Inhibition in Cognition. In Benjamin, A.S. (Ed.), *Successful Remembering and Successful Forgetting: A Festschrift in Honor of Robert A. Bjork.* (pp. 107-132). Taylor and Francis.
180. Gross, J. (1998). Antecedent and Response-Focused Emotion Regulation: Divergent Consequences for Experience, Expression, and Physiology. *Journal of Personality and Social Psychology, 74*(1), 224-237.
181. Garssen, B. (2007). Repression: Finding Our Way in the Maze of Concepts. *Journal of behavioural, 30*(6), 471-481.
182. Gross, J., & Levenson, R. (1997). Hiding Feelings: The Acute Effects of Inhibiting Negative and Positive Emotion. *Journal of Abnormal Psychology, 106*(1), 95–103.
183. Teicher, M. (2002). Scars That Won't Heal: The Neurobiology of Child Abuse. *Scientific American, 286*(3), 68-75.
184. Beblo, T., Fernando, S., Klocke, S., Griepenstroh, J., Aschenbrenner, S., & Driessen, M. (2012). Increased

Suppression of Negative and Positive Emotions in Major Depression. *Journal of Affective Disorders, 141*(2), 474-479.
185. Joormann, J., & Gotlib, I.H. (2010). Emotion Regulation in Depression: Relation to Cognitive Inhibition. *Cognition and Emotion: Emotional States, Attention, and Working Memory, 24*(2), 281-298.
186. Machunsky, M., Toma, C., Yzerbyt, V., & Corneille, O. (2014). Social Projection Increases for Positive Targets: Ascertaining the Effect and Exploring Its Antecedents. *Personality and Social Psychology Bulletin, 40*(10), 1373-1388.
187. Üstün, S., Kale, E., & Çiçek, M. (2017). Neural Networks for Time Perception and Working Memory. *Frontiers in Human Neuroscience, 11*(83), 1-11.
188. Craig, A. (2009). Emotional Moments Across Time: A Possible Neural Basis for Time Perception in the Anterior Insula. *Philosophical Transactions of The Royal Society B-Biological Sciences, 364*(1525), 1933-1942.
189. Kalisch, R. (2009). The Functional Neuroanatomy of Reappraisal: Time Matters. *Neuroscience and Biobehavioral Reviews, 33*(8), 1215-1226.
190. Alberini, C., & Ledoux, J. (2013). Memory Reconsolidation. *Current Biology, 23*(17), 746-750.
191. de Jonge, P., Roest, A.M., Lim, C.C,W., Levinson, D., & Scott, K.M. (2017). In Scott, K.M. (Ed.), *Panic Disorder and Panic Attacks. Mental Disorders Around the World.* (pp. 93-105). Cambridge University Press.
192. Verplanken, B., Friborg, O., Wang, C.E., Trafimow, D., & Woolf, K. (2007). Mental Habits: Metacognitive Reflection on Negative Self-Thinking. *Journal of Personality and Social Psychology, 92*(3), 526-541.
193. Wood, W., & Rünger, D. (2016). Psychology of Habit. *Annual Review of Psychology, 67*(1), 289-314.
194. Marien H., Custers R., & Aarts H. (2018). Understanding the Formation of Human Habits: An Analysis of Mechanisms of Habitual Behaviour. In Verplanken, B. (Ed.), *The Psychology of Habit* (pp. 51-69). Springer.
195. Tugade, M. M., & Fredrickson, B. L. (2004). Resilient Individuals Use Positive Emotions to Bounce Back From Negative Emotional Experiences. *Journal of Personality and Social Psychology, 86*(2), 320-333.
196. Fredrickson, B. L. (2001). The Role of Positive Emotions in Positive Psychology: The Broaden-and-Build Theory of Positive Emotions. *American Psychologist, 56*(3), 218–226.
197. Longe, O., Maratos, F.A., Gilbert, P., Evans, G., Volker, F., Rockliff, H., & Rippon, G. (2010). Having a Word with Yourself: Neural Correlates of Self-Criticism and Self-Reassurance. *NeuroImage, 49*(2), 1849-1856.
198. Lamblin, M., Murawski, C., Whittle, S., & Fornito, A. (2017). Social Connectedness, Mental Health and the Adolescent Brain. *Neuroscience and Biobehavioral Reviews, 80*, 57-68.
199. Baumeister, R. F., & Leary, M.R. (1995). The Need to Belong: Desire for Interpersonal Attachments as a Fundamental Human Motivation. *Psychological Bulletin, 117*(3), 497-529.
200. Lee, R. M., & Robbins, S. B. (1998). The Relationship Between Social Connectedness and Anxiety, Self-Esteem, and Social Identity. *Journal of Counseling Psychology, 45*(3), 338–345.
201. Lingawi, N., & Balleine, B. (2012). Amygdala Central Nucleus Interacts With Dorsolateral Striatum to Regulate the Acquisition of Habits. *The Journal of Neuroscience, 32*(3), 1073-1081.
202. Wingard, J.C., & Packard, M.G. (2008). The Amygdala and Emotional Modulation of Competition Between Cognitive and Habit Memory. *Behavioural Brain Research, 193*(1), 126-131.
203. Wang, M., & Saudino, K.J. (2011). Emotion Regulation and Stress. *Journal of Adult Development, 18*(2), 95-103.
204. Zelazo, P.D. (2015). Executive Function: Reflection, Iterative Reprocessing, Complexity, and the Developing Brain. *Developmental Review, 38*, 55-68.
205. Dyck, M., Loughead, J., Kellermann, T., Boers, F., Gur, R.C., & Mathiak, K. (2011). Cognitive Versus Automatic Mechanisms of Mood Induction Differentially Activate Left and Right Amygdala. *NeuroImage, 54*(3), 2503-2513.
206. Ehring, T., Tuschen-Caffier, B., Schnülle, J., Fischer, S., & Gross, J.J. (2010). Emotion Regulation and Vulnerability to Depression: Spontaneous Versus Instructed Use of Emotion Suppression and Reappraisal. *Emotion, 10*(4), 563-572.
207. Smith, K.S., & Graybiel, A.M. (2016). Habit Formation. *Dialogues in Clinical Neuroscience, 18*(1), 33-43.
208. Berridge, K. C., & Kringelbach, M. L. (2015). Pleasure Systems in the Brain. *Neuron, 86*(3), 646–664.
209. Schott, B.H., Minuzzi, L., Krebs, R.M., Elmenhorst, D., Lang, M., Winz, O.H., Seidenbecher, C.I., Coenen, H.H., Heinze, H.J., Zilles, K., Düzel, E., & Bauer, A. (2008). Mesolimbic Functional Magnetic Resonance Imaging Activations During Reward Anticipation Correlate With Reward-Related Ventral Striatal Dopamine Release. *The Journal of Neuroscience, 28*(52), 14311-14319.
210. Everitt, B.J., & Robbins, T.W. (2005). Neural Systems of Reinforcement for Drug Addiction: From Actions to Habits to Compulsion. *Nature Neuroscience, 8*(11), 1481-1489.
211. Koob, G.F., & Volkow, N.D. (2010). Neurocircuitry of Addiction. *Neuropsychopharmacology, 35*(1), 217-238.
212. Amaya, K.A., & Smith, K.S. (2018). Neurobiology of Habit Formation. *Current Opinion in Behavioral Sciences, 20*, 145-152.
213. Schoenbaum, G., & Roesch, M. (2005). Orbitofrontal Cortex, Associative Learning, and Expectancies. *Neuron, 47*(5), 633-636.
214. Hornak, J., O'Doherty, J., Bramham, J., Rolls, E.T., Morris, R.G., Bullock, P.R., & Polkey, C.E. (2004). Reward-Related Reversal Learning After Surgical Excisions in Orbito-frontal or Dorsolateral Prefrontal Cortex in Humans. *Journal of Cognitive Neuroscience, 16*(3), 463-478.
215. Verplanken, B. (Ed.). (2018). *The Psychology of Habit: Theory, Mechanisms, Change, and Contexts.* Springer.
216. Mischel, W., & Ebbesen, E.B. (1970). Attention in Delay of Gratification. *Journal of Personality and Social Psychology, 16*(2), 329–337.
217. Wallis, J.D. (2007). Orbitofrontal Cortex and Its Contribution to Decision-Making. *Annual Review of Neuroscience, 30*(1), 31-56.
218. Casey, B. J., Somerville, L. H., Gotlib, I. H., Ayduk, O., Franklin, N. T., Askren, M. K., Jonides, J., Berman, M. G., Wilson, N. L., Teslovich, T., Glover, G., Zayas, V., Mischel, W., & Shoda, Y. (2011). Behavioral and Neural Correlates of Delay of Gratification 40 Years Later. *Proceedings of the National Academy of Sciences, 108*(36), 14998–15003.
219. Staudinger, M.R., Erk, S., & Walter, H. (2011). Dorsolateral Prefrontal Cortex Modulates Striatal Reward Encoding During Reappraisal of Reward Anticipation. *Cerebral Cortex, 21*(11), 2578–2588.
220. Verplanken, B., & Sui, J. (2019). Habit and Identity: Behavioral, Cognitive, Affective, and Motivational Facets of an Integrated Self. *Frontiers in Psychology, 10*, 1504.
221. Ouellette, J.A., & Wood, W. (1998). Habit and Intention in Everyday Life. *Psychological Bulletin, 124*(1), 54-74.
222. Verplanken, B., & Faes, S. (1999). Good Intentions, Bad Habits, and Effects of Forming Implementation

Intentions on Healthy Eating. *European Journal of Social Psychology, 29*(5–6), 591-604.
223. Li, Z., Yan, C., Xie, W. Z., Li, K., Zeng, Y. W., Jin, Z., Cheung, E. F., & Chan, R. C. (2015). Anticipatory Pleasure Predicts Effective Connectivity in the Mesolimbic System. *Frontiers in Behavioral Neuroscience, 9*(217), 1-8.
224. Gu, R., Huang, W., Camilleri, J., Xu, P., Wei, P., Eickhoff, S.B., & Feng, C. (2019). Love is Analogous to Money in Human Brain: Coordinate-Based and Functional Connectivity Meta-Analyses of Social and Monetary Reward Anticipation. *Neuroscience and Biobehavioral Reviews, 100*, 108-128.
225. Carver, C. S. (2012). Self-awareness. In Leary. M. R., & Tangney, J. P. (Eds.), *Handbook of Self and Identity* (pp. 50–68). The Guilford Press.
226. Cardinal, R.N., Parkinson, J.A., Hall, J., & Everitt, B.J. (2002). Emotion and Motivation: The Role of the Amygdala, Ventral Striatum, and Prefrontal Cortex. *Neuroscience and Biobehavioral Reviews, 26*(3), 321-352.
227. Davis, M., & Whalen, P. (2001). The Amygdala: Vigilance and Emotion. *Molecular Psychiatry, 6*, 13–34.
228. Healy, J.M. (1989). Emotional Adaptation to Life Transitions: Early Impact on Integrative Cognitive Processes. In Buss D.M. & Cantor N. (Eds.), *Personality Psychology*. Springer.
229. Koole, S. (2009). The Psychology of Emotion Regulation: An Integrative Review. *Cognition and Emotion, 23*(1), 4-41.
230. Santos, C., Oliveira, P., Assunção, S., Catarina Almeida, T., & Ramos, C. (2019). Coping Strategies in the Management of Traumatic Events and Cognitive and Emotional Processing from Disclosure. *Annals of Medicine, 51*(1), 192.
231. Teixeira, C., Santos, C., Diogo, R., Gonçalves, A., Freire, F., Catarina Almeida, T., & Ramos, C. (2019). Posttraumatic Growth and Cognitive and Emotional Processing from Disclosure in the Aftermath of a Traumatic Event. *Annals of Medicine, 51*(1), 193.
232. Hayes, S. (2019). Acceptance and Commitment Therapy: Towards a Unified Model of Behavior Change. *World Psychiatry, 18*(2), 226-227.
233. Cordova, J.V. (2001). Acceptance in Behavior Therapy: Understanding the Process of Change. *The Behavior Analyst, 24*(2), 213-226.
234. Hayes, S.C. (2001). Acceptance and Change. In Smelser, N. J. & Baltes, P. B. (Eds.), *International Encyclopedia of the Social and Behavioral Sciences.* (pp. 27-30). Pergamon.
235. Kanske, P., Heissler, J., Schönfelder, S., Bongers, A., & Wessa, M. (2011). How to Regulate Emotion? Neural Networks for Reappraisal and Distraction. *Cerebral Cortex, 21*(6), 1379-1388.
236. Ochsner, K., Gross, J., Poldrack, R., & Wagner, A. (2008). Cognitive Emotion Regulation: Insights From Social Cognitive and Affective Neuroscience. *Current Directions in Psychological Science, 17*(2), 153-158.
237. Brown, V., & Morey, R. (2012). Neural Systems for Cognitive and Emotional Processing in Posttraumatic Stress Disorder. *Frontiers in Psychology, 3*(449), 1-14.
238. MacDonald, A., Cohen, J., Stenger, V., & Carter, C. (2000). Dissociating the Role of the Dorsolateral Prefrontal and Anterior Cingulate Cortex in Cognitive Control. *Science, 288*(5472), 1835-1838.
239. LeDoux, J.E. (1996). *The Emotional Brain: The Mysterious Underpinnings of Emotional Life.* Simon & Schuster.
240. Humphreys, G.W., & Price, C.J. (2001). Cognitive Neuropsychology and Functional Brain Imaging: Implications for Functional and Anatomical Models of Cognition. *Acta Psychologica, 107*(1),119-153.
241. Mahon, B.Z., & Cantlon, J.F. (2011). The Specialization of Function: Cognitive and Neural Perspectives. *Cognitive Neuropsychology, 28*(3), 147–155.
242. Wikenheiser, A.M., & Schoenbaum, G. (2016). Over the River, Through the Woods: Cognitive Maps in the Hippocampus and Orbitofrontal Cortex. *Nature Reviews Neuroscience, 17*(8), 513–523.
243. Rosenbloom, M.H., Schmahmann, J.D., & Price, B.H. (2012). The Functional Neuroanatomy of Decision-Making. *Journal of Neuropsychiatry and Clinical Neurosciences, 24*(3), 266-277.
244. Williams, R. (2017). Anger as a Basic Emotion and Its Role in Personality Building and Pathological Growth: The Neuroscientific, Developmental and Clinical Perspectives. *Frontiers in Psychology, 8*(1950), 1-9.
245. Pulcu, E., Zahn, R., & Elliott, R. (2013). The Role of Self-Blaming Moral Emotions in Major Depression and Their Impact on Social-Economical Decision Making. *Frontiers in Psychology, 4*(310), 1-17.
246. Freed, P.J., Yanagihara, T.K., Hirsch, J., & Mann, J.J. (2009). Neural Mechanisms of Grief Regulation. *Biological Psychiatry, 66*(1), 33-40.
247. Saleh, A., Potter, G., McQuoid, D., Boyd, B., Turner, R., MacFall, J., & Taylor, W. (2017). Effects of Early Life Stress on Depression, Cognitive Performance and Brain Morphology. *Psychological Medicine, 47*(1), 171-181.
248. Zelviene, P., & Kazlauskas, E. (2018). Adjustment Disorder: Current Perspectives. *Neuropsychiatric Disease and Treatment. 14*, 375–381.
249. Courtois, C. (2008). Complex Trauma, Complex Reactions: Assessment and Treatment. *Psychotherapy: Theory, Research, Practice, Training. Psychotherapy, 41*(4), 412-425.
250. Lundorff, M., Holmgren, H., Zachariae, R., Farver-Vestergaard, I., & O'connor, M. (2017). Prevalence of Prolonged Grief Disorder in Adult Bereavement: A Systematic Review and Meta-Analysis. *Journal of Affective Disorders. 212*, 138–149.
251. Aarts, H., Ruys, K., Veling, H., Renes, R., De Groot, J., Van Nunen, A., & Geertjes, S. (2010). The Art of Anger: Reward Context Turns Avoidance Responses to Anger-Related Objects into Approach. *Psychological Science, 21*(10), 1406 – 1410.
252. Garfinkel, S.N., Zorab, E., Navaratnam, N., Engels, M., Mallorquí-Bagué, N., Minati, L., Dowell, N.G., Brosschot, J.F., Thayer, J.F., Critchley, H.D. (2015). Anger in Brain and Body: The Neural and Physiological Perturbation of Decision-Making by Emotion. *Social Cognitive and Affective Neuroscience, 11*(1), 150-158.
253. Waugh, C., Shing, E., & Avery, B. (2015). Temporal Dynamics of Emotional Processing in the Brain. *Emotion Review, 7*(4), 323–329.
254. Fernandez-Alcantara, M., Verdejo-Roman, J., Cruz-Quintana, F., Perez-Garcia, M., Catena-Martinez, A., Inmaculada Fernandez-Avalos, M., & Nieves Perez-Marfil, M. (2020). Increased Amygdala Activations during the Emotional Experience of Death-Related Pictures in Complicated Grief: An fMRI Study. *Journal of Clinical Medicine, 9*(3), 851.
255. Gündel, H., O'Connor, M., Littrell, L., Fort, C., & Lane, R. (2003). Functional Neuroanatomy of Grief: An FMRI Study. *American Journal of Psychiatry, 160*(11), 1946-1953.
256. Zubieta, J., Ketter, T., Bueller, J., Xu, Y., Kilbourn, M., Young, E., & Koeppe, R. (2003). Regulation of Human Affective Responses by Anterior Cingulate and Limbic μ-Opioid Neurotransmission. *Archives of General Psychiatry, 60*(11), 1145–1153.
257. Worden, J. W. (2018). *Grief Counseling and Grief Therapy: A Handbook for the Mental Health Practitioner*

(5th ed.). Springer Publishing Co.
258. Jacobs, S. (1993). *Pathologic Grief: Maladaptation to Loss*. American Psychiatric Association.
259. Shear, K., & Shair, H. (2005). Attachment, Loss, and Complicated Grief. *Developmental Psychobiology*, 47(3), 253-267.
260. Lieberman, M., Eisenberger, N., Crockett, M., Tom, S., Pfeifer, J., & Way, B. (2007). Putting Feelings into Words Affect Labeling Disrupts Amygdala Activity in Response to Affective Stimuli. *Psychological Science*, 18(5), 421-428.
261. Mendolia, M., & Kleck, R.E. (1993) Effects of Talking About a Stressful Event on Arousal: Does What We Talk About Make a Difference? *Journal of Personality and Social Psychology*, 64(2), 283-92.
262. Zisook, S., & Reynolds, C. (2017). Complicated Grief. *Focus*, 15(4), 12–13.
263. Arizmendi, B., Kaszniak, A., & O'Connor, M. (2016). Disrupted Prefrontal Activity During Emotion Processing in Complicated Grief: An fMRI Investigation. *NeuroImage*, 124, 968-976.
264. Lerner, J.S., Li, Y., Valdesolo, P., & Kassam, K.S. (2015). Emotion and Decision Making. *Annual Review of Psychology*, 66(1), 799-823.
265. Palombo, D.J., Keane, M.M. & Verfaellie, M. (2015). How Does The Hippocampus Shape Decisions?. *Neurobiology of Learning and Memory*, 125, 93-97.
266. Gosling, P., Denizeau, M., & Oberlé, D. (2006). Denial of responsibility: A New Mode of Dissonance Reduction. *Journal of Personality and Social Psychology*. 90(5), 722–733.
267. Blumenthal-Barby, J., & Ubel, P. (2018). In Defense of "Denial": Difficulty Knowing When Beliefs Are Unrealistic and Whether Unrealistic Beliefs Are Bad. *The American Journal of Bioethics*, 18(9), 4-15.
268. Sharot, T, Korn, C.W., & Dolan, R.J. (2011). How Unrealistic Optimism is Maintained in The Face of Reality". *Nature neuroscience*, 14(11), 1475–1479.
269. Kuzmanovic, B., Rigoux, L., & Tittgemeyer, M. (2018). Influence of vmPFC on dmPFC Predicts Valence-Guided Belief Formation. *Journal of Neuroscience*, 38(37), 7996-8010.
270. Rief, W., Glombiewski, J. A., Gollwitzer, M., Schubö, A., Schwarting, R., & Thorwart, A. (2015). Expectancies as Core Features of Mental Disorders, *Current Opinion in Psychiatry*, 28(5), 378-385.
271. Dumontheil, I. (2016). Adolescent Brain Development. *Current Opinion in Behavioral Sciences*, 10, 39-44.
272. Johnson, S., Blum, R., & Giedd, J. (2009). Adolescent Maturity and the Brain: The Promise and Pitfalls of Neuroscience Research in Adolescent Health Policy. *Journal of Adolescent Health*, 45(3), 216-221.
273. Olesen, P., Macoveanu, J., Tegnér, J., & Klingberg, T. (2007). Brain Activity Related to Working Memory and Distraction in Children and Adults. *Cerebral Cortex*, 17(5), 1047-1054.
274. Williams, S. (2002). Strategic Planning and Organizational Values: Links to Alignment. *Journal Human Resource Development International*. 5(2), 217-233.
275. Sullivan, W., Sullivan, R., & Buffton, B. (2001). Aligning Individual and Organisational Values to Support Change. *Journal of Change Management*, 2(3), 247-254.
276. Berns, G., Bell, E., Capra, C., Prietula, M., Moore, S., Anderson, B., Ginges, J., & Atran, S. (2012). The Price of Your Soul: Neural Evidence for the Non-Utilitarian Representation of Sacred Values. *Philosophical Transactions. Biological Sciences*, 367(1589), 754-762.
277. Luo, Q., Nakic, M., Wheatley, T., Richell, R., Martin, A., & Blair, R. (2006). The Neural Basis of Implicit Moral Attitude—An IAT Study Using Event-Related fMRI. *Neuroimage*, 30(4), 1449-1457.

278. Baron, J., & Spranca, M. (1997). Protected Values. *Organizational Behavior and Human Decision Processes*, 70(1), 1-16.
279. Harmon-Jones, E., & Mills, J. (2019). An Introduction to Cognitive Dissonance Theory and an Overview of Current Perspectives on the Theory. In Harmon-Jones, E. (Eds.), *Cognitive Dissonance: Reexamining a Pivotal Theory in Psychology* (pp.3-24). American Psychological Association.
280. Izuma, K., & Murayama, K. (2019). Neural Basis of Cognitive Dissonance. In Harmon-Jones, E. (Eds.) *Cognitive Dissonance: Reexamining a Pivotal Theory in Psychology* (pp.227-245). American Psychological Association.
281. Oyserman, D. (2015). Values, Psychology of. In Wright J.D. (Ed.), *International Encyclopedia of the Social & Behavioral Sciences* (2nd ed., pp. 36-40). Elsevier.
282. Jarcho, J. M., Berkman, E. T., & Lieberman, M. D. (2011). The Neural Basis of Rationalization: Cognitive Dissonance Reduction During Decision-Making. *Social Cognitive and Affective Neuroscience*, 6(4), 460–467.
283. Baumeister, R. F. , Heatherton, T. F. , & Tice, D. M (1994). *Losing Control: How and Why People Fail at Self Regulation*. Academic Press.
284. Müller-Pinzler, L., Gazzola, Keysers, Sommer, Jansen Frässle, Einhäuser, W., & Krach, S. (2015). Neural Pathways of Embarrassment and Their Modulation by Social Anxiety. *NeuroImage*, 119, 252-261.
285. Hariri, A., Mattay, V., Tessitore, A., Fera, F., & Weinberger, D. (2003). Neocortical Modulation of the Amygdala Response to Fearful Stimuli. *Biological Psychiatry*, 53(16), 494-501.
286. Üstün, S., Kale, E., & Çiçek, M. (2017). Neural Network for Time Perception and Working Memory. *Frontiers in Human Neuroscience*, 11(83), 1-11.
287. Rao, H., Betancourt, L., Giannetta, J., Brodsky, N., Korczykowski, M., Avants, B., Gee, J., Wang, J., Hurt H., Detre, J., & Farah, M. (2010). Early Parental Care Important for Hippocampal Maturation: Evidence from Brain Morphology in Humans. *NeuroImage*, 49(1), 114-1150.
288. Lenroot, R., & Giedd, J. (2006). Brain Development in Children and Adolescents: Insights from Anatomical Magnetic Resonance Imaging. *Neuroscience and Biobehavioral Reviews*, 30(6), 718-729.
289. Yim J. (2016). Therapeutic Benefits of Laughter in Mental Health: A Theoretical Review. *The Tohoku Journal of Experimental Medicine*, 239(3), 243–249.
290. Bartolo, A., Benuzzi, F., Nocetti, L., Baraldi, P., & Nichelli, P. (2006). Humor Comprehension and Appreciation: An fMRI Study. *Journal of Cognitive Neuroscience*, 18(11), 1789-1798.
291. Wild, B., Rodden, F., Rapp, A., Erb, M., Grodd,W., Ruch,W. (2006). Humor and Smiling: Cortical Regions Selective for Cognitive, Affective, and Volitional Components. *Neurology*, 66(6), 887-893.
292. Fonzi, L., Matteucci, G., & Bersani, G. (2010). Laughter and Depression: Hypothesis of Pathogenic and Therapeutic Correlation. *Rivista di Psichiatria*, 45(1), 1–6.
293. Berk, L. S., Tan, S. A., Fry, W. F., Napier, B. J., Lee, W., Hubbard, R. W., Lewis, J. E., & Eby, W. C. (1989). Neuroendocrine and Stress Hormone Changes During Mirthful Laughter. *The American Journal of the Medical Sciences*, 298(6), 390–396.
294. Itani, J., & Suzuki, A. (1967). The Social Unit of Chimpanzees. *Primates*, 8(4), 355–381.
295. Harlow, H. F. (1958). The Nature of Love. *American Psychologist*, 13(12), 673–685.
296. Feldman, R. (2017). The Neurobiology of Human Attachments. *Trends in Cognitive Sciences*, 21(2), 80-99.
297. Maslow, A. H. (1943). A Theory of Human Motivation

Psychological Review, 50(4), 370–396.

298. Young, S. N. (2008). The Neurobiology of Human Social Behaviour: An Important but Neglected Topic. *Journal of Psychiatry & Neuroscience*, 33(5), 391–392.
299. Morris, A.S, Silk, J.S., Steinberg, L., Myers, S.S., & Robinson, L.R. (2007). The Role of the Family Context in the Development of Emotion Regulation. *Social Development*, 16(2), 361-388.
300. Newman, L., Sivaratnam, C & Komiti, A. (2015). Attachment and Early Brain Development Neuroprotective Interventions in Infant–Caregiver Therapy. *Translational Developmental Psychiatry* 3(1), 28647.
301. Laurita, A.C., Hazan, C., & Spreng, R.N. (2019). An Attachment Theoretical Perspective for The Neural Representation of Close Others. *Social Cognitive and Affective Neuroscience*, 14(3), 237-251.
302. Allen, J. P. & Miga, E. M. (2010). Attachment in Adolescence: A Move to The Level of Emotion Regulation. *Journal of Social and Personal Relationships*, 27(2), 181-190.
303. Sutcliffe, A., Dunbar, R., Binder, J. & Arrow, H. (2012). Relationships and The Social Brain: Integrating Psychological and Evolutionary Perspectives. *British Journal of Psychology*, 103(2), 149-168.
304. Lea, M. & Duck, S. (1982), A Model for The Role of Similarity of Values in Friendship Development. *British Journal of Social Psychology*, 21(4), 301-310.
305. Massen, J., Sterck, E., & de Vos, H. (2010). Close Social Associations in Animals and Humans: Functions and Mechanisms of Friendship. *Behaviour*, 147(11), 1379-1412.
306. Patrick, C. (2005). *Handbook of Psychopathy* (2nd ed.). Guilford Press.
307. Ogloff, J.R.P. (2006). Psychopathy/Antisocial Personality Disorder Conundrum. *Australian & New Zealand Journal of Psychiatry*, 40(6-7), 519-528.
308. Coid, J., Yang, M., Ullrich, S., Roberts, A., & Hare, R.D. (2009). Prevalence and Correlates of Psychopathic Traits in The Household Population of Great Britain. *International Journal of Law and Psychiatry*, 32(2), 65-73.
309. World Health Organization. (1992). *The ICD-10 Classification of Mental and Behavioural Disorders: Clinical Descriptions and Diagnostic Guidelines*. World Health Organization.
310. Schulkin, J., & Sterling, P. (2019). Allostasis: A Brain-Centered, Predictive Mode of Physiological Regulation, *Trends in Neurosciences*, 42(10), 740-752.
311. McEwen, B. (2004). Protection and Damage from Acute and Chronic Stress: Allostasis and Allostatic Overload and Relevance to The Pathophysiology of Psychiatric Disorders. *Annals of the New York Academy of Sciences*, 1032(1), 1–7.
312. McEwen, B., Bowles, N., Gray, J., Hill, M., Hunter, R., Karatsoreos, I., & Nasca, C. (2015). Mechanisms of Stress in The Brain. *Nature Neuroscience*. 18(10), 1353–1363.
313. McEwen, B., Gray, J., & Nasca, C. (2015) 60 Years of Neuroendocrinology: Redefining Neuroendocrinology: Stress, Sex and Cognitive and Emotional Regulation. *Journal of Endocrinology*, 226(2), 67–83.
314. McEwen, B. (2007). Physiology and Neurobiology of Stress and Adaptation: Central Role of the Brain. *Physiological Reviews*, 87(3), 873-904.
315. Smith, S., & Vale, W. (2006). The Role of The Hypothalamic-Pituitary-Adrenal Axis in Neuroendocrine Responses to Stress. *Dialogues in clinical neuroscience*, 8(4), 383–395.
316. Kumar, A., Rinwa, P., Kaur, G., & Machawal, L. (2013). Stress: Neurobiology, Consequences and Management.

Journal of Pharmacy & Bioallied Science, 5(2), 91-97.

317. Osório, C., Probert, T., Jones, E., Young, A.H., & Robbins, I. (2017). Adapting to Stress: Understanding the Neurobiology of Resilience, *Behavioral Medicine*, 43(4), 307-322.
318. Lam, J., Shields, G., Trainor, B., Slavich, G., & Yonelinas, A. (2018). Greater Lifetime Stress Exposure Predicts Blunted Cortisol but Heightened DHEA Responses to Acute Stress. *Stress & Health*, 35(1), 15-26.
319. King, A. (2016). Neurobiology: Rise of Resilience. *Nature*, 531(7592), 18-19.
320. Wu, G., Feder, A., Cohen, H., Kim, J., Calderon, S., Charney, S., & Mathé, A. (2013). Understanding Resilience. *Frontiers in Behavioral Neuroscience*, 7(10), 1-15.
321. Godoy, L., Rossignoli, M., Delfino-Pereira, P., Garcia-Cairasco, N., & De Lima Umeoka, E. (2018). A Comprehensive Overview on Stress Neurobiology: Basic Concepts and Clinical Implications. *Frontiers in Behavioral Neuroscience*, 12(127), 1-23.
322. Lupien, S., Juster, R., Raymond, C., & Marin, M. (2018). The Effects of Chronic Stress on The Human Brain: From Neurotoxicity, to Vulnerability, to Opportunity. *Frontiers in Neuroendocrinology*, 49, 91-105.
323. McEwen, B. (2017). Neurobiological and Systemic Effects of Chronic Stress. *Chronic Stress*, 1, 1-11.
324. Bollini, A.M., Walker, E.F., Hamman, S., & Kestler, L. (2004). The influence of perceived control and locus of control on the cortisol and subjective responses to stress. *Biological Psychology*, 67(3), 245-260. https://doi.org/10.1016/j.biopsycho.2003.11.002
325. Boudarene, M., Legros, J. J., & Timsit-Berthier, M. (2002). Etude de la réponse de stress: rôle de l'anxiété, du cortisol et des DHEAs [Study of the stress response: role of anxiety, cortisol and DHEAs]. *L'Encephale*, 28(2), 139–146.
326. Contrada, R.J., & Baum, A. (Eds.) (2011). *The Handbook of Stress Science: Biology, Psychology & Health*. Springer Publishing Co.
327. Shields, G., Lam, J., Trainor, B., & Yonelinas, A. (2016). Exposure to Acute Stress Enhances Decision-Making Competence: Evidence for The Role of DHEA. *Sychoneuroendocrinology*, 67, 51-60.
328. Boudarene, M., Legros, J.J., & Timsit-Berthier, M. (2002). Study of the Stress Response: Role of Anxiety, Cortisol and DHEAs. *Encephale*, 28(2),139-146.
329. Jamieson, J., Crum, A., Goyer, J., Marotta, M. & Akinola, M. (2018). Optimizing Stress Responses with Reappraisal and Mindset Interventions: An Integrated Model. *Anxiety, Stress & Coping*, 31(3), 245-261.
330. Casper, A., Sonnentag, S. & Tremmel, S. (2017). Mindset Matters: The Role of Employees' Stress Mindset for Day-Specific Reactions to Workload Anticipation. *European Journal of Work and Organizational Psychology*, 26(6), 798-810.
331. Papousek, I., Weiss, E., Perchtold, C., Weber, H., De Assunção, V.L., Schulter, G., Lackner. H.K, & Fink, A. (2017). The Capacity for Generating Cognitive Reappraisals Is Reflected in Asymmetric Activation of Frontal Brain Regions. *Brain Imaging and Behavior*, 11(2), 577-90.
332. Morgan, C., Southwick, S, Hazlett, G, Rasmusson, A., Hoyt, G., Zimolo, Z., & Charney, D. (2004). Relationships Among Plasma Dehydroepiandrosterone Sulfate and Cortisol Levels, Symptoms of Dissociation, and Objective Performance in Humans Exposed to Acute Stress. *Archives of General Psychiatry*, 61(8), 819–825.
333. Kamin, H.S., & Kertes, D.A. (2017). Cortisol and DHEA in Development and Psychopathology. *Hormones and Behavior*, 89, 69-85.
334. Everly, G.S., & Lating, J.M. (2019). *The Link from Stress*

Arousal to Disease. In: *A Clinical Guide to the Treatment of the Human Stress Response*. Springer.
335. McEwen, B. & Karatsoreos, I. (2015). Sleep Deprivation and Circadian Disruption: Stress, Allostasis, and Allostatic Load. *Sleep Medicine Clinics, 10*(1), 1-10.
336. Van Reeth, O., Weibel, L., K. Spiegel, K., Leproult, R., Dugovic, C. & Maccari, S. (2000). PHYSIOLOGY OF SLEEP (REVIEW): Interactions between stress and sleep: from basic research to clinical situations. *Sleep Medicine Reviews, 4*(2), 201–219.
337. Almeida, D., & Kessler, R. (1998). Everyday Stressors and Gender Differences in Daily Distress. *Journal of Personality and Social Psychology, 75*(3), 670-680.
338. Almeida, D. (2005). Resilience and Vulnerability to Daily Stressors Assessed via Diary Methods. *Current Directions in Psychological Science, 14*(2), 64-68.
339. Shapiro, D.H., Schwartz, C.E., & Astin, J.A. (1996). Controlling Ourselves, Controlling Our World. *The American Psychologist, 51*(12), 1213-1230.
340. Leotti, L.A., Iyengar, S.S., & Ochsner, K.N. (2010). Born to Choose: The Origins and Value of the Need for Control. *Trends in Cognitive Sciences, 14*(10), 457-463.
341. Lachman, M.E., Neupert, S.D., & Agrigoroaei, S. (2011). The Relevance of Control Beliefs for Health and Aging. In Schaie, K.W., & Willis, S.L. (Eds.), *Handbooks of Aging. Handbook of the Psychology of Aging.* (7th ed, pp. 175-190). Academic Press.
342. Chanda, M.L., & Levitin, D.J. (2013). The Neurochemistry of Music. *Trends in Cognitive Sciences, 17*(4), 179-193.
343. De Witte, M., Spruit, A., Van Hooren, S., Moonen, X., & Stams, G.J. (2020). Effects of Music Interventions on Stress-Related Outcomes: A Systematic Review and Two Meta-Analyses. *Health Psychology Review, 14*(2), 294-324.
344. Nagata, K., Iida, N., Kanazawa, H., Fujiwara, M., Mogi, T., Mitsushima, T., Lefor, A.T., & Sugimoto, H. (2014). Effect of Listening to Music and Essential Oil Inhalation on Patients Undergoing Screening CT Colonography: A Randomized Controlled Trial. *European Journal of Radiology, 83*(12), 2172-2176.
345. Tyrer, P. (2019). Nidotherapy: A Cost–Effective Systematic Environmental Intervention. *World Psychiatry, 18*(2), 144-145.
346. Tyrer, P., & Tyrer, H. (2018). *Nidotherapy: Harmonising the Environment with the Patient* (2nd ed.). Royal College of Psychiatrists.
347. Tyrer, P., & Kramo, K. (2009). Nidotherapy in Practice. *Journal of Mental Health, 16*(1), 117-129.
348. Tyrer, P. (2003). Editorial: Nidotherapy as a Treatment Strategy in Stress. *Stress and Health, 19*(3), 127-128.
349. Bilgili, B., Ozkul, E., & Koc, E. (2020). The Influence of Colour of Lighting on Customers' Waiting Time Perceptions. *Total Quality Management & Business Excellence, 31*(9-10), 1098-1111.
350. Yildirim, K., Akalin-Baskaya, A., & Hidayetoglu, M.L. (2007). Effects of Indoor Color on Mood and Cognitive Performance. *Building and Environment, 42*(9), 3233-3240.
351. Elliot, A.J., & Maier, M.A. (2014). Color Psychology: Effects of Perceiving Color on Psychological Functioning in Humans. *Annual Review of Psychology, 65*(1), 95-120.
352. Holzman, D. C. (2010). What's in a Color? The Unique Human Health Effects of Blue Light. *Environmental Health Perspectives, 118*(1), 22-27.
353. Hill, R.A., & Barton, R.A. (2005). Psychology Red Enhances Human Performance in Contests. *Nature, 435*(7040), 293.
354. Farah, M.J., Betancourt, L., Shera, D.M., Savage, J.H., Giannetta, J.M., Brodsky, N. L., Malmud, E.K., & Hurt, H. (2008). Environmental Stimulation, Parental Nurturance and Cognitive Development in Humans. *Developmental Science, 11*(5), 793-801.
355. Avants, B.B., Hackman, D.A., Betancourt, L.M., Lawson, G.M., Hurt, H., & Farah, M.J. (2015). Relation of Childhood Home Environment to Cortical Thickness in Late Adolescence: Specificity of Experience and Timing. *PloS One, 10*(10), 1-10.
356. Kelly, M.E., Loughrey, D., Lawlor, B.A., Robertson, I.H., Walsh, C., & Brennan, S. (2014). The Impact of Cognitive Training and Mental Stimulation on Cognitive and Everyday Functioning of Healthy Older Adults: A Systematic Review and Meta-Analysis. *Ageing Research Reviews, 15*(1), 28-43.
357. Kramer, A.F., Bherer, L., Colcombe, S.J., Dong, W., & Greenough, W.T. (2004). Environmental Influences on Cognitive and Brain Plasticity During Aging. *The Journals of Gerontology. Series A, Biological Sciences and Medical Sciences, 59*(9), 940-957.
358. Schooler, C., Mulatu, M.S., & Oates, G. (1999). The Continuing Effects of Substantively Complex Work in the Intellectual Functioning of Older Workers. *Psychology and Aging, 14*(3), 483-506.
359. Hernandez, R., Bassett, S., Boughton, S., Schuette, S., Shiu, E., & Moskowitz, J. (2018). Psychological Well-Being and Physical Health: Associations, Mechanisms, and Future Directions. *Emotion Review, 10*(1), 18–29.
360. Diener, E., & Ryan, K. (2009). Subjective Well-Being: A General Overview. *South African Journal of Psychology, 39*(4), 391–406.
361. Halson, S. (2005). Enhancing Recovery: Strategies and Recent Research. *Journal of Science and Medicine in Sport, 8*(4), 214.
362. Halson, S. (2008). Nutrition, Sleep and Recovery. *European Journal of Sport Science, 8*(2), 119-126.
363. Medic, G., Wille, M., & Hemels, M. E. (2017). Short- and Long-Term Health Consequences of Sleep Disruption. *Nature and Science of Sleep, 9,* 151–161.
364. Van Dongen, H., Maislin, G., Mullington, J., & Dinges, D. (2003). The Cumulative Cost of Additional Wakefulness: Dose-Response Effects on Neurobehavioral Functions and Sleep Physiology from Chronic Sleep Restriction and Total Sleep Deprivation. *Sleep, 26*(2),117–26.
365. Whitney, P., Hinson, J., Chee, M., Honn, K., & Van Dongen, H. (Eds.). (2019). *Sleep Deprivation and Cognition. Progress in Brain Research.* Academic Press.
366. Fullagar, H., Skorski, S., Duffield, R., Hammes, D., Coutts, A.J., & Meyer, T. (2015). Sleep and Athletic Performance: The Effects of Sleep Loss on Exercise Performance, and Physiological and Cognitive Responses to Exercise. *Sports Medicine, 45*(2), 161-186.
367. Schmitt, A., Belschak, F., & Den Hartog, D. (2017) Feeling Vital After a Good Night's Sleep: The Interplay of Energetic Resources and Self-Efficacy for Daily Proactivity. *Journal of Occupational Health Psychology 22*(4), 443–454.
368. van Duinen, H., Renken, R., Maurits, N., & Zijdewind I. (2007). Effects of Motor Fatigue on Human Brain Activity, an fMRI Study. *NeuroImage, 35*(4), 1438-1449
369. Yoo, S., Gujar, N., Hu, P., Jolesz, F., & Walker, M. (2007) The Human Emotional Brain Without Sleep--a Prefrontal Amygdala Disconnect. *Current Biology, 17*(20), 877-878.
370. Kühnel, J., Zacher, H., De Bloom, J., & Bledow, R. (2017). Take a Break! Benefits of sleep and Short Breaks for Daily Work Engagement, *European Journal of Work and Organizational Psychology, 26*(4), 481-491.
371. Ariga, A., & Lleras, A. (2011). Brief and Rare Mental "Breaks" Keep You Focused: Deactivation and Reactivation of Task Goals Preempt Vigilance Decrements. *Cognition, 118*(3), 439-443.
372. Helton, W., & Russell, P. (2017). Rest Is Still Best: The Role of the Qualitative and Quantitative Load of

Interruptions on Vigilance. *Human Factors: The Journal of Human Factors and Ergonomics Society, 59*(1), 91-100.

373. Smith, A. P., Brockman, P., Flynn, R., Maben, A., & Thomas, M. (1993). Investigation of The Effects of Coffee on Alertness and Performance During the Day and Night. *Neuropsychobiology, 27*(4), 217–223.

374. Hindmarch, I., Rigney, U., Stanley, N., Quinlan, P., Rycroft, J., & Lane, J. (2000). A Naturalistic Investigation of The Effects of Day-Long Consumption of Tea, Coffee and Water on Alertness, Sleep Onset and Sleep Quality. *Psychopharmacology, 149*(3), 203-216.

375. Cappelletti, S., Piacentino, D., Sani, G., & Aromatario, M. (2015). Caffeine: Cognitive and Physical Performance Enhancer or Psychoactive Drug?. *Current neuropharmacology, 13*(1), 71–88.

376. de Jonge, J. (2020) What Makes a Good Work Break? Off-Job and On-Job Recovery as Predictors of Employee Health. *Industrial Health, 58*(2), 142-152.

377. Gluschkoff, K., Elovainio, M., Kinnunen, U., Mullolal, S., Hintsanen, M., Keltikangas-Järvinen, L., & Hintsa, T. (2016). Work Stress, Poor Recovery and Burnout in Teachers. *Occupational Medicine, 66*(7), 564-570.

378. Sonnentag, S., Arbeus, H., Mahn, C., & Fritz, C. (2014). Exhaustion and Lack of Psychological Detachment from Work During Off-Job Time: Moderator Effects of Time Pressure and Leisure Experiences. *Journal of Occupational Health Psychology, 19*(2), 206-216.

379. Gilbert, D., & Abdullah, J. (2004). Holidaytaking and The Sense of Well-Being. *Annals of Tourism Research. 31*(1), 103-121.

380. Bloom, J., Geurts, S., & Kompier, M. (2012). Effects of Short Vacations, Vacation Activities and Experiences on Employee Health and Well–Being. *Stress and Health, 28*(4), 305-318.

381. Bloom, J., Geurts, S., & Kompier, M. (2010) Vacation from work as prototypical recovery opportunity, *Gedrag en Organisatie, 23*(4), 333-349.

382. Weigelt, O., Gierer, P., & Syrek, C.J. (2019). My Mind is Working Overtime-Towards an Integrative Perspective of Psychological Detachment, Work-Related Rumination, and Work Reflection. *International Journal of Environmental Research and Public Health, 16*(16), 2987.

383. Sonnentag, S., Venz, L., & Casper, A. (2017). Advances in Recovery Research: What Have We Learned? What Should Be Done Next?. *Journal of Occupational Health Psychology, 22*(3), 365-380.

384. Blank, C., Gatterer, K., Leichtfried, V., Pollhammer, D., Mair-Raggautz, M., Duschek, S., Humpeler, E., & Schobersberger, W. (2018). Short Vacation Improves Stress-Level and Well-Being in German-Speaking Middle-Managers-A Randomized Controlled Trial. *International Journal of Environmental Research and Public Health, 15*(1), 130.

385. Siltaloppi, M., Kinnunen, U., Feldt, T., & Tolvanen, A. (2011). Identifying Patterns of Recovery Experiences and Their Links to Psychological Outcomes Across One Year. *International Archives of Occupational and Environmental Health. 84*(8), 877-888.

386. Tempesta, D., Socci, V., De Gennaro, L., & Ferrara, M. (2018). Sleep and Emotional Processing. *Sleep Medicine Reviews, 40*, 183-95.

387. Walker, M. (2009). The Role of Sleep in Cognition and Emotion. *Annals of the New York Academy of Sciences, 1156*(1), 168-197.

388. Fultz, N., Bonmassar, G., Setsompop, K., Stickgold, R., Rosen, B., Polimeni, J., & Lewis, L. (2019). Coupled Electrophysiological, Hemodynamic, and Cerebrospinal Fluid Oscillations in Human Sleep. *Science, 366*(6465), 628-631.

389. Raven, F., Van Der Zee, E., Meerlo, P., & Havekes, R. (2018). The Role of Sleep in Regulating Structural Plasticity and Synaptic Strength: Implications for Memory and Cognitive Function. *Sleep Medicine Reviews, 39*, 3-11.

390. Tononi, G. (2012). Sleep Function and Synaptic Homeostasis. *Journal of Molecular Neuroscience, 48*, 118.

391. Nofzinger, E., & Maquet, P. (2013). *Neuroimaging of Sleep and Sleep Disorders.* Cambridge University Press.

392. Kreutzmann, J., Havekes, R., Abel, T., & Meerlo, P. (2015). Sleep deprivation and Hippocampal Vulnerability: Changes in Neuronal Plasticity, Neurogenesis and Cognitive Function. *Neuroscience, 309*, 173-190.

393. Goldstein, A., & Walker, M. (2014). The Role of Sleep in Emotional Brain Function. *Annual Review of Clinical Psychology, 10*(1), 679-708.

394. Waterhouse, J., Fukuda, Y., & Morita, T. (2012). Daily Rhythms of The Sleep-Wake Cycle. *Journal of Physiological Anthropology, 31*(1), 5.

395. Mander, B., Winer, J., & Walker, M. (2017). Sleep and Human Aging. *Neuron, 94*(1), 19-36.

396. Moore, R. Y. (2007). Suprachiasmatic Nucleus in Sleep–Wake Regulation. *Sleep Medicine, 8*(3), 27-33.

397. Blume, C., Garbazza, C., & Spitschan, M. (2019). Effects of light on Human Circadian Rhythms, Sleep and Mood", *Somnologie, 23*(3), 147 – 156.

398. Bedont J L., & Blackshaw S. (2015). Constructing the Suprachiasmatic Nucleus: A Watchmaker's Perspective on The Central Clockworks. *Frontiers in Systems Neuroscience,* (9), 1-21.

399. Jones, B.E. (2017). Neurobiology of Waking and Sleeping. *Handbook of Clinical Neurology, 98*, 131–149.

400. Daniel, C., & Mason, O.J. (2015). Predicting Psychotic-Like Experiences During Sensory Deprivation. *BioMed Research International, 2015*(439379), 1-10.

401. Daniel, C., Lovatt, A., & Mason, O.J. (2014). Psychotic-Like Experiences and Their Cognitive Appraisal Under Short-Term Sensory Deprivation. *Frontiers in Psychiatry, 5*(106), 1-8.

402. Hayashi, M., Morikawa, T., & Hori, T. (1992). EEG Alpha Activity and Hallucinatory Experience During Sensory Deprivation. *Perceptual and Motor Skills, 75*(2), 403-12.

403. Morgenthaler, T., Kramer, M., Alessi, C., Friedman, L., Boehlecke, B., Brown, T., Coleman, J., Kapur, V., Lee-Chiong, T., Owens, J., Pancer, J., & Swick, T. (2006). Practice Parameters for The Psychological and Behavioral Treatment of Insomnia: An Update. An American Academy of Sleep Medicine Report. *Sleep, 29*(11), 1415-1419.

404. Taylor, D.J., & Roane, B.M. (2010). Treatment of Insomnia in Adults and Children: A Practice–Friendly Review of Research. *Journal of Clinical Psychology, 66*(11), 1137-1147.

405. Engel-Yeger, B., & Shochat, T. (2012). The Relationship between Sensory Processing Patterns and Sleep Quality in Healthy Adults. *Canadian Journal of Occupational Therapy, 79*(3), 134-141.

406. Landolt, H. P. (2008). Sleep Homeostasis: A Role for Adenosine in Humans? *Biochemical Pharmacology, 75*(11), 2070-2079.

407. Schwartz, J., & Roth, T. (2008). Neurophysiology of Sleep and Wakefulness: Basic Science and Clinical Implications. *Current Neuropharmacology, 6*(4), 367-378.

408. Fredholm, B. (1995) Adenosine, Adenosine Receptors and the Actions of Caffeine. *Basic & Clinical Pharmacology & Toxicology, 76*, 93–101.

409. Lieberman, H., Wurtman, R., Emde, J., Roberts, G., & Coviella, C. (1987). The Effects of low

Doses of Caffeine on Human Performance and Mood. *Psychopharmacology, 92*(3), 308-312.
410. Nehlig, A., Daval, J., & Debry, G. (1992). Caffeine and the Central Nervous System: Mechanisms of Action, Biochemical, Metabolic and Psychostimulant Effects. *Brain Research Reviews, 17*(2), 139-170.
411. Elmenhorst, D., Meyer, P., Winz, O., Matusch, A., Ermert, J., Coenen, H., Basheer, R., Haas, H., Zilles, K., & Bauer, A. (2007). Sleep Deprivation Increases A_1 Adenosine Receptor Binding in the Human Brain: A Positron Emission Tomography Study. *Journal of Neuroscience, 27*(9), 2410-2415.
412. Colrain, I.M., Nicholas, C.L., & Baker, F.C. (2014). Chapter 24 - Alcohol and The Sleeping Brain. *Handbook of Clinical Neurology, 125*, 415-431.
413. Muzur, A., Pace-Schott, E., & Hobson, J. (2002). The Prefrontal Cortex in Sleep. *Trends in Cognitive Sciences, 6*(11), 475-481.
414. Nishida M, Uchida, S., Hirai, N., Miwakeichi, F., Maehara, T., Kawai K., Shimizu, H., & Kato, S. (2005). High Frequency Activities in The Human Orbitofrontal Cortex in Sleep–Wake Cycle, *Neuroscience Letters, 379*(2), 110-115.
415. Parker, J.D.A., Summerfeldt, L.J., Walmsley, C., O'Byrne, R., Dave, H.P., & Crane, A.G. (2021). Trait Emotional Intelligence and Interpersonal Relationships: Results From A 15-Year Longitudinal Study. *Personality and Individual Differences, 169*, 110013.
416. Finkel, E.J., Simpson, J.A., & Eastwick, P.W. (2017). The Psychology of Close Relationships: Fourteen Core Principles. *Annual Review of Psychology, 68*(1), 383-411.
417. Harris, K., & Vazire, S. (2016). On Friendship Development and the Big Five Personality Traits. *Social and Personality Psychology Compass, 10*(11), 647– 667.
418. Saxe, R., & Baron-Cohen, S. (2006). The Neuroscience of Theory of Mind. *Social Neuroscience, 1*(3-4), 1-9.
419. Banakou, D., Kishore, S., & Slater, M. (2018). Virtually Being Einstein Results in an Improvement in Cognitive Task Performance and a Decrease in Age Bias. *Frontiers in psychology, 9*, 917. https://doi.org/10.3389/fpsyg.2018.00917
420. Kirlic, N., Aupperle, R. L., Misaki, M., Kuplicki, R., & Alvarez, R. P. (2017). Recruitment of Orbitofrontal Cortex During Unpredictable Threat Among Adults at Risk for Affective Disorders. *Brain and Behavior, 7*(8), 1-11.
421. Rogers, C. R., & Farson, R. E. (1957). *Active Listening*. USA University of Chicago.
422. Jones, S.M. (2011) Supportive Listening. *International Journal of Listening, 25*(1-2), 85-103.
423. Watt, D.F. (2005). Social Bonds and the Nature of Empathy. *Journal of Consciousness Studies, 12*(8-10), 185-209.
424. Niediek, J., & Bain, J. (2014). Human Single-Unit Recordings Reveal a Link Between Place-Cells and Episodic Memory. *Frontiers in Systems Neuroscience, 8*(158), 1-2.
425. Amaral, D., & Lavenex, P. (2007). Hippocampal Neuroanatomy. In Andersen, P., Morris, R., Amaral, D., Bliss, T., & O'Keefe, J. (Eds.). *The Hippocampus Book* (p. 37–114). Oxford University Press.
426. Goode, T.D., Tanaka, K.Z., Sahay, A., & McHugh, T.J. (2020). An Integrated Index: Engrams, Place Cells, and Hippocampal Memory. *Neuron, 107*(5), 805-820.
427. Lind, S., Hall, L., Breidegard, B., Balkenius, C., & Johansson, P. (2014). Speakers' Acceptance of Real-Time Speech Exchange Indicates That We Use Auditory Feedback to Specify the Meaning of What We Say. *Psychological Science, 25*(6), 1198-1205.
428. Casey, B.J. (2015). Beyond Simple Models of Self-Control to Circuit-Based Accounts of Adolescent Behavior. *Annual Review of Psychology, 66*(1), 295-319
429. Luna, B., Padmanabhan, A., & O'hearn, K. (2010). What Has fMRI Told Us About the Development of Cognitive Control Through Adolescence?. *Brain and Cognition 72*(1), 101-113.
430. Arain, M., Haque, M., Johal, L., Mathur, P., Nel, W., Rais, A., Sandhu, R., & Sharma, S. (2013). Maturation of the Adolescent Brain. *Neuropsychiatric Disease and Treatment, 9*, 449-461.
431. Smith, D., Xiao, L., & Bechara, A. (2012). Decision Making in Children and Adolescents: Impaired Iowa Gambling Task Performance in Early Adolescence. *Developmental Psychology, 48*(4), 1180-1187.
432. Bowlby, J. (2008). *A Secure Base: Parent-Child Attachment and Healthy Human Development*. Basic books.
433. Schore, A.N. (2018). The Right Brain Implicit Self: A Central Mechanism of the Psychotherapy Change Process. In Craparo, G., & Mucci, C. (Eds.), *Unrepressed Unconscious, Implicit Memory, and Clinical Work* (99-129). Routledge.
434. Beauregard, M. (2007). Mind Does Really Matter. Evidence From Neuroimaging Studies Of Emotion-Self-Regulation, Psychotherapy, And Placebo Effect. *Progress in Neurobiology, 81*(4), 218-236.
435. Beck, L., Pietromonaco, P., DeBuse, C., Powers, S., & Sayer, A. (2013). Spouses' Attachment Pairings Predict Neuroendocrine, Behavioral, and Psychological Responses to Marital Conflict. *Journal of Personality and Social Psychology, 105*(3), 388-424.
436. Edgar-Bailey, M., & Kress, V. (2010). Resolving Child and Adolescent Traumatic Grief: Creative Techniques and Interventions. *Journal of Creativity in Mental Health, 5*(2), 158-176.
437. Miller, D.T., & Ross, M. (1975). Self-serving Biases in the Attribution of Causality: Fact or Fiction?. *Psychological Bulletin, 82*(2), 213–225
438. Shepperd, J., Malone, W., & Sweeny, K. (2008). Exploring Causes of the Self–serving Bias. *Social and Personality Psychology Compass, 2*(2), 895-908.
439. Lindquist, K., Satpute, A., & Gendron, M. (2015). Do Language Do More Than Communicate Emotion? *Current Directions in Psychological Science, 24*(2), 9-108.
440. Brooks, J., Shablack, H., Gendron, M., Satpute, A., Parrish, M., & Lindquist, K. (2016). The Role of Language in the Experience and Perception of Emotion: A Neuroimaging Meta-Analysis. *Social Cognitive and Affective Neuroscience, 12*(2), 169-183.
441. Reidy, D., Zeichner, A., Hunnicutt-Ferguson, K., & Lilienfeld, S. (2008). Psychopathy Traits and the Processing of Emotion Words Results of a Lexical Decision Task. *Cognition and Emotion, 20*(6), 117-1186.
442. Anderson, N.E., & Kiehl, K.A. (2014). Psychopathy Developmental Perspectives and Their Implications for Treatment. *Restorative Neurology and Neuroscience, 32*(1), 103-117.
443. Blanch-Hartigan, D., Andrzejewski, S.A., & Hill, K.M. (2012). The Effectiveness of Training to Improve Person Perception Accuracy: A Meta-Analysis. *Basic and Applied Social Psychology, 34*(6), 483-498.
444. Boutet, I., LeBlanc, M., Chamberland, J.A., & Collin, C.A. (2021). Emojis influence emotional communication, social attributions, and information processing. *Computers in Human Behaviour, 119*, 106722. https://doi.org/10.1016/j.chb.2021.106722
445. Kernis, M.H., Brown, A.C., & Brody, G. (2000). Fragile Self–Esteem in Children and its Associations With Perceived Patterns of Parent–Child Communication. *Journal of Personality, 68*(2), 225-25.

46. Cramer, P. (2011). Young Adult Narcissism: A 20 Year Longitudinal Study of the Contribution of Parenting Styles, Preschool Precursors of Narcissism, and Denial. *Journal of Research in Personality, 45*(1), 19-28.
47. Aunola, K., & Nurmi, J.E. (2005). The Role of Parenting Styles in Children's Problem Behavior. *Child Development, 76*(6), 1144-1159.
48. Brummelman, E., & Sedikides, C. (2020). Raising Children With High Self–Esteem (But Not Narcissism). *Child Development Perspectives, 14*(2), 83-89.
49. Sciaraffa, M.A., Zeanah, P.D., & Zeanah, C.H. (2018). Understanding and Promoting Resilience in the Context of Adverse Childhood Experiences. *Early Childhood Education Journal, 46*(3), 343-353.
50. Masten, A. S., Cutuli, J. J., Herbers, J. E., & Reed, M.G.J. (2009). Resilience in Development. In Lopez, S. J., & Snyder, C. R. (Eds.). *Oxford Library of Psychology: Oxford Handbook of Positive Psychology.* (pp. 117–131). Oxford University Press.
51. Eisenberg, N., Spinrad, T.L., & Eggum, N.D. (2010). Emotion-Related Self-Regulation and Its Relation To Children's Maladjustment. *Annual Review of Clinical Psychology, 6*(1), 495-525.
52. Steinberg, L., & Morris, A. (2001). Adolescent Development. *Journal of Cognitive Education and Psychology, 2*(1), 55-87.
53. Rubin, K. H., Bukowski, W. M., & Parker, J. G. (2006). Peer Interactions, Relationships, and Groups. In Damon, W., Lerner, R. M., & Eisenberg, N. (Eds.). *Handbook of Child Psychology: Vol. 3. Social, Emotional, and Personality Development.* (6th ed., pp. 571-645). Wiley.
54. Etherington, L. (2020). *Melanie Klein and Object Relations Theory.* Simply Psychology. https://www.simplypsychology.org/Melanie-Klein.html
55. GoodTherapy. (2016, September 5). *Object Relations.* https://www.goodtherapy.org/learn-about-therapy/types/object-relations
56. Motzkin, J., Philippi, C., Wolf, R., Baskaya, M., & Koenigs, M. (2015). Ventromedial Prefrontal Cortex Is Critical for the Regulation of Amygdala Activity in Humans. *Biological Psychiatry, 77*(3), 276-284.
57. Rolls, E., Grabenhorst, F., & Deco, G. (2010). Choice, Difficulty, and Confidence in the Brain. *Neuroimage, 53*(2), 694-706.
58. Kurgansky, A. (2019). Functional Organization of the Human Brain in the Resting State. *Neuroscience and Behavioral Physiology, 49*(9), 1135–1144.
59. Buckner, R. & DiNicola, L. (2019). The Brain's Default Network: Updated Anatomy, Physiology and Evolving Insights. *Nature Reviews Neuroscience, 20*(10), 593–608.
60. Amati, V., Meggiolaro, S., Rivellini, G., & Zaccarin, S. (2018). Social Relations and Life Satisfaction: The Role of Friends. *Genus, 74*(1), 1-18.
61. Eisenberger, N. & Cole, S. (2012). Social Neuroscience and Health: Neurophysiological Mechanisms Linking Social Ties with Physical Health. *Nature Neuroscience, 15*(5), 669–674.
62. Van Dongen, H.P., Whitney, P., Hinson, J.M., Honn, K.A. & Chee, M.W.L. (2019). *Progress in Brain Research: Sleep Deprivation and Cognition.* Academic Press.
63. Owen, L. & Corfe, B. (2017). The Role of Diet and Nutrition n Mental Health and Wellbeing. *Proceedings of the Nutrition Society, 76*(4), 425-426.
64. Gómez-Pinilla F. (2008). Brain Foods: The Effects of Nutrients on Brain Function. *Nature reviews. Neuroscience, 9*(7), 568–578.
65. Jorm, A.F., Morgan, A.J. & Hetrick, S.E. (Eds.). (2010). *Relaxation for Depression.* John Wiley & Sons Ltd.
66. Klainin-Yobas, P., Oo, W., Suzanne Yew, P. & Lau, Y. (2015). Effects of Relaxation Interventions on Depression and Anxiety Among Older Adults: A Systematic Review. *Aging & Mental Health, 19*(12), 1043–1055.
467. Stangl, D. & Thuret, S. (2009). Impact of Diet on Adult Hippocampal Neurogenesis. *Genes & Nutrition, 4*(4), 271-282.
468. Sasinthiran. (2016, February 4). *The Neuroscience of Love.* NUS Neuroscience Interest Group. https://nusneurointerest.wordpress.com/2016/02/14/the-neuroscience-of-love/
469. Killgore, W. (2010). Effects of Sleep Deprivation on Cognition. *Progress in Brain Research, 185*, 105-129.
470. Saghir, Z., Syeda, J., Muhammad, A., & Balla Abdalla, T. (2018). The Amygdala, Sleep Debt, Sleep Deprivation, and the Emotion of Anger: A Possible Connection?. *Cureus, 10*(7), 1-5.
471. Motomura, Y., Kitamura, S., Oba, K., Terasawa, Y., Enomoto, M., Katayose, Y., Hida, A., Moriguchi, Y., Higuchi, S., & Mishima, K. (2013). Sleep Debt Elicits Negative Emotional Reaction through Diminished Amygdala-anterior Cingulate Functional Connectivity. *PLoS ONE, 8*(2), 1-10.
472. Hoehl S, Hellmer K, Johansson M and Gredebäck G (2017) Itsy Bitsy Spider…: Infants React with Increased Arousal to Spiders and Snakes. *Front. Psychol.* 8:1710. doi: 10.3389/fpsyg.2017.01710

Index

- for searching out key areas

Acceptance 186, 196
Addressing
   - concerns 65
   - five important points 74
Anger in grief 206
Approval by others need for 113
Assertiveness 343
Autopilots
   - definition 15
   - Grade A hits, Truths of life, differences 229

Bananas 76
Bargaining 203
Basic needs list 289, 290
Bigger picture 252
   - with long-term vision 254
Blue square experiment 7

Change of mind and a change of food 112
Chimp Model 8
Chimp
   - accepting nature of 80
   - befriending using computer 94
   - boxing 65
   - definition 8
   - exercising 64
   - exercising with insight and change 158
   - fulfilling its drives 106
   - harnessing power 91
   - managing negative aspects 82
   - managing using external support 266
   - managing with bananas 76
   - managing with pause button 76
   - managing with three-step structured process 66
   - positive aspects 86
   - provoking in others 156
   - recognising 11, 22
   - speaking its language 133, 136
   - system developing 8

Chronic stress 285,
   - checking for and removing 286, 288
Colour and sleep – scientific points 303
Commitment 176-178
Commitment and motivation 101
   - scientific points 102
Communication
   - checking skill 373
   - differences in approach Human and Chimp 372
   - perceiving 367
   - verbal interactions 371
   - words and emojis 369
Comparing to others 97
Computer
   - as adviser 190
   - contents 15
   - definition 9
   - programming 25, 190
   - system scientific points 9, 189
   - tidying up 39
CORE Principle 102

Death bed advice scenario 249
Denial 202
Depression 157, 208
Developing self 52
Disorganisation 207
Distressed person
   - Chimp to Human 326
   - EUAR 327
Drives
   - inbuilt 3, 109
   - managing by drawing a line 107
   - reviewing your 123
   - scientific points 3

Eating drive
   - change of mind scientific point 112
   - managing 110

Index

Efficient versus effective 102
Emotion
- bubbling under the surface 163
- displaced 140
- express effectively and constructively 153
- managing projected 146
- mixed 142
- over emphasised 139
- real and Ghost 146
- replaced or altered emotion 138
- suppressed and repressed 143

Emotional messages
- Chimp's to us 130
- managing negative 131
- that keep on being sent 129
- scars 296

Emotional response Programming 192
Environment to manage mood 303
Establishing your options 49
EUAR 327
External support 266
- employing 271
- establishing 267
- maintaining 270

Feelings of missing out 120
Finding the real you: exercise 21
Fridge door syndrome 388

Ghost emotions 146
Goblins
- definition 15
- examples 17

Grade A and Grade B hits 65
Gremlins
- and autopilots in action 16
- definition 15
- finding and replacing 28, 38, 42

Grief reaction 200
- pathological 211
- typical chimp emotions 201-202

Habits
- changing with Triangle of Change 174, 179
- how they are formed 25, 167, 171
- managing moaning 154
- scientific points 167, 168
- self image 173

Happiness and Peace of mind lists 236-242, 248
Human
- advantages 45
- and Chimp working together 53, 55
- definition 8
- system scientific points 9

Ice cream scenario 259
Information entering the mind 33
Instincts
- interaction with the world 3
- scientific points 4

Judging or using your judgement 332, 344

Learning to say 'no' 99
Life Force 249
Lifeline Exercise 214
Links
- formed in mind 335
- turning to advantage 335

Logic and emotion bases 13

Marshmallow experiment 170
Memory storage 6
Mind processes information 187
Minnows and Whale 60
Most crucial Key Point of them all 261
Mushroom syndrome 298

NEAT 58
Needs and wants Human Chimp Computer 330-331
Nidotherapy 303

One in five rule 361, 365
Options - establishing them 49
Orbitofrontal cortex
- definition of success Scientific point 170
- team leader 5
- order of action 31
- motivation 102

Order of interpreting experiences 31

Panic attacks 148
Past experiences dealing with 347
Pause button 76
Peace of mind list - forming 242
Perfectionism 87
Perspective
- and confidence 256- 257
- and long-term vision 254
- defining successful life 255
- how to keep it 249, 251, 259
- resetting the brain with laughter 262
Processing
- complex 197, 199
- simple 186-189
Psychological mindedness 175
Psychopath 269, 368

React or respond 57, 63
Reality
- computer programming 223, 226
- working with 220
Reasonable and realistic 224, 233
Reasonable and realistic difference 18
Recuperation 308
- brief rest periods 309
Redrawing the line 118
Relationships
- building and maintaining 360
- dealing with past 362
- dependency 356
- foundation for successful 354
- with parents 345
Reorganisation 209
Resilience
- becoming 385
- building in children 387
- by losing self 391
- definition 376
- preventing common problems 401
Robustness
- checklist for 383-384
- definition 376
- how to achieve 377

Security - Chimp and Human differences 113
Self-serving bias 357, 364

Sleep 311
- activating system 314
- alcohol use 317
- brain activity during 311
- diary 321
- light dark system 313
- sleep cycles 312 -313
- summary 321
- tiredness system 316
Speaker thinkers and thinker speakers 339
Starting from where you are 51, 59, 63
Stone of life
- deeper level of operating 18
- definition and components 220
- reasons why not working 392
- summary of reasons why not working 401
Stress
- prevention 289
- reasons for persistent 295
Stress reaction
- alerting stage 276
- managing 275
- resilience stage 278
- return to resilience stage 284
- scientific points 274
- stress stage 283
- three stages 273

Three teams 5
Trigger points - finding personal 291, 299
Truths of Life 220, 227, 234

Ultimate mind stabiliser 220
Uncooperative person - building relationship with 331

Values
- basis for enduring relationship 247
- defining 235
- in conflict 245
Ventromedial prefrontal cortex (VMPFC) 17
Verbal interactions 371

World – Human and Chimp views 301

Yearning 205
You – finding self 21

and acknowledgements

There are so many people that I would like to thank during the making of this book, specifically those patients and clients who have trusted and shared their lives with me and taught me so much. There are too many people to name that have read passages but I am very grateful to all of them. From a practical perspective, I am indebted to the following four people who read, re-read, found references and generally kept me on my toes:

Andy Varns   Hazel Barker   Dean Coomer   Jess Radburn

I would like to say a special thank you to Jeff Battista, who has painstakingly worked through and designed the layout, graphics and pin-people for the book, bringing the science and key points to life.

Lastly, a big thank you for working through the book. I hope the book has helped to bring about thought and behaviours that will give you a better quality of life as you take your path through the jungle. I hope our paths cross.

Also Available

- by Prof Steve Peters

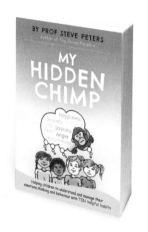

The Chimp Paradox

The Chimp Paradox is an introduction to the Chimp Model. It describes the struggle that takes place within our minds, with examples of how to apply this understanding, so that you can:

- Recognise how your mind is working
- Understand and manage your emotions and thoughts
- Manage yourself and become the person you would like to be.

MY HIDDEN CHIMP

My Hidden Chimp is an educational book for children to work through with an adult or by themselves. The book offers parents, teachers and carers some ideas and thoughts on how to help children to develop healthy habits for life.

The science behind the habits is discussed in a practical way with exercises and activities. The neuroscience of the mind is simplified for children to understand and then use to their advantage.

The Silent Guides

The Silent Guides is a companion book to My Hidden Chimp. Prof Steve Peters uses his Chimp Mind Management Model to help parents, teachers and carers understand the neuroscience behind unconscious beliefs and habits that may be silently guiding children's emotions, thinking and behaviours.

During our childhood, we learn to manage emotions and thinking by developing coping strategies. These strategies, whether helpful or unhelpful, often progress into habits for life – our Silent Guides.

It offers practical ideas for implementing into the lives of children the 10 helpful habits described in the book, My Hidden Chimp